AF388916

STEAM Education and the Innovative Pedagogies in the Intelligence Era

STEAM Education and the Innovative Pedagogies in the Intelligence Era

Editors

Zehui Zhan
Qintai Hu
Shan Wang
Xuan Liu

MDPI • Basel • Beijing • Wuhan • Barcelona • Belgrade • Manchester • Tokyo • Cluj • Tianjin

Editors
Zehui Zhan
South China Normal
University
China

Qintai Hu
Guangdong University of
Technology
China

Shan Wang
University of Saskatchewan
Canada

Xuan Liu
The Open University of
Sichuan
China

Editorial Office
MDPI
St. Alban-Anlage 66
4052 Basel, Switzerland

This is a reprint of articles from the Special Issue published online in the open access journal *Applied Sciences* (ISSN 2076-3417) (available at: https://www.mdpi.com/journal/applsci/special_issues/STEAM_Education_Intelligence_Era).

For citation purposes, cite each article independently as indicated on the article page online and as indicated below:

LastName, A.A.; LastName, B.B.; LastName, C.C. Article Title. *Journal Name* **Year**, *Volume Number*, Page Range.

ISBN 978-3-0365-7624-4 (Hbk)
ISBN 978-3-0365-7625-1 (PDF)

Contents

Editorial

STEAM Education and the Innovative Pedagogies in the Intelligence Era

Zehui Zhan [1], Qintai Hu [2,*], Xuan Liu [3] and Shan Wang [4]

[1] School of Information Technology in Education, South China Normal University, Guangzhou 510631, China; zhanzehui@m.scnu.edu.cn
[2] School of Computing, Guangdong University of Technology, Guangzhou 510006, China
[3] Editorial Department of Modern Distance Education Research, The Open University of Sichuan, Chengdu 610073, China; liuxuan123@126.com
[4] Department of Finance and Management Science, University of Saskatchewan, Saskatoon, SK S7N 2A5, Canada; wang@edwards.usask.ca
* Correspondence: huqt8@gdut.edu.cn

1. Introduction

As we delve into the era of intelligence, the importance of STEAM (Science, Technology, Engineering, Arts, and Mathematics) education has become increasingly evident. The rapid advancement of technologies such as Artificial Intelligence (AI) and the Internet of Things (IoT) also offers new opportunities for enhancing the educational experience. However, despite the increasing awareness of the value of STEAM education, there remains a scarcity of research on how to effectively incorporate these emerging technologies and cultivate essential 21st-century competencies in students.

This Special Issue aims to address the pressing need for transformative teaching and learning methods that integrate intelligent technologies and innovative pedagogies in STEAM education. It brings together a collection of original research articles and reviews, providing valuable insights into the current state of STEAM education and the innovative pedagogies in the intelligence era. In this editorial, we will introduce articles from the Special Issue, with the aim of providing a concise overview of the research content and results presented in these articles. We believe that this collection of articles will inspire further exploration and collaboration within the field, ultimately contributing to the ongoing advancement of STEAM education for the benefit of future generations.

With a total of 12 articles, including both empirical research and review articles, the collection offers valuable insights into the current state of STEAM education and its future prospects. The articles can be broadly categorized into two themes: technological application innovations in STEAM education and innovative pedagogies in STEAM educational contexts.

2. Technological Application Innovations in STEAM Education

By embracing technology and leveraging new methodologies, educators can create more engaging and effective learning experiences, better-preparing students for the challenges of the future.

Wu, Liang, and Zhan [1] in their paper on Course Recommendation Based on Enhancement of Meta-Path Embedding in Heterogeneous Graph, address the issue of student dropout in online courses, often resulting from a loss of interest during the learning process. The authors propose a novel course recommendation method called HGE-CRec, which leverages context formation for heterogeneous graphs to model students and courses. This approach employs meta-path embedding simulation and meta-path weight fusion to enhance the meta-path embedding set, thereby improving the representation ability of meta-path embedding and avoiding the manual setting of the meta-path. The article

Citation: Zhan, Z.; Hu, Q.; Liu, X.; Wang, S. STEAM Education and the Innovative Pedagogies in the Intelligence Era. *Appl. Sci.* **2023**, 13, 5381. https://doi.org/10.3390/app13095381

Received: 17 April 2023
Accepted: 20 April 2023
Published: 25 April 2023

demonstrates that the HGE-CRec method outperforms several existing baseline methods, providing more effective course recommendations for online learners.

Another article, A Novel Method for Cross-Modal Collaborative Analysis and Evaluation in the Intelligence Era by Wu et al. [2], explores the use of intelligent information technology in cross-modal learning analytics to facilitate procedural and scientific educational evaluation. The authors focus on assessing learners' emotional status during the learning process and propose an intelligent analysis model for this purpose. This model aims to accurately capture the emotional changes of learners, providing effective technical solutions for cross-modal learning analytics. The article showcases the effectiveness and superiority of the proposed method through experimental results, highlighting its potential to innovate classroom teaching evaluation in the intelligence era and improve the overall quality of modern teaching.

Moreover, this Special Issue highlights the potential of digital technology in various educational contexts.

Ke and Lin [3] present a study on the Dynamic Generation of Knowledge Graph Supporting STEAM Learning Theme Design. This research proposes a dynamic completion model for knowledge graphs based on subject semantic tensor decomposition. This model aims to provide more reasonable STEAM project-based learning themes for teachers by calculating multidisciplinary curriculum standard knowledge semantics. The study demonstrates the effectiveness of the model through an application experiment, generating STEAM learning themes.

López-Nores et al. [4] focus on the topic of Digital Technology in Managing Erasmus+ Mobilities: Efficiency Gains and Impact Analysis from Spanish, Italian, and Turkish Universities, assesses the efficiency gains that can be achieved through the ongoing digital transformation in managing the Erasmus+ program. The authors analyze the workload, resources, and expenses associated with Erasmus+ proceedings at four universities in Spain, Italy, and Turkey. The study reveals significant savings in terms of paper wastage and administrative time, potentially enabling the management of up to 80% more mobilities with the same resources and staff.

In their paper, Practice Promotes Learning: Analyzing Students' Acceptance of a Learning-by-Doing Online Programming Learning Tool, Iftikhar, Guerrero-Roldán, and Mor [5], investigate the factors that influence students' acceptance of learning-by-doing tools, such as CodeLab, in the context of an introductory programming course. The authors employed the Unified Theory of Acceptance and the Use of Technology (UTAUT) model, which they extended by adding the factor of motivation. The results reveal a strong relationship between acceptance and motivation, suggesting that students are more likely to use online learning-by-doing tools if they feel motivated and engaged in the learning activities.

By focusing on gamification, Parody and coworkers [6] in their paper on Gamification in Engineering Education: The Use of Classcraft Platform to Improve Motivation and Academic Performance, present a gamification teaching experience using the Classcraft platform in a first-year university mathematics course. The authors hypothesized that using Classcraft could enhance learning and promote the development of the Four C's: critical thinking, communication, collaboration, and creativity. A comparison between a gamification group and a control group demonstrated that the gamification group achieved higher mean marks and improved Four C's development, supporting the effectiveness of the gamification platform.

3. Innovative Pedagogies in STEAM Education

Several articles in this Special Issue emphasize the importance of interdisciplinary and learner-centered approaches.

For instance, the study by Wang et al. [7], Developing Computational Thinking: Design-Based Learning and Interdisciplinary Activity Design, explores the concept of design-based learning (DBL) and its connection with computational thinking (CT) teaching.

The study establishes an interdisciplinary activity design model by analyzing existing design-based scientific cycle models and research into STEAM education. The researchers design specific activities using Scratch to teach graphical programming to fifth-grade students, comparing the promotion effects of interdisciplinary activity design and traditional programming activities on students' CT development. The results indicate that the proposed interdisciplinary activity design is more effective in promoting students' CT levels than traditional programming activities.

Similarly, In the study presented by Montés et al. [8], titled "EXPLORIA, STEAM Education at University Level as a New Way to Teach Engineering Mechanics in an Integrated Learning Process", focuses on implementing STEAM learning in the Bachelor of Engineering in Industrial Design and Product Development at CEU Cardenal Herrera University through the EXPLORIA project. The authors describe the integration of STEAM learning within the EXPLORIA project to improve the learning of the physics subject, particularly regarding the part of the syllabus related to mechanical engineering. The study shows the adaptation made in the physical part to teach the integrated mechanics part of this learning process. The complete learning process is carried out through several challenges and milestones that students must overcome through the application of the physical knowledge learned in class. An ad hoc questionnaire validates the effectiveness of the proposed methodology, highlighting the students' assessment regarding the new teaching methodology.

Kučera et al. [9] introduce Educational Case Studies for Pilot Engineer 4.0 Programme: Monitoring and Control of Discrete-Event Systems Using OPC UA and Cloud Applications, focusing on the development of case studies for educational purposes. These case studies aim to address the modeling and control of a virtual discrete-event system using a PLC program and its subsequent interfacing with a cloud application. The prepared case studies are suitable for use in the education of engineers for the digitalization of production processes and can also be helpful in research on creating digital twins.

The growing popularity of design thinking in K-12 education, as shown in Li and Zhan's [10] systematic review paper, A Systematic Review on Design Thinking Integrated Learning in K-12 Education, indicates a shift toward more creative and human-centered problem-solving approaches in education. The authors find a growing popularity of integrating design thinking into K-12 education, particularly in STEM-related curricula. They also identify several core concepts of design thinking frequently valued and pursued in K-12 education, such as prototype, ideate, define, test, explore, empathize, evaluate, and optimize. The review reveals that while design thinking shows great educational potential in K-12 education, empirical evidence supporting the effectiveness of DTIL is still limited. This trend aligns with the broader goals of STEAM education, which aims to foster collaboration, innovation, and adaptability in students.

Zhao et al. [11] explore the Factors Influencing Student Satisfaction toward STEM Education: Exploratory Study Using Structural Equation Modeling. By extending the planned behavior theory, the study aims to predict high school students' learning satisfaction with STEM education. The results indicate that subjective norms and playfulness factors of STEM education positively relate to students' attitudes toward STEM education, with attitude being the most important factor influencing student satisfaction and acceptance.

Wu and coworkers [12] in their paper, How K12 Teachers' Readiness Influences Their Intention to Implement STEM Education: Exploratory Study Based on Decomposed Theory of Planned Behavior, examine the factors that affect K12 teachers' intentions to implement STEM education in China. Using a decomposed theory of planned behavior combined with teacher readiness, the authors developed an assumption model of the factors influencing teachers' STEM education intentions. Their findings indicate that teachers' intentions are significantly influenced by attitudes, perceived behavioral control, perceived usefulness, self-efficacy, and behavioral readiness. Additionally, emotional readiness directly impacts teachers' intentions to implement STEM education, while behavioral and cognitive readiness indirectly affect intentions through self-efficacy.

4. Conclusions

These articles collectively showcase the potential of various innovative pedagogies, technologies, and methodologies in STEAM education from primary and secondary schools to universities. The articles cover topics such as design thinking, digital technologies, student satisfaction, educational case studies, and integrated learning processes. As we continue to explore and develop these innovative approaches, it is crucial to maintain a focus on creating engaging, effective, and inclusive learning environments that prepare students for the challenges and opportunities of the future. We hope that this Special Issue serves as a valuable resource for educators, researchers, and policymakers interested in understanding and implementing innovative pedagogies and technologies in the context of STEAM education and the Intelligence Era.

Funding: This research was financially supported by the National Natural Science Foundation in China (62237001; 62277018), Ministry of Education in China Project of Humanities and Social Sciences (22YJC880106), the Major Project of Social Science in South China Normal University (ZDPY2208).

Conflicts of Interest: The authors declare no conflict of interest.

References

1. Wu, Z.; Liang, Q.; Zhan, Z. Course Recommendation Based on Enhancement of Meta-Path Embedding in Heterogeneous Graph. *Appl. Sci.* **2023**, *13*, 2404. [CrossRef]
2. Wu, W.; Hu, Q.; Feng, G.; He, Y. A Novel Method for Cross-Modal Collaborative Analysis and Evaluation in the Intelligence Era. *Appl. Sci.* **2023**, *13*, 163. [CrossRef]
3. Ke, Q.; Lin, J. Dynamic Generation of Knowledge Graph Supporting STEAM Learning Theme Design. *Appl. Sci.* **2022**, *12*, 11001. [CrossRef]
4. López-Nores, M.; Pazos-Arias, J.; Gölcü, A.; Kavrar, Ö. Digital Technology in Managing Erasmus+ Mobilities: Efficiency Gains and Impact Analysis from Spanish, Italian, and Turkish Universities. *Appl. Sci.* **2022**, *12*, 9804. [CrossRef]
5. Iftikhar, S.; Guerrero-Roldán, A.; Mor, E. Practice Promotes Learning: Analyzing Students' Acceptance of a Learning-by-Doing Online Programming Learning Tool. *Appl. Sci.* **2022**, *12*, 12613. [CrossRef]
6. Parody, L.; Santos, J.; Trujillo-Cayado, L.; Ceballos, M. Gamification in Engineering Education: The Use of Classcraft Platform to Improve Motivation and Academic Performance. *Appl. Sci.* **2022**, *12*, 11832. [CrossRef]
7. Wang, D.; Luo, L.; Luo, J.; Lin, S.; Ren, G. Developing Computational Thinking: Design-Based Learning and Interdisciplinary Activity Design. *Appl. Sci.* **2022**, *12*, 11033. [CrossRef]
8. Montés, N.; Aloy, P.; Ferrer, T.; Romero, P.; Barquero, S.; Carbonell, A. EXPLORIA, STEAM Education at University Level as a New Way to Teach Engineering Mechanics in an Integrated Learning Process. *Appl. Sci.* **2022**, *12*, 5105. [CrossRef]
9. Kučera, E.; Haffner, O.; Drahoš, P.; Cigánek, J. Educational Case Studies for Pilot Engineer 4.0 Programme: Monitoring and Control of Discrete-Event Systems Using OPC UA and Cloud Applications. *Appl. Sci.* **2022**, *12*, 8802. [CrossRef]
10. Li, T.; Zhan, Z. A Systematic Review on Design Thinking Integrated Learning in K-12 Education. *Appl. Sci.* **2022**, *12*, 8077. [CrossRef]
11. Zhao, J.; Wijaya, T.; Mailizar, M.; Habibi, A. Factors Influencing Student Satisfaction toward STEM Education: Exploratory Study Using Structural Equation Modeling. *Appl. Sci.* **2022**, *12*, 9717. [CrossRef]
12. Wu, P.; Yang, L.; Hu, X.; Li, B.; Liu, Q.; Wang, Y.; Huang, J. How K12 Teachers' Readiness Influences Their Intention to Implement STEM Education: Exploratory Study Based on Decomposed Theory of Planned Behavior. *Appl. Sci.* **2022**, *12*, 11989. [CrossRef]

Article

Course Recommendation Based on Enhancement of Meta-Path Embedding in Heterogeneous Graph

Zhengyang Wu [1,†], Qingyu Liang [1,†] and Zehui Zhan [2,*,†]

1   School of Computer Science, South China Normal University, Guangzhou 510631, China
2   School of Information Technology in Education, South China Normal University, Guangzhou 510631, China
*   Correspondence: zhanzehui@m.scnu.edu.cn
†   These authors contributed equally to this work.

Abstract: The main reason students drop out of online courses is often that they lose interest during learning. Moreover, it is not easy for students to choose an appropriate course before actually learning it. Course recommendation is necessary to address this problem. Most existing course recommendation methods depend on the interaction result (e.g., completion rate, grades, etc.). However, the long period required to complete a course, especially large-scale online courses in higher education, can lead to serious sparsity of interaction results. In view of this, we propose a novel course recommendation method named HGE-CRec, which utilizes context formation for heterogeneous graphs to model students and courses. HGE-CRec develops meta-path embedding simulation and meta-path weight fusion to enhance the meta-path embedding set, which can expand the learning space of the prediction model and improve the representation ability of meta-path embedding, thereby avoiding tedious manual setting of the meta-path and improving the effectiveness of the resulting recommendations. Extensive experiments show that the proposed approach has advantages over a number of existing baseline methods.

Keywords: online learning; heterogeneous graph; graph neural networks; course recommendation system

Citation: Wu, Z.; Liang, Q.; Zhan, Z. Course Recommendation Based on Enhancement of Meta-Path Embedding in Heterogeneous Graph. *Appl. Sci.* **2023**, *13*, 2404. https://doi.org/10.3390/app13042404

Academic Editor: Rafael Valencia-Garcia

Received: 31 October 2022
Revised: 9 February 2023
Accepted: 10 February 2023
Published: 13 February 2023

1. Introduction

In recent ten years the growth rate of distance education has been impressive, enabling students from all over the world to study courses at low cost. Many distance learning platforms, such as Coursera and edX, provide a large number of online courses and attract millions of registered users. The goal of users (students) is to master knowledge through course learning. The ongoing surge in the number of courses requires effective recommendation methods to help learners find suitable courses [1]. The recommendation accuracy is very important in regard to improving the service quality of online course platforms [2,3].

The purpose of course learning is to master new knowledge or improve the level of mastery of previously learned knowledge. Course learning is different from reading or browsing news, newsletters, and documents. The learning process is continuous (usually a semester), and the interactive objects (e.g., teacher, course video, subject) are diverse. For example, a higher education course on the XuetangX platform (xuetangx.com), e.g., Introduction to Machine Learning, lasts for 12 weeks and has 6–7 sections per week, while Introduction to Psychology is a 12-week course with 5–7 sections per week. Another example can be found in higher education courses on the edX (edx.org) platform, such as Big Data and Education, each of which are 8 weeks long and require 6–12 h per week. The Architectural Imagination course takes 10 weeks and 3–5 h per week. However, massive online course resources place students in danger of becoming lost in the large amount of information present at any time during the learning process [1]. In addition, the diversity of interactive objects enables relationships between students and students, students and courses, and courses and courses, even if these relationships are indirect or potential. The network in the real world is complex, with maany different elements affecting each other.

The traditional methods of modeling the real world with homogeneous networks neglect the impact of multi-source heterogeneous elements, while heterogeneous graphs (abbr. HG) have proved to be an effective method for modeling graph structures composed of multiple entities and relationships [4,5]. The advantage of HG is that it fully and intuitively uses the heterogeneous network structure in the dataset [6]. However, the main disadvantage is that it needs to manually design the meta-path, which is difficult to achieve optimally in real practice. Unfortunately, most existing methods do not solve these problems.

Faced with the aforementioned problems and issues, in this paper we propose a course recommendation method based on an enhancement of HG meta-path embedding (abbr. HGE-CRec). First, we use the skip-gram algorithm to generate the original meta-path embedding. Then, Graph Convolutional Networks (abbr. GCN) are utilized to generate simulated meta-path embedding based on the construction of the original meta-paths, enriching the dataset used for HG meta-path embedding. Third, we adopt a Graph Attention Network (abbr.GAT) to aggregate the neighboring node information of each node on the meta-path in order to enrich the semantics of node embedding. Fourth, a nonlinear method is adopted to fuse all information in order to enhance the node embedding of the meta-paths. Finally, the enhanced embedding is combined with a Matrix Factorization-based model to predict ratings for students and courses. We compare and analyze HGE-CRec with other existing Matrix Factorization-based recommendation models using two real-world online course datasets and two general open datasets dealing with HIN embedding-based recommendation research.

In brief, the contributions of this paper can be summarized based on the following three perspectives:

- We propose a novel course recommendation method based on heterogeneous graph embedding, and our experiments prove that the performance of this method is better than existing methods.
- We propose a novel solution for enhancing the embedding of the meta-path in HG.
- Extensive experiments on four real-world datasets demonstrate the effectiveness of the proposed approach. In addition, we show that the proposed approach can maintain good performance even in the absence of meta-path data.

The rest of this paper is organized as follows. We briefly review related works in Section 2. In Section 3, we introduce preliminaries and important notation. In Section 4, we describe the framework and implementation of HGE-CRec. The extensive experimental studies we conducted are described in Section 5. Finally, we provide our conclusions in Section 6.

2. Related Work

In this section, we first review related research about course recommendation. Then, we define and explain the meaning of exercise difficulty and revisit existing methods for predicting exercise difficulty.

2.1. Course Recommendation System

In the educational domain, personalized learning recommender system have been a research hotspot. Traditional recommendation methods have applications in personalized course recommendation, such as content-based filtering [7,8], collaborative filtering [9,10], and Hybrid-based course recommendation systems [11]. Due to the hierarchical structure of knowledge concepts in curricula, research into ontology-based course recommendations [1,12,13] have received extensive attention. Generally, an administrator is required to operate an operational course recommender system. The administrator can adjust the algorithm parameters to generate more accurate learning log-based recommendations, monitor user feedback on the courses recommended by the system, and adjust the algorithm parameters according to actual situation to ensure the normal operation of the system.

Traditional recommendation methods mostly come from the field of e-commerce. They regard courses as commodities, students as customers, and course learning behaviors as commodity purchase behaviors. However, course selection behavior is essentially different from commodity purchase behavior. When comparing learning content to movies or books, the cognitive state of the student and the learning content may change over time and context [14–16]. Learning a course generally takes a relatively long period of time, and constantly consumes students' attention and energy. In addition, students intend to achieve good grades through a course, which strongly affects students' level of satisfaction towards the course. Furthermore, due to changes in their level of knowledge mastery, students' preference for courses is not persistent; that is, after completing a course and mastering relevant knowledge, their preference for courses with similar knowledge concepts decreases. Student satisfaction with a course should not be equated with preference, as course recommendation methods based on student preference have certain limitations. In [17], the authors pointed out that student satisfaction with course learning is significantly affected by teachers, course content, curriculum, and grades. Recently, the satisfaction of students has become an important factor in research on course recommendation. In [18], course recommendation based on learning performance prediction was proposed.

A number of course recommendation methods focus on the characteristics and behavior data of students in the learning system. For example, [19] proposed a recommendation method for learning objects based on the integration of social signals, interests, and learner user preferences in an e-learning system. The authors of [20] proposed a course recommender system that uses extracted rules to find suitable courses according to student behaviors and preferences. In [21], a path that satisfies students' limited time while maximizing their grades is used to make lesson-by-lesson recommendations.

Similar to other recommendation systems, course recommendation systems must face the cold start problem, meaning that it is difficult to recommend courses for new students. The literature [22] points out that when there are insufficient data in one domain and relatively rich data in another, transfer learning can be used to overcome the cold-start problem when the two domains are explicitly or implicitly related. In addition, with sufficient data from the source domain, transfer learning and collaborative filtering can be used in combination to extract knowledge in order to improve the accuracy of recommendations in the target domain [23].

2.2. Heterogeneous Information Network Embedding-Based Recommendation

In natural and social systems, interacting components form interconnected networks, which can be referred to as complex information networks. These networks are homogeneous, containing the same type of objects and links; however, most real systems contain multiple types of interacting components, and as such can be modeled as heterogeneous information networks (abbr. HIN). HIN consist of multiple types of nodes and links, and have been proposed as a powerful information modeling method [24] able to naturally model complex objects and their rich relations. Figure 1 shows the HIN of two course datasets. HIN embedding, which aims to embed multiple types of nodes into a low-dimensional space, has received growing attention as well.

Heterogeneous information networks can naturally model complex objects and their rich relations in recommender systems, as in such systems objects are of different types and links among objects represent different relations. In [25], the authors propose an approach to evaluate the similarity of items or users in HIN. In [26], the authors proposed an MF-based recommendation method that uses the entity similarity calculated on a heterogeneous information network based on a meta-path algorithm. In [27], a collaborative filtering method on a weighted heterogeneous information network was proposed. This method is constructed by connecting users and items with the same rating. It can flexibly integrate heterogeneous information to make recommendations through weighted meta-paths and weighted integration methods. In order to take full advantage of the relationship heterogeneity in information networks, the authors of [28] introduced metapath-based latent features to represent the connectivity between users and items along different types of paths.

The network embedding approach is more resistant to sparse and noisy data. Considerable research has been done on representation learning for HIN. Broadly speaking, the existing works on HIN embedding can be categorized into four types: random walk-based methods, decomposition-based methods, deep neural network-based methods, and task-specific methods. In [6], an HIN embedding-based approach for recommendation was proposed. In [29], the authors exploited the attention-guided walk model in a heterogeneous information network to selectively sample discriminative attributes and representative explanation meta-paths to explain the recommendations.

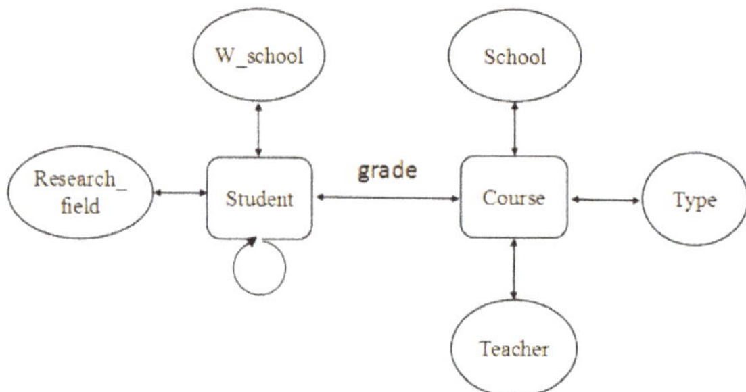

(**a**) Network schemas of HIN for a Scholat dataset (2020–2021)

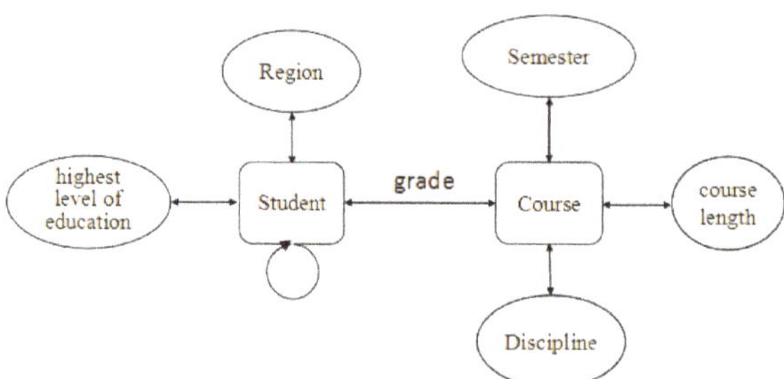

(**b**) Network schemas of HIN for a CNPC dataset (2014–2015)

Figure 1. Network schemas of heterogeneous information networks for the two course datasets.

3. Preliminary

The term *Heterogeneous Graph* is used uniformly in the descriptions in this paper. Because the graph structure means that it contains more comprehensive information and rich semantics, it has been widely used in many data mining tasks. In this section, we first introduce several definitions used in this article, including the HG and the meta-path of the HG. Then, we illustrate the working process of performance prediction using meta-path embedding. Several important notations are listed in Table 1, and we present a more detailed explanation of their role in this context.

Definition 1 (Heterogeneous Graph). *(1) Given a graph $G = \{V, E\}$, where V and E are the node set and relation set, respectively, if there is a mapping function $\phi(\cdot)$ of node type and a mapping function $\theta(\cdot)$ of edge type which map the nodes $v \in V$ and edges $e \in E$ to the specific types A and C, respectively, then this type graph is a heterogeneous graph. (2) If $|A| + |C| > 1$, then G*

is a heterogeneous graph and $G = \{A, C\}$ is called heterogeneous graph (abbr. HG) schema. The schematic of a heterogeneous graph of a course is shown in Figure 1.

Definition 2 (Meta-path of HG). *A meta-path is defined as a sub-path that links two different nodes in a heterogeneous graph. In this study, a meta-path is denoted as a path in the form of $B_1 \xrightarrow{C_1} B_2 \xrightarrow{C_2} \ldots \xrightarrow{C_l} B_l$ which describes a composite relation $C = C_1 \circ C_2 \circ \cdots \circ C_l$ between entity B_1 and B_{i+1}, where $\circ$ denotes the composition operator on relations.*

Definition 3 (Reciprocal Relationship of HG Embedding). *Given the HG embedding of a course, denoted by $\vec{c}$, and a student, denoted by $\vec{s}$, the reciprocal relationship between them is an interactive module Θ, e.g., $\mathcal{R} = \Theta(\vec{c}, \vec{s})$, where $\mathcal{R}$ is the representation of the interactive result.*

Table 1. Important notations used in this paper.

Notation	Description
$G = (V, E)$	The heterogeneous graph
A	The set of the types of node
P	A meta-path
Γ	The set of the meta-path
A_t	The type of node t
$|N^{A_t}(v)|$	The number of nodes of type A_t in the neighbor of node v.
M^a	An adjacency matrix of HG.
$e^{(l)}$	A node embedding on the l meta-path.
$e_u^{(U)}, e_i^{(I)}$	The finally embedding of user u and item i.
$\hat{r}_{u,i}$	The rating predicted by user u on item i.
u_u, v_i	The potential factors for user u and item i

4. Proposed Approach for Course Recommendation

In this section, we propose a novel method for course recommendation based on HG meta-path embedding. For simplicity, in the following presentation we use the short name "HGE-CRec" for our proposed method. The framework of HGE-CRec is shown in Figure 2. As shown in the figure, HGE-CRec is a triple-layer architecture model that takes students' ratings of each course as the output. The first layer is named the *Meta-path Embedding Layer* (MEL), the second layer is named the *Embedding Weight Aggregation Layer* (EWAL), and the third layer is named the *Matrix Factorization based Prediction Layer* (MFPL). The MEL generates meta-path embeddings for heterogeneous graphs and employs a GCN to simulate more meta-path embeddings. The EWAL employs GAT to aggregate neighbor node embeddings and concatenate them with the original embeddings. The MFPL uses a matrix factorization framework to predict students' ratings on courses. The details of MEL, EWAL, and MFPL are illustrated in Sections 4.1–4.3.

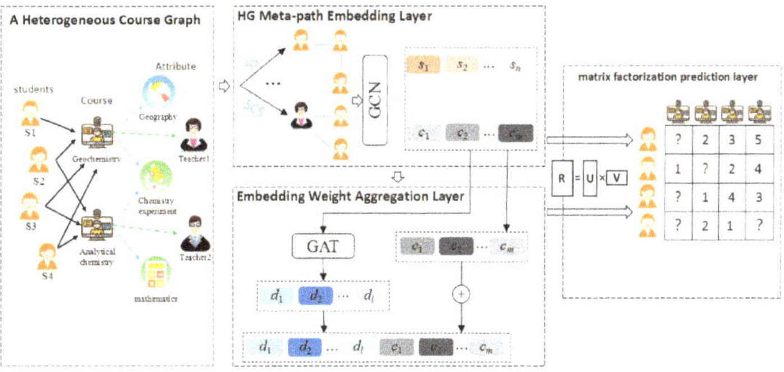

Figure 2. Framework of HGE-CRec.

4.1. Meta-Path Embedding Layer

4.1.1. Original Meta-Path Embedding

This module is used to generate the embedding of nodes on meta-paths of the heterogeneous graph. For this task, the meta-paths of the heterogeneous graph are generally sampled first, then the nodes on the sampled meta-paths are embedded. This module uses a meta-path based Deep Walk (abbr. MDK) to generate node embedding on each meta-path. MDK combines the random walk and word2vec algorithms. First, the heterogeneous graph is sampled and filtered for node sequences of various meta-paths, then the skip-gram algorithm in word2vec is used to map the node sequences into low-dimensional space:

$$R(n_{t+1} = y | n_t = x, P) = \begin{cases} \frac{1}{|N_{A_{t+1}}(x)|}, & (x,y) \in E, \Phi(y) = A_{t+1} \\ 0, & (x,y) \in E, \Phi(y) \neq A_{t+1} \\ 0, & (x,y) \notin E \end{cases}$$

where n_t represents the t-th node of the walk, x is a set of the node type which includes A_t, $|N_{A_{t+1}}(x)|$ represents the number of nodes with node type A_{t+1} among the neighbors of node x, and $R(n_{t+1} = y | n_t = x, P)$ represents the probability for a given node x of selecting a node y which represents the constraints of meta-path P. This regular sampling can ensure uniform sampling of various heterogeneous types of nodes under the constraint of the meta-path P to obtain the node sequence $X = [x_1, x_2, x_3, ..., x_n]$ of various meta-paths. Next, in order to improve the training efficiency of meta-path nodes, we filter X, that is, we keep nodes of the same type as the starting node, map nodes of the same type to the same space, form a new homogeneous node sequence $\hat{X}$, and then pass embedding methods to generate embedded representations of the filtered nodes.

Next, the skip-gram algorithm is used to generate node embeddings of $\hat{x}$. The skip-gram algorithm adopts the principle of similarity of adjacent nodes, uses the central node to predict the context node, and obtains the embedding of the target node after continuous iterative optimization. Skip-gram is a three-layer network structure consisting of an input layer, hidden layer, and output layer. It first generates a node sequence according to the size of the window, then takes the central node of the sequence as input and continuously maximizes the occurrence probability of its neighbor nodes, and finally obtains the embeddings of all nodes. The optimization objective of this algorithm is as follows:

$$L = \max_{f} \sum_{u \in V} \log Pr(N_u | f(u)), \tag{1}$$

where f is a function that maps each node u to a d-dimensional space vector, e.g., $f = V \rightarrow R^d$, u is the current node, N_u is the neighbor node of the u node obtained through the meta-path, and *Stochastic Gradient Descent* is used to optimize L.

4.1.2. Simulated Meta-Path Embedding

Using the method in the previous section, the embeddings generated under each meta-path in the heterogeneous graph can be obtained. In order to gather more neighbor information to improve the learning effect of node feature representation, this module uses GCN to aggregate the information of neighbor nodes to generate more meta-path node embeddings. As Figure 2 shows, a GCN model inside MEL is used to achieve this task; that is, more meta-path node embeddings are simulated by using GCN to aggregate neighbor embeddings for the node embeddings of the initial meta-path, thereby generating new node embeddings. The schematic of this process is shown in Figure 3.

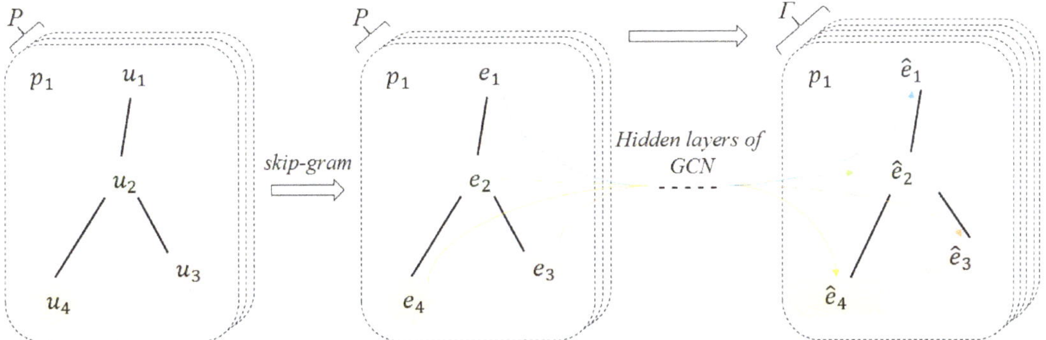

Figure 3. Schematic of GCN simulated embedding.

For a heterogeneous graph $G = (V, E)$, M^a denotes an adjacency matrix of G generated from the set of filtered meta-path sampling sequences $\hat{X}$; M^a is added as an identity matrix I for extension, e.g., $M^a = M^a + I$. Then, we have a matrix D

$$D_{ii} = \sum_j M^a_{ij}, \tag{2}$$

where D denotes a diagonal matrix of the node degree. Then, the GCN is used to generate the embeddings of the simulated meta-path following the process in Equation (3):

$$\hat{e}^P = softmax(D^{-\frac{1}{2}} M^a D^{-\frac{1}{2}} LeakyReLU(D^{-\frac{1}{2}} M^a D^{-\frac{1}{2}} e^P W_0^P) W_1^P), \tag{3}$$

where $\hat{e}^P$ denotes the simulated meta-path embeddings, e^P denotes the original ones, and W_0^P and W_1^P represent the trainable weight matrix.

4.2. Embedding Weight Aggregation Layer

4.2.1. Aggregation of Meta-Path Embedding Based on Attention Mechanism

In order to aggregate the neighbor information of each node, EWAL adopts a GAT mechanism to update the embedding representation of each node to form a new node embedding vector. For each meta-path embedding, including the original meta-path embedding and the simulated meta-path embedding, Equation (4) can be used to generate a new node embedding for each node under each meta-path:

$$t_{v,j}^{(l)} = LeakyReLU(\eta^{(l)T} \cdot [W^{(l)} e_v^{(l)} || W^{(l)} e_j^{(l)}]), \tag{4}$$

where $e_v^{(l)}$ represents the embedding of node v under the l-th meta-path, $t_{v,j}^{(l)}$ represents the contribution of features of neighbor j to node v under the l-th meta-path, $\eta^{(l)T}$ represents the attention parameter vector under the lth meta-path, $W^{(l)}$ represents the trainable weight matrix, and $||$ represents the splicing operation between vectors. Then, we can use Equation (5) for normalization:

$$a_{i,j}^{(l)} = \frac{exp(t_{vj}^{(l)})}{\sum_{k=1}^{k \in N_v} exp(t_{vk}^{(l)})} \tag{5}$$

After the weighted average of the neighborhood features of node v is obtained according to contribution $a_{i,j}^{(l)}$, a nonlinear conversion function σ is added to obtain the new features of node v, i.e., $\tilde{e}_v^{(l)}$. Equation (6) shows this process:

$$\tilde{e}_v^{(l)} = \sigma(\sum_{j \in N_v} a_{vj}^{(l)} W^{(l)} e_j^{(l)}) \tag{6}$$

In order to better learn the correlation between nodes and neighbors, the module uses the multi-header attention mechanism to perform K times attention operations, then splices the K times of the results, as shown in Equation (7):

$$\tilde{e}_v^{(l)}(K) = \overset{K}{\underset{k=1}{\|}} \sigma(\sum_{j \in N_v} [a_{vj}^{(l)}]^k [W^{(l)}]^k e_j^{(l)}), \tag{7}$$

where $[a_{vj}^{(l)}]^k$ represents the normalized neighbor contribution weight to node v on the k attention header under the l meta-path, and $[W^{(l)}]^k$ represents the trainable weight matrix on the k attention header under the l meta-path.

In order to retain the embedded information before and after updating the node, the embedded information before and after the update is spliced, meaning that the final embedded dimension after splicing is twice the previously embedded dimension. Both the user node and the project node need to undergo graph attention network aggregation embedding.

The framework of GAT Embedding is shown in Figure 4.

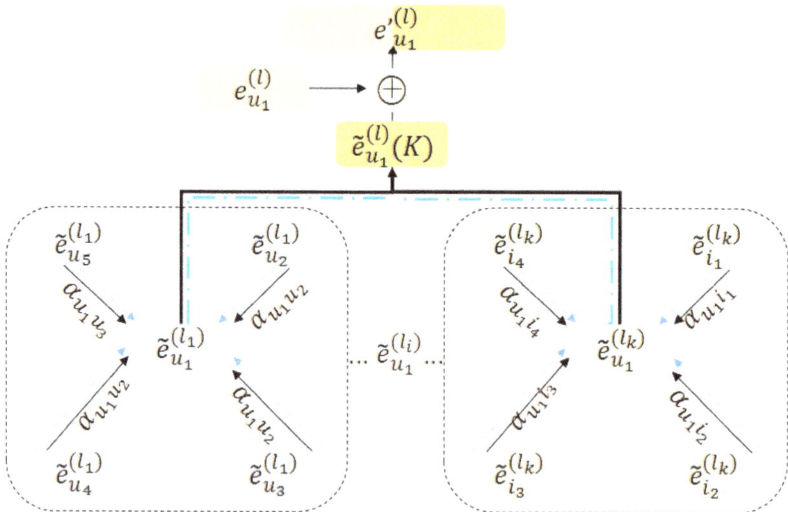

Figure 4. Framework of GAT Embedding.

4.2.2. Fusion of Meta-Path Embedding

Next, we use nonlinear depth fusion method to fuse all of the different kinds of embedded information of the meta-path into the embedded information targeted at a single node. This embedding fusion transformation can transform the embedding into a more appropriate form, improving the recommendation performance.

With a node $v \in V$, we can find the embedded representation set of v nodes $\{e_v^{(l)}\}_{l=1}^{|\Gamma|}$, where Γ is a set of meta-paths that include the original meta-paths and the simulated meta-paths, and $e_v^{(l)}$ represents the embedded information of node v under meta-path l. Taking the user node as an example, the nonlinear depth fusion function $\mathcal{M}(\cdot)$ provided below can be used to fuse the embedded information of each user node u, as shown in Equation (8):

$$\mathcal{M}(\{e_u^{(l)}\}) = \sigma(\sum_{l=1}^{|\Gamma|} W_u^{(l)} \sigma(M^{(l)} e_u^{(l)} + b^{(l)})), \tag{8}$$

where $M^{(l)} \in R^{D \times d}$ is the conversion matrix under path l, $b^{(l)} \in R^D$ is the offset vector under path l, and $W_u^{(l)}$ is the preference weight of user v under meta-path l. The function $\sigma(\cdot)$ is a nonlinear function. Here, the *sigmoid* function is used. For the fusion of embedded information of project nodes, a fusion method consistent with the embedded information of the user nodes is adopted. The algorithm for enhancing meta-path embedding is shown in Algorithm 1.

Algorithm 1: Algorithm for enhancing meta-path embedding

Input: the heterogeneous graph G; the adjacency matrix M^a; the meta-path sets $P^{(U)}$ for users and $P^{(I)}$ for items.

Output: the enhanced meta-path embedding set of users and items: $\{e_u^{(l)}\}_l^{|\Gamma^{(U)}|}, \{e_i^{(l)}\}_l^{|\Gamma^{(I)}|}$.

1: Get original meta-path embedding sets $\{e_i^{(l)}\}_l^{|P^{(I)}|}$ and $\{e_u^{(l)}\}_l^{|P^{(U)}|}$ of G by using *skip-gram*

2: **for** $e_i^{(l)}$ in $\{e_i^{(l)}\}_l^{|P^{(I)}|}$ **do**

3:     $\tilde{e}_i^{(l)} \leftarrow GCN(e_i^{(l)}, M^a)$. {Equation (3)}

4:     $\Gamma^{(I)} \leftarrow \{e_i^{(l)}, \tilde{e}_i^{(l)}\}$. {Get a set including two types meta-path embedding.}

5: **end for**

6: **for** $\bar{e}_i^{(l)}$ in $\Gamma^{(I)}$ **do**

7:     $t_{i,j}^{(l)} \leftarrow GAT(\bar{e}_i^{(l)}, \bar{e}_j^{(l)})$. {$\bar{e}_j^{(l)}$ is the embedding of the neighbor node j. Equation (4)}

8:     $a_{i,j}^{(l)} \leftarrow normalize(t_{i,j}^{(l)})$. {Equation (5)}

9:     $\hat{e}_i^{(l)} \leftarrow F(a_{i,j}^{(l)}, e_j^{(l)})$. {Equation (6), Equation (7)}

10:    $e_i^{(l)} \leftarrow Con(e_i^{(l)}, \hat{e}_i^{(l)})$. Update $e_i^{(l)}$ by connecting $\hat{e}_i^{(l)}$.

11: **end for**

12: **for** $e_u^{(l)}$ in $\{e_u^{(l)}\}_l^{|P^{(U)}|}$ **do**

13:    $\tilde{e}_u^{(l)} \leftarrow GCN(e_u^{(l)}, M^a)$. {Equation (3)}

14:    $\Gamma^{(I)} \leftarrow \{e_u^{(l)}, \tilde{e}_u^{(l)}\}$. {Get a set including two types meta-path embedding.}

15: **end for**

16: **for** $\bar{e}_u^{(l)}$ in $\Gamma^{(U)}$ **do**

17:    $t_{u,j}^{(l)} \leftarrow GAT(\bar{e}_u^{(l)}, \bar{e}_j^{(l)})$. {$\bar{e}_j^{(l)}$ is the embedding of the neighbor node j. Equation (4)}

18:    $a_{u,j}^{(l)} \leftarrow normalize(t_{u,j}^{(l)})$. {Equation (5)}

19:    $\hat{e}_u^{(l)} \leftarrow F(a_{u,j}^{(l)}, e_j^{(l)})$. {Equation (6), Equation (7)}

20:    $e_u^{(l)} \leftarrow Con(e_u^{(l)}, \hat{e}_u^{(l)})$. Update $e_u^{(l)}$ by connecting $\hat{e}_u^{(l)}$.

21: **end for**

22: **return** $\{e_u^{(l)}\}_l^{|\Gamma^{(U)}|}, \{e_i^{(l)}\}_l^{|\Gamma^{(I)}|}$.

4.3. Matrix Factorization-Based Prediction Layer

The *MFPL* module uses Matrix Factorization to predict the grade of students on courses. The HGE-CRec model proposed in this paper combines HG meta-path embedding with Matrix Factorization, and further combines the embedded vectors of users u and items i obtained from MEL and EWAL into the Matrix Factorization model. The prediction process of the rating is shown in Eqution (9):

$$\hat{r}_{u,i} = u_u^T \cdot v_i + \alpha \cdot e_u^T \cdot \gamma_i + \beta \cdot \gamma_u^T \cdot e_i, \tag{9}$$

where $\hat{r}_{u,i}$, u, and i belong to the rating matrix $\mathcal{R}$, e_u is the embedding of the user after merging the embedding vectors of different meta-paths, and e_i is the embedding of the item after merging the embedding vectors of different meta-paths, that is, the result of merging function $\mathcal{M}(\cdot)$ in Equation (8). In order to keep the form consistent, the hidden factors γ_i and γ_u are introduced to multiply the embedding vectors e_u and e_i, respectively, while α and β are adjustable parameters.

In order to provide the model with better generalization ability and prevent overfitting, regular constraints can be added to the loss function. For modeling scenarios of heterogeneous information networks, heterogeneous regular terms need to be added. The final optimization goal is shown in Eqution (10):

$$\ell = \sum_{<u,i,r_{u,i}>\in\mathcal{R}} (r_{u,i} - \hat{r}_{u,i})^2 +$$
$$\lambda \sum_u (\parallel U_u \parallel_2 + \parallel V_i \parallel_2 + \parallel \gamma_u \parallel_2 + \parallel \gamma_i \parallel_2 + \parallel \mathbb{E}_u \parallel_2 + \parallel \mathbb{E}_i \parallel_2), \tag{10}$$

where $\hat{r}_{u,i}$ is the prediction score calculated using Equation (9), λ is the regularization parameter, $\mathbb{E}_u$ is the parameter set of the function $\mathcal{M}(\cdot)$ for users, and $\mathbb{E}_i$ is the parameter set of the function $\mathcal{M}(\cdot)$ for items. The model training uses *Gradient Descent* to obtain the partial derivative, then updates the variables according to the direction of the negative gradient. After multiple iterations, the low-dimensional matrix is updated continuously until the algorithm finally converges; the parameters of Equation (10) are updated according to the following formula:

$$U_u \leftarrow U_u - \eta((r_{u,i} - \hat{r}_{u,i})V_i + \lambda_U U_u) \tag{11}$$

$$\mathbb{E}_{u,l} \leftarrow \mathbb{E}_{u,l} - \eta(-\alpha(r_{u,i} - \hat{r}_{u,i})\gamma_u \nabla^{(U)} + \lambda_{\mathbb{E}} \mathbb{E}_{u,l}), \tag{12}$$

$$V_i \leftarrow V_i - \eta((r_{u,i} - \hat{r}_{u,i})U_u + \lambda_V V_i) \tag{13}$$

$$\mathbb{E}_{i,l} \leftarrow \mathbb{E}_{i,l} - \eta(-\alpha(r_{u,i} - \hat{r}_{u,i})\gamma_i \nabla^{(I)} + \lambda_{\mathbb{E}} \mathbb{E}_{i,l}), \tag{14}$$

$$\frac{\partial e_i^{(l)}}{\partial \mathbb{E}_{i,l}} = \begin{cases} W_i^{(l)}\sigma(X_a)\sigma(X_f)(1-\sigma(X_a))(1-\sigma(X_f))e_i^{(l)}, & \mathbb{E}=M; \\ W_i^{(l)}\sigma(X_a)\sigma(X_f)(1-\sigma(X_a))(1-\sigma(X_f)), & \mathbb{E}=b; \\ \sigma(X_a)\sigma(X_f)(1-\sigma(X_a)), & \mathbb{E}=W, \end{cases} \tag{15}$$

The algorithm for training HGE-CRec is shown in Algorithm 2.

Algorithm 2: HGE-CRec training algorithm

Input: the heterogeneous graph G; the adjacency matrix M^a; the rating matrix $\mathcal{R}$; the adjustable parameters α, β; the regularization parameter λ; the learning rate coefficients for integrating embedding features; the enhanced meta-path embedding sets $\{e_u^{(l)}\}_l^{|\Gamma^{(U)}|}$ for users and $\{e_i^{(l)}\}_l^{|\Gamma^{(I)}|}$ for items.

Output: $\hat{r}_{u,i}$, the users and items feature matrices U and V; the weights of users and items HG embedding; the weights of feature interaction matrix and; the parameters in the fusion function of embedding

1: Get enhanced meta-path embedding sets $\{e_u^{(l)}\}_l^{|\Gamma^{(U)}|}$ and $\{e_i^{(l)}\}_l^{|\Gamma^{(I)}|}$ by using Algorithm 1

2: Initialize $U, V, \gamma^{(U)}, \gamma^{(I)}, \mathbb{E}_u, \mathbb{E}_i$ by standard normal distribution;
3: **while** termination criterion is not satisfied **do**
4:    $r_{u,i} \leftarrow$ Randomly select a tuple $< u, i > \in R$
5:    Update U_u and V_i by a typical MF model;
6:    **for** $e_u^{(l)}$ in $\Gamma^{(U)}$ **do**
7:       $\nabla^{(U)} \leftarrow \frac{\partial e_u^{(l)}}{\partial \mathbb{E}_{u,l}}$ {Equation (15)};
8:       Update $\mathbb{E}_{u,l}$ by Equation (12);
9:    **end for**
10:    Update $\gamma_u^{(U)}$ by Equation (11);
11:    **for** $e_i^{(l)}$ in $\Gamma^{(I)}$ **do**
12:       $\nabla^{(I)} \leftarrow \frac{\partial e_i^{(l)}}{\partial \mathbb{E}_{i,l}}$ {Equation (15)};
13:       Update $\mathbb{E}_{i,l}$ by Equation (14);
14:    **end for**
15:    Update $\gamma_i^{(I)}$ by Equation (13);
16: **end while**
17: **return** $U, V, \gamma^{(U)}, \gamma^{(I)}, \mathbb{E}_u, \mathbb{E}_i$

---

5. Experiments

In this section, we conduct extensive experimental studies to verify the effectiveness and advantages of our approach. To this end, we compare the recommendation performance produced by HGE-CRec with several baselines on four datasets.

5.1. Datasets

In order to verify the effective of our proposed approach, the experiments are carried on four real datasets: two large-scale online courses datasets, Scholat and CNPC, and two other datasets, Yelp and Movielens, which are commonly used in research on HIN embedding-based recommendation. CNPC, Yelp, and Movielens are public datasets. Table 2 presents statistical information about the nodes and relationships in these datasets.

- Scholat

  This dataset is from a real academic social course platform (scholat.com) which provides courses offered by Chinese universities, including undergraduate and graduate courses. The courses involve computer science, economics, pedagogy, and other disciplines. The student profiles include the school, grade, major, courses learned, etc. The dataset used in our experiment contains 3168 courses, 150,563 users, and 1,237,485 course visit records for the 2020–2021 academic year. The frequency of students' attendance of courses represents information about student interest in the course. In this experiment, we scaled the attendance frequency to an interval.

- CNPC (https://dataverse.harvard.edu/dataset.xhtml?persistentId=doi:10.7910/DVN/26147, accessed on 28 September 2022)
  This dataset consists of the Canvas Network Open Course (canvas.net), which hosts open online courses, including Massive Open Online Courses (MOOCs) that are freely available to participants around the world. The dataset used in our experiments is from January 2014 to September 2015, including 224,914 users and 238 courses as well as various attribute information on users' social relations, forums, users, and courses. The courses include ten disciplines, e.g., mathematics, statistics, and education.
- Yelp (https://www.yelp.com/dataset/documentation/main, accessed on 30 September 2022)
  This dataset comes from the largest merchant rating website in the United States, yelp.com. The dataset records user ratings of merchants, the users' social relationship, and attribute information on users and merchants, including 16,239 users, 14,282 merchants, and 198,397 ratings.
- Movielens (http://files.grouplens.org/datasets/movielens/ml-100k-README.txt, accessed on 30 September 2022)
  The Movielens dataset is a classic movie recommendation dataset from movielens.org. Movielens-100k was selected for this experiment. This dataset has 943 users, 1682 movies, and 100,000 scores, and contains social relationship and attribute information between users and movies.

Table 2. Statistics for nodes and relationships in the datasets.

Datasets	Relations (A-B)	Number (A)	Number (B)	Number (A-B)
Scholat	Student-Course	25,293	1670	53,988
	Student-Unit_of_study	150,563	5753	150,563
	Student-Research_field	150,563	6458	150,563
	Course-School	3168	344	3168
	Course-Type	3168	13	3168
	Course-Teacher	3168	1060	7846
CNPC	User-Course	224,914	238	325,199
	User-Learner_type	32,719	7	32,719
	User-Age	224,914	4	224,914
	Course-Discipline	238	10	238
	Course-Course_length	238	79	238
Yelp	User-Business	16,239	14,284	198,397
	User-User	10,580	10,580	158,590
	User-Compliment	14,411	11	76,875
	Business-City	14,267	47	14,267
	Business-Category	14,180	511	40,009
Movielens	User-Movie	943	1682	100,000
	User-User	943	943	47,150
	User-Occupation	943	21	943
	User-Age	943	8	943
	Movie-Movie	1682	1682	82,798
	Movie-Genre	1682	18	2891

For the proposed model, the interactive data between students and courses is the most important. In addition, we are concerned about the characteristics that are conducive to the formation of the rich meta-paths from datasets, such as course type and teacher or student school and type. Based on the statistical analysis in Table 2, we defined the meta-paths in all four datasets, as shown in Table 3.

Table 3. Meta-paths of the four datasets.

Scholat	CNPC	Yelp	Movielens
S-C-S, C-S-C,	U-C-U, C-U-C,	U-B-U, B-U-B,	U-M-U, M-U-M,
S-C-Te-C-S,	U-C-D-C-U,	U-B-Ci-B-U,	U-M-G-M-U,
C-Te-C, C-Ty-C,	C-D-C,	B-Ci-B,	M-G-M,
S-C-Ty-C-S,	U-C-Co-C-U,	U-B-Ca-B-U,	M-M,
S-C-Sc-C-S,	C-Co-C	B-Ca-B	U-M-M-U
C-Sc-C			

5.2. Experimental Setup

In this subsection, we first clarify the implementation details of the experimental setup used for HGE-CRec. Then, we introduce the comparison baselines and evaluation metrics used in the experiments.

5.2.1. Baselines

This experiment selects three baseline methods to carry out comparative experiments with the proposed approach: PMF, SoMF, and HERec.

- PMF [30]: This is a recommended algorithm for classical probability matrix factorization models which decomposes the scoring matrix into two low-dimensional matrices.
- SoMF [31]: In this algorithm, social relations have the characteristics of social regularization items, helping to integrate social relations into basic recommendations in the matrix factorization model.
- HERec [6]: This classical recommendation algorithm based on heterogeneous information network embedding adopts the random walk strategy based on the meta-path to generate the embedding, then integrates the embedded fusion into the matrix factorization model for recommendation.

5.2.2. Evaluation Metrics

This experiment selects two indicators commonly used in research on recommendation systems, namely, the Mean Absolute Error (MAE) and Root Mean Square Error (RMSE), as the evaluation indicators of the model. MAE and RMSE calculate the prediction accuracy by calculating the deviation between the predicted user score and the actual user score. The smaller the MAE and RMSE, the higher the recommendation quality.

The *MAE* can be expressed by Equation (16):

$$MAE = \frac{1}{|D_{test}|} \sum_{(i,j) \in D_{test}} |r_{i,j} - \hat{r}_{i,j}|. \tag{16}$$

The *RMSE* can be expressed by Equation (17):

$$RMSE = \sqrt{\frac{1}{|D_{test}|} \sum_{(i,j) \in D_{test}} (r_{i,j} - \hat{r}_{i,j})^2}. \tag{17}$$

where $r_{i,j}$ is the real rating of item j by user i, $\hat{r}_{i,j}$ is the predicted score of user i on item j, and D_{test} is the test set of the scoring records.

5.3. Results of the Comparative Experiment

This subsection illustrates the comparative experimental results using the proposed approach and baseline methods on four datasets.

For each dataset, the preference (or score) record data are divided into a training set and a test set. For the Scholat, CNPC, and Movielens datasets, four training segmentation ratios are set, namely, 80%, 60%, 40%, and 20%; For the Yelp dataset, because it is more sparse than the other three datasets, four larger training ratios are set, namely, 90%, 80%, 70%, and 60%. For each ratio, teen evaluation sets are randomly generated, then the results are averaged to the final results, which as shown in Tables 4–7.

Table 4. Results on Scholat.

Training Rate	Metrics	PMF	SoMF	HERec	HGE-CRec
80%	MAE	0.4732	0.4685	0.4529	0.4435
	RMSE	0.7199	0.7087	0.6596	0.6294
60%	MAE	0.5023	0.4832	0.4627	0.4486
	RMSE	0.7693	0.7303	0.6908	0.6516
40%	MAE	0.5758	0.5090	0.4801	0.4608
	RMSE	0.8988	0.7748	0.7185	0.6683
20%	MAE	0.8302	0.5856	0.5677	0.5516
	RMSE	1.4199	0.8048	0.7866	0.6976

Table 5. Results on CNPC.

Training Rate	Metrics	PMF	SoMF	HERec	HGE-CRec
80%	MAE	0.8998	0.9074	0.8775	0.8658
	RMSE	1.2254	1.2293	1.1666	1.1361
60%	MAE	0.9124	0.9248	0.8843	0.8695
	RMSE	1.2504	1.2563	1.1761	1.1374
40%	MAE	0.9335	0.9585	0.8955	0.8754
	RMSE	1.2934	1.3092	1.1928	1.1403
20%	MAE	1.0504	1.0236	0.9156	0.8805
	RMSE	1.4053	1.4465	1.2273	1.1428

Table 6. Results on Yelp.

Training Rate	Metrics	PMF	SoMF	HERec	HGE-CRec
90%	MAE	1.0412	1.0095	0.8395	0.7723
	RMSE	1.4268	1.3392	1.0907	0.9787
80%	MAE	1.0791	1.0373	0.8475	0.7804
	RMSE	1.4816	1.3782	1.1117	0.9884
70%	MAE	1.1170	1.0694	0.8580	0.7817
	RMSE	1.5387	1.4201	1.1256	0.9899
60%	MAE	1.1778	1.1135	0.8759	0.7853
	RMSE	1.6167	1.4748	1.1488	0.9928

Table 7. Results on Movielens.

Training Rate	Metrics	PMF	SoMF	HERec	HGE-CRec
80%	MAE	0.7324	0.7289	0.7103	0.6992
	RMSE	0.9862	0.9851	0.9274	0.8980
60%	MAE	0.7463	0.7450	0.7181	0.7041
	RMSE	1.0121	1.0112	0.9369	0.8998
40%	MAE	0.7661	0.7784	0.7293	0.7104
	RMSE	1.0542	1.0650	0.9536	0.9033
20%	MAE	0.8527	0.8451	0.7495	0.7179
	RMSE	1.1641	1.1423	0.9881	0.9086

Tables 4–7 show a comparison of the two indicators, *MAE* and *RMSE*, with respect to the recommended results using the proposed approach and baseline methods on the four datasets.

Figure 5 shows the results on the four datasets in terms of *MAE*. Compared with the baselines, the line when using HGE-CRec is gentler in terms of MAE as the training ratio changes from 80% to 20%. In other words, the MAE growth of HGE-CRec is lower than the other three baselines. The same can be observed for all of the datasets.

Figure 6 shows the results on the four datasets in terms of *RMSE*. Compared with the baselines, the line when using HGE-CRec is gentler in terms of RMSE as the training ratio changes from 80% to 20%. In other words, as for MAE, the growth in RMS for EHGE-CRec is lower than for the other three baselines. Again, this can be observed for all of the datasets.

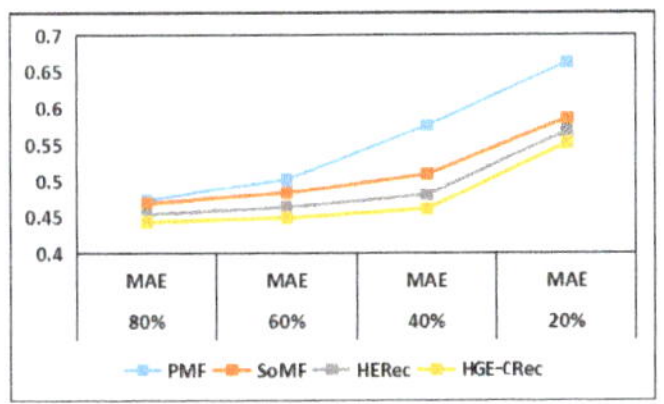

(**a**) Comparison of MAE with the change in the training rate on dataset Scholat

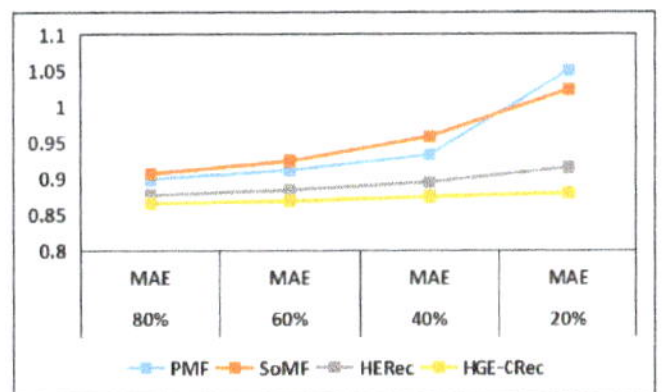

(**b**) Comparison of MAE with the change in the training rate on dataset CNPC

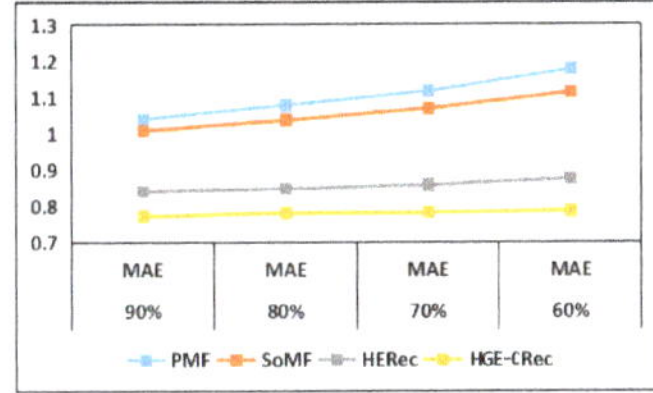

(**c**) Comparison of MAE with the change in the training rate on dataset Yelp

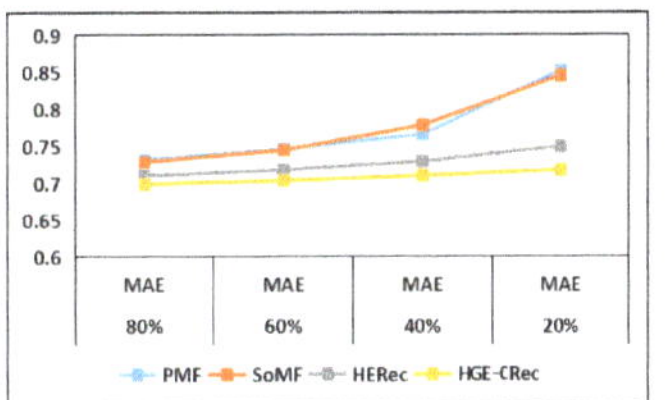

(**d**) Comparison of MAE with the change in the training rate on dataset Movielens

Figure 5. Comparison of MAE with the change in the training rate on four datasets.

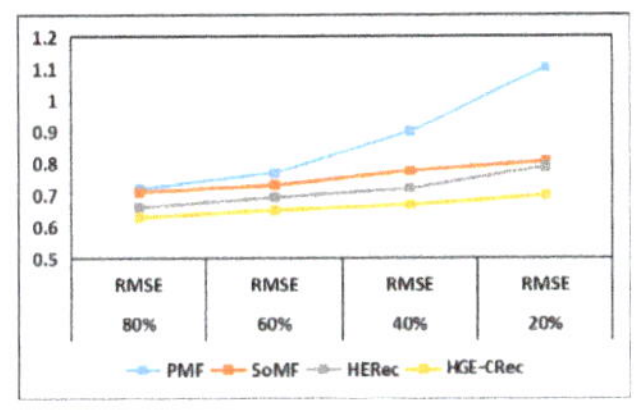

(**a**) Comparison of RMSE with the change in the training rate on dataset Scholat

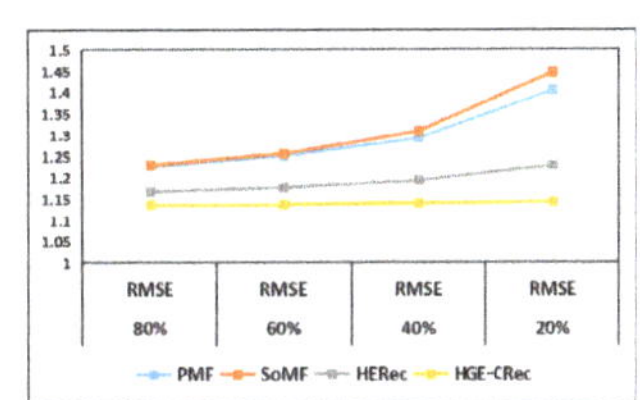

(**b**) Comparison of RMSE with the change in the training rate on dataset CNPC

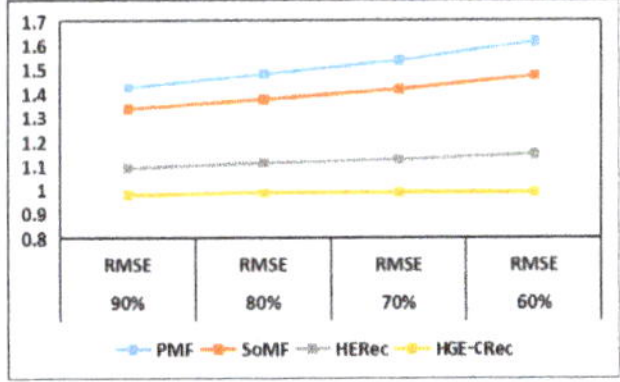

(**c**) Comparison of RMSE with the change in the training rate on dataset Yelp

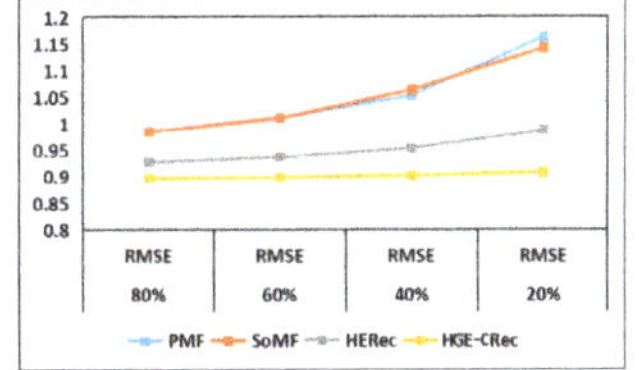

(**d**) Comparison of RMSE with the change in the training rate on dataset Movielens

Figure 6. Comparison of RMSE with the change in the training rate on four datasets.

Moreover, from Figures 5 and 6 it can be seen that the experimental results curves of the proposed approach is flat compared to the baselines, indicating that the effectiveness of the proposed approach is more stable than the baselines.

5.4. Ablation Study

In this section, we conduct an ablation study of the proposed approach through two experiments.

5.4.1. Component Adjustment

The first experiment observes the changes in the performance of the various approaches by adjusting the model's components. The HGE-CRec proposed in this paper uses GCN and GAT to improve the meta-path embedding after generating the original meta-path embedding. The first source of improvement is realized by using GCN to generate the simulated meta-path embedding, which enhances the meta-path embedding and expands the range of sample feature distribution. The second improvement is to use GAT to aggregate the various kinds of meta-path embedding based on neighbor weights. In order to explore the effect of these two improvements on the model, an ablation study of the component was set up using the following models for comparison:

- $HGE\text{-}CRec_{GCN}$
  Only the first part of the HGE-CRec model's meta-path embedding is improved, that is, while GCN is used to generate analog meta-path embedding, GAT is not used to aggregate various kinds of meta-path embedding based on neighbor weight.
- $HGE\text{-}CRec_{GAT}$
  Only the second part of the HGE-CRec model's meta-path embedding is improved, that is, while GAT is used to aggregate the original meta-path embedding based on the neighbor weights, GCN is not used to generate simulated meta-path embeddings.

The training ratio was uniformly 80%. The experimental results of the above models on the four datasets are shown in Table 8.

It can be seen from the experimental results that $HGE\text{-}CRec_{GCN}$ has the weakest effect, which means that only the first part of the improvement has no significant impact on the model; slight jitter is observed during the training process, with a slight increase or decrease. This further indicates that, based on the premise of there being sufficient of original meta-path embeddings, more analog meta-path embeddings are generated by aggregating neighborhood information, and there is little improvement on the model. That is, when meta-path embedding reaches a sufficient amount, saturation occurs and it is of little significance to add meta-path embedding. The effect of $HGE\text{-}CRec_{GAT}$ is significantly improved, showing that the improvement in the second part has a good impact on the model. Furthermore, this shows that using GAT to aggregate the neighbor embedding information of each node based on weight can effectively improve the performance of the model. That is, by learning the influence weight of different neighbors on nodes in the meta-path, the node representation can be enriched and the accuracy of prediction can be improved.

Table 8. Results of ablation study on component adjustment.

Dataset	Metrics	$HGE\text{-}CRec_{GCN}$	$HGE\text{-}CRec_{GAT}$	HGE-CRec
Scholat	MAE	0.4528	0.4435	0.4435
	RMSE	0.6598	0.6295	0.6294
CNPC	MAE	0.8775	0.8658	0.8658
	RMSE	1.1667	1.1362	1.1361
Yelp	MAE	0.8479	0.7807	0.7804
	RMSE	1.1111	0.9880	0.9884
Movielens	MAE	0.7103	0.6992	0.6992
	RMSE	0.9272	0.8981	0.8980

5.4.2. Meta-Path Embedding Adjustment

From the above component adjustment experiment, it can be observed that the effect of the first source of model improvement does not result in a significant effect. Our analysis indicates that the reason for this is that the embedding of the selected meta-path from the dataset becomes saturated. In order to verify this hypothesis, the simulation experiment described in this section was designed to simulate whether the GCN can simulate the embedding of meta-paths in the case of insufficient meta-paths in the dataset (i.e., the embedding of meta-paths is not saturated), expand the range of feature distribution for the meta-paths embedded in the sample, and improve the recommendation effect. The design of the analog element path ablation experiment is described below.

Assuming that each dataset only provides one meta-path, a comparative experiment was conducted to investigate whether GCN can be used to simulate the embedding of meta-paths without considering the weight aggregation of GAT embedding. This experiment adopted a unified meta-path; the type of meta-path was User-Item-User and the training rate was 80%. The model that does not use GCN to simulate the embedding of meta-paths is named HGE-CRec$_{None-OneMP}$, the model embedded by GCN analog meta-path is named HGE-CRec$_{GCN-OneMP}$, and the results of the analog element path ablation experiment for the four datasets are shown in Table 9.

Table 9. Results of meta-path embedding enhancement.

Datasets	Metrics	HGE-CRec$_{None-OneMP}$	HGE-CRec$_{GCN-OneMP}$
Scholat	MAE	0.4559	0.4533
	RMSE	0.6719	0.6604
CNPC	MAE	0.8808	0.8795
	RMSE	1.1784	1.1682
Yelp	MAE	0.8804	0.8529
	RMSE	1.1607	1.1283
Movielens	MAE	0.7136	0.7112
	RMSE	0.9360	0.9298

The results of this experiment verify that in the case of insufficient meta-paths the use of a graph convolution network to aggregate neighbor information to generate more analog meta-path embedding can improve the experimental results. The method of adding small sample data to GCN can generate more embedded data for recommendation, improving the accuracy of small-sample training data in the final recommendation process. This further shows that the HGE-CRec proposed in this paper can be applied to datasets with insufficient meta-paths. In addition, the traditional predefined meta-path depends on the manual definition provided by researchers. In order to find a better predefined meta-path, a large number of experiments need to be carried out. HGE-CRec effectively expands the range of the feature distribution of the meta-path embedding sample by using GCN to generate more simulated meta-path embedding. For cases with only a small number of meta-paths, the best effect that can be achieved by manually predefined meta-paths can be obtained. From this point of view, the proposed model is conducive to relieving the complexity and uncertainty of manually predefined meta-paths.

6. Conclusions

This paper proposes a course recommendation model, HGE-CRec, based on heterogeneous graph and Graph Neural Network and adopts a heterogeneous graph approach to model multi-source heterogeneous information in the scenario of course recommendation for distance education. The proposed HGE-CRec model uses GCN and GAT to improve the meta-path embedding process in HG. Specifically, after using the skip-gram algorithm to generate the original meta-path embeddings, GCN is used to generate new embeddings to simulate more meta-path embeddings, effectively expanding the range of feature distribution of the meta-path embedded samples. Next, GAT is used to learn the neighborhood contribution degree, aggregate the neighbor embedding information of nodes, and obtain the node embedding vector with importance semantics. The node embedding vectors of different meta-paths are weighted nonlinearly and aggregated to form the low dimensional embedding vectors of each node, and the recommendation is implemented under the framework of matrix decomposition.

Data sparsity is a difficult problem in large-scale online course recommendation. The model proposed in this paper uses a heterogeneous graph structure to represent online course learning and expands the relationship domain of entities (e.g., students, courses) through the meta-paths, thereby alleviating the problem of data sparsity. This work can expand recommendation methods for large-scale online courses from a local perspective to a global perspective, providing a reference for further optimization of online course platforms. On the whole, our proposed model fully considers the different contributions of different neighbors and distinguishes the weight of different neighbor embeddings in order to better model the heterogeneous graph and improve the recommendation effect. Moreover, traditional meta-path generation methods rely on researchers manually predefining the meta-path. In order to find a better meta-path, repeated experiments need to be carried out. Using graph convolution network to generate new embeddings for simulation of more meta-paths might be a good way to solve this problem, as defining only a few meta-paths can already achieve a better effect than that achieved by manually predefining the meta-path. In this way, the problem of tedious and uncertain manual adjustment of predefined meta-paths can be solved. In our future work, we intend to make our model more flexible and extensible. For example, by considering the prevalence of streaming education as a global trend [32], this model can be adjusted to recommend interdisciplinary and cross-domain courses by adding attribute variables to generate meta-paths in the model. In addition, the extension of our framework into an advanced version with further consideration of the multi-modal information of courses and the social network behavior of learners [33] could potentially enrich the knowledge concept representation.

Author Contributions: Conceptualization, Z.W. and Z.Z.; Software, Q.L.; Data curation, Q.L.; Writing—original draft, Z.W. and Q.L.; Writing—review and editing, Z.Z.; Funding acquisition, Z.Z. All authors have read and agreed to the published version of the manuscript.

Funding: This work is supported by the National Natural Science Foundation of China (62277018; 62237001), the China Ministry of Education Project in the Humanities and Social Sciences (22YJC880106), and the Major Project in the Social Sciences of South China Normal University (ZDPY2208).

Institutional Review Board Statement: Not applicable.

Informed Consent Statement: Not applicable.

Data Availability Statement: Not applicable.

Conflicts of Interest: The authors declare no conflict of interest.

References

1. Li, Y.; Chen, D.; Zhan, Z. Research on personalized recommendation of MOOC resources based on ontology. *Interact. Technol. Smart Educ.* **2022**, *19*, 422–440. [CrossRef]
2. Lin, Y.; Feng, S.; Lin, F.; Zeng, W.; Liu, Y.; Wu, P. Adaptive course recommendation in MOOCs. *Knowl. Based Syst.* **2021**, *224*, 107085. [CrossRef]
3. Tian, X.; Liu, F. Capacity Tracing-Enhanced Course Recommendation in MOOCs. *IEEE Trans. Learn. Technol.* **2021**, *14*, 313–321. [CrossRef]
4. Zhu, Y.; Lu, H.; Qiu, P.; Shi, K.; Chambua, J.; Niu, Z. Heterogeneous teaching evaluation network based offline course recommendation with graph learning and tensor factorization. *Neurocomputing* **2020**, *415*, 84–95. [CrossRef]
5. Wang, C.; Peng, C.; Wang, M.; Yang, R.; Wu, W.; Rui, Q.; Xiong, N.N. CTHGAT: Category-aware and Time-aware Next Point-of-Interest via Heterogeneous Graph Attention Network. In Proceedings of the 2021 IEEE International Conference on Systems, Man, and Cybernetics, SMC 2021, Melbourne, Australia, 17–20 October 2021; pp. 2420–2426.
6. Shi, C.; Hu, B.; Zhao, W.X.; Yu, P.S. Heterogeneous Information Network Embedding for Recommendation. *IEEE Trans. Knowl. Data Eng.* **2019**, *31*, 357–370. [CrossRef]

7. Morsomme, R.; Alferez, S.V. Content-based Course Recommender System for Liberal Arts Education. In Proceedings of the 12th International Conference on Educational Data Mining, EDM 2019, Montréal, QC, Canada, 2–5 July 2019.

8. Chau, H.; Barria-Pineda, J.; Brusilovsky, P. Content Wizard: Concept-Based Recommender System for Instructors of Programming Courses. In Proceedings of the Adjunct Publication of the 25th Conference on User Modeling, Adaptation and Personalization, UMAP 2017, Bratislava, Slovakia, 9–12 July 2017; pp. 135–140.

9. Li, X.; Li, X.; Tang, J.; Wang, T.; Zhang, Y.; Chen, H. Improving Deep Item-Based Collaborative Filtering with Bayesian Personalized Ranking for MOOC Course Recommendation. In Proceedings of the Knowledge Science, Engineering and Management-13th International Conference, KSEM 2020, Hangzhou, China, 28–30 August 2020; Proceedings, Part I; Li, G., Shen, H.T., Yuan, Y., Wang, X., Liu, H., Zhao, X., Eds.; Springer: Berlin/Heidelberg, Germany, 2020; Volume 12274, pp. 247–258.

10. Madani, Y.; Erritali, M.; Bengourram, J.; Sailhan, F. Social Collaborative Filtering Approach for Recommending Courses in an E-learning Platform. In *Proceedings of the 10th International Conference on Ambient Systems, Networks and Technologies (ANT 2019)/The 2nd International Conference on Emerging Data and Industry 4.0 (EDI40 2019)/Affiliated Workshops, Leuven, Belgium, 29 April–2 May 2019*; Shakshuki, E.M., Yasar, A., Eds.; Elsevier: Amsterdam, The Netherlands, 2019; Volume 151, pp. 1164–1169.

11. Chang, P.; Lin, C.; Chen, M. A Hybrid Course Recommendation System by Integrating Collaborative Filtering and Artificial Immune Systems. *Algorithms* **2016**, *9*, 47. [CrossRef]

12. Ibrahim, M.E.; Yang, Y.; Ndzi, D.L.; Yang, G.; Al-Maliki, M. Ontology-Based Personalized Course Recommendation Framework. *IEEE Access* **2019**, *7*, 5180–5199. [CrossRef]

13. Huang, C.; Chen, R.; Chen, L. Course-recommendation system based on ontology. In Proceedings of the International Conference on Machine Learning and Cybernetics, ICMLC 2013, Tianjin, China, 14–17 July 2013; pp. 1168–1173.

14. George, G.; Lal, A.M. Review of ontology-based recommender systems in e-learning. *Comput. Educ.* **2019**, *142*, 103642. [CrossRef]

15. Núñez-Valdéz, E.R.; Lovelle, J.M.C.; Martínez, O.S.; García-Díaz, V.; de Pablos, P.O.; Marín, C.E.M. Implicit feedback techniques on recommender systems applied to electronic books. *Comput. Hum. Behav.* **2012**, *28*, 1186–1193. [CrossRef]

16. Tong, Y.; Zhan, Z. An evaluation model based on procedural behaviors for predicting MOOC learning performance: Students' online learning behavior analytics and algorithms construction. *Interact. Technol. Smart Educ.* **2023** , *1*, 1–22. [CrossRef]

17. Hew, K.F.; Hu, X.; Qiao, C.; Tang, Y. What predicts student satisfaction with MOOCs: A gradient boosting trees supervised machine learning and sentiment analysis approach. *Comput. Educ.* **2020**, *145*, 103724. [CrossRef]

18. Huang, L.; Wang, C.; Chao, H.; Lai, J.; Yu, P.S. A Score Prediction Approach for Optional Course Recommendation via Cross-User-Domain Collaborative Filtering. *IEEE Access* **2019**, *7*, 19550–19563. [CrossRef]

19. da Silveira Dias, A.; Wives, L.K. Recommender system for learning objects based in the fusion of social signals, interests, and preferences of learner users in ubiquitous e-learning systems. *Pers. Ubiquitous Comput.* **2019**, *23*, 249–268. [CrossRef]

20. Dahdouh, K.; Oughdir, L.; Dakkak, A.; Ibriz, A. Smart Courses Recommender System for Online Learning Platform. In Proceedings of the 5th IEEE International Congress on Information Science and Technology, CiSt 2018, Marrakech, Morocco, 21–27 October 2018; pp. 328–333.

21. Nabizadeh, A.H.; Gonçalves, D.; Gama, S.; Jorge, J.A.; Rafsanjani, H.N. Adaptive learning path recommender approach using auxiliary learning objects. *Comput. Educ.* **2020**, *147*, 103777. [CrossRef]

22. Lu, J.; Behbood, V.; Hao, P.; Zuo, H.; Xue, S.; Zhang, G. Transfer learning using computational intelligence: A survey. *Knowl. Based Syst.* **2015**, *80*, 14–23. [CrossRef]

23. Zhang, Q.; Wu, D.; Lu, J.; Liu, F.; Zhang, G. A cross-domain recommender system with consistent information transfer. *Decis. Support Syst.* **2017**, *104*, 49–63. [CrossRef]

24. Shi, C.; Li, Y.; Zhang, J.; Sun, Y.; Yu, P.S. A Survey of Heterogeneous Information Network Analysis. *IEEE Trans. Knowl. Data Eng.* **2017**, *29*, 17–37. [CrossRef]

25. Shi, C.; Kong, X.; Huang, Y.; Yu, P.S.; Wu, B. HeteSim: A General Framework for Relevance Measure in Heterogeneous Networks. *IEEE Trans. Knowl. Data Eng.* **2014**, *26*, 2479–2492. [CrossRef]

26. Yu, X.; Ren, X.; Sun, Y.; Sturt, B.; Khandelwal, U.; Gu, Q.; Norick, B.; Han, J. Recommendation in heterogeneous information networks with implicit user feedback. In Proceedings of the 7th ACM International Conference on Recommender Systems, Hong Kong, China, 12–16 October 2013; pp. 347–350.

27. Shi, C.; Zhang, Z.; Luo, P.; Yu, P.S.; Yue, Y.; Wu, B. Semantic Path based Personalized Recommendation on Weighted Heterogeneous Information Networks. In Proceedings of the 24th ACM International Conference on Information and Knowledge Management, CIKM 2015, Melbourne, Australia, 19–23 October 2015; pp. 453–462.

28. Yu, X.; Ren, X.; Sun, Y.; Gu, Q.; Sturt, B.; Khandelwal, U.; Norick, B.; Han, J. Personalized entity recommendation: A heterogeneous information network approach. In Proceedings of the 7th ACM International Conference on Web Search and Data Mining, New York, NY, USA, 24–28 February 2014; pp. 283–292.

29. Wang, X.; Wang, Y.; Ling, Y. Attention-Guide Walk Model in Heterogeneous Information Network for Multi-Style Recommenda-tion Explanation. In Proceedings of the 34th AAAI Conference on Artificial Intelligence, New York, NY, USA, 7–12 February 2020; pp. 6275–6282.

30. Salakhutdinov, R.; Mnih, A. Probabilistic Matrix Factorization. In Proceedings of the 21st Advances in Neural Information Processing Systems, Vancouver, BC, Canada, 3–6 December 2007; pp. 1257–1264.

31. Ma, H.; Zhou, D.; Liu, C.; Lyu, M.R. Recommender systems with social regularization. In Proceedings of the 4th International Conference on Web Search and Web Data Mining, Hong Kong, China, 9–12 February 2011; pp. 287–296.

32. Zhan, Z.; Shen, W.; Xu, Z.; Niu, S.; You, G. A bibliometric analysis of the global landscape on STEM education (2004–2021): Towards global distribution, subject integration, and research trends. *Asia Pac. J. Innov. Entrep.* **2022**, *16*, 171–203. [CrossRef]

33. Zhan, Z.; Mei, H.; Liang, T.; Huo, L.; Bonk, C.; Hu, Q. A longitudinal study into the effects of material incentives on knowledge-sharing networks and information lifecycles in an online forum. *Interact. Learn. Environ.* **2021**, *3*, 1–14. [CrossRef]

Article

A Novel Method for Cross-Modal Collaborative Analysis and Evaluation in the Intelligence Era

Wenyan Wu [1], Qintai Hu [1,*], Guang Feng [1] and Yaxuan He [2]

[1] School of Computer Science and Technology, Guangdong University of Technology, Guangzhou 510006, China
[2] Center of Campus Network and Modern Educational Technology, Guangdong University of Technology, Guangzhou 510006, China
* Correspondence: huqt8@gdut.edu.cn

Abstract: The development of intelligent information technology provides an effective way for cross-modal learning analytics and promotes the realization of procedural and scientific educational evaluation. To accurately capture the emotional changes of learners and make an accurate evaluation of the learning process, this paper takes the evaluation of learners' emotional status in the behavior process as an example to construct an intelligent analysis model of learners' emotional status in the behavior process, which provides effective technical solutions for the construction of cross-modal learning analytics. The effectiveness and superiority of the proposed method are verified by experiments. Through the analysis of learners' cross-mode learning behavior, it innovates the evaluation method of classroom teaching in the intelligent age and improves the quality of modern teaching.

Keywords: educational evaluation; artificial intelligence; cross-modal collaborative analysis

Citation: Wu, W.; Hu, Q.; Feng, G.; He, Y. A Novel Method for Cross-Modal Collaborative Analysis and Evaluation in the Intelligence Era. *Appl. Sci.* **2023**, *13*, 163. https://doi.org/10.3390/app13010163

Academic Editor: Gaetano Zizzo

Received: 19 October 2022
Revised: 18 December 2022
Accepted: 20 December 2022
Published: 23 December 2022

1. Introduction

1.1. Educational Evaluation in the Age of Intelligence

With the development of science and information technology, contemporary education has undergone a series of reforms and innovations with a focus on improving the quality of education rather than increasing the level of its scale and speed. Establishing a high-quality educational system has become the aim for the country and society as a whole.

In recent years, technologies such as big data, the Internet of Things, artificial intelligence, short video, virtual reality (VR), augmented reality (AR), and the metaverse have been gradually applied to the field of education. The upgraded level of information technology accelerates reform and innovation in education and profoundly affects the connotation, method, and path of educational evaluation, making educational evaluation transfer from a result-oriented, subjective direction to a process-oriented, concomitant, and scientific direction. There is an emphasis on the innovation of evaluation tools and the application of modern information technologies such as artificial intelligence and big data to explore and carry out the longitudinal evaluation of students' learning at all grades, such as the horizontal evaluation of moral, intellectual, physical, aesthetic, and labor factors.

In accordance with the national aim and requirements of education, which lie in the scientific, procedural, and intelligent standards of educational evaluation, and with the use of advanced technologies in information and artificial intelligence, this paper adopts an innovative teaching and learning evaluation model that focuses on intelligence education evaluation methods. It is deemed essential for the development of education reform in terms of the integration of information technology, intelligent technology, and educational philosophy.

1.2. Proposition of Research Questions

Education in the age of intelligence emphasizes the cultivation of students' abilities in terms of innovation, interdisciplinary problem-solving, teamwork, and other comprehensive qualities through the process of hands-on practice. The biggest difference compared with traditional education is that it focuses more on the process of learning than the ultimate results. The emotional status generated by students during the learning process can, to some extent, reflect their learning interests and the learning effects, and therefore form the mechanism of learning eventually. The status can be recorded and analyzed through cross-modal data such as facial expressions, actions, language, and heartbeat.

Cross-modal collaborative analysis and evaluation can provide an effective technical solution for the construction of a teaching evaluation system in the Age of Intelligence. Thus, the data analysis and conclusions from the evaluation system can shed some light on research questions such as how to improve the quality of intelligent education.

In the smart education environment, based on technologies such as the Internet of Things and artificial intelligence, as well as intelligent sensing devices such as eye trackers and electroencephalographs, it is possible to monitor and record complex teaching processes in real-time and form cross-modal data such as audios, texts, images, videos, etc. Based on cross-modal learning analysis technology, collaborative analysis of student learning behavior data is carried out to mine and restore the students' learning emotional status information implied by the cross-modal and accompanying behavioral information in the classroom. A scientific model further reveals the implicit relationship between teaching and learning process information and classroom education quality, and a new evaluation model of classroom teaching and learning has been constructed. This evaluation method is more comprehensive and scientific by eliminating the weaknesses of the traditional educational evaluation model, which emphasized tests and grades, and by optimizing teaching methods based on evaluation and improving the quality of classroom teaching. The purposed method focuses on evaluating and developing students' innovative thinking ability, comprehensive quality, personality, and mental health.

Taking the cross-modal learning behavior data as the carrier, with technologies such as artificial intelligence, big data, etc., this research collects students' emotional status through the whole process of learning, conducts a collaborative analysis of learning behavior, and thus constructs a new evaluation method of teaching that links learning behavior with learners' emotional status. This research indicates significance in research and real-world practice in terms of achieving a comprehensive and accurate assessment of a student's ability, cognitive level, personality traits, mental health, etc., explaining and exploring cross-modal data-driven educational phenomena and educational laws, offering guidance on the optimization of the teaching process, and improving the goal of classroom teaching quality.

2. Literature Review

Research on cross-modal learning analysis could be dated back a decade, and the academic community continues to focus on its development. In 2012, scholars raised the idea of applying cross-modal data such as text, video, and audio in the field of cross-modal interaction research, a new direction for the research of cross-modal learning interaction, which put cross-modal learning as the core status of educational evaluation. In 2012, the International Conference on Cross-Modal Interaction (ICMI) established the frontier status of cross-modal learning analysis by organizing relevant scholars to participate in the seminar on "Cross-Modal Learning Analysis" for the first time in the form of workshops [1], declaring that the aim of this research field is to combine cross-modal analysis technology with learning science research to help researchers better understand the mechanism of learning. Since 2014, with the gradual development of blended learning, learning analysis, and big data technologies, a large amount of online and offline learning data have become available for identifying patterns such as learning emotions by extracting and analyzing cross-modal feature data. In 2016, the International Conference on Learning Analysis and Knowledge (LAK) set up a data challenge workshop for cross-modal learning and analysis, organized relevant scholars to participate in the practical work of cross-modal data analysis, and explored the development direction of learning analysis research supported by cross-modal data [2]. The application of artificial intelligence technology has been widely spread in cross-modal learning analysis since 2018. By applying artificial intelligence to education, Li Songqing [3] proposed to use cross-modal machine learning to create and replicate human cognition, analyze and design artificial artifacts based on artificial intelligence technology through cross-modal learning, and ultimately support, help, and expand human cognitive abilities. The American Educational Research Association (AERA) annual meeting in 2019 proposed to use cross-modal data such as text, charts, audio, and video to construct a research paradigm based on cross-modal narratives to seek evidence of fairness in educational research [4].

In terms of specific research content, Hu Qintai et al. [5] analyzed the interpretability problem of cross-modal learning behavior and used the HDRBM (Hybrid Deep Restricted Boltzmann Machine), a deep neural network model, to obtain the implicit physiological and emotional characteristics of students. This research used the Bayesian network and the junction tree algorithm to analyze the interpretability of the results, which has a certain effect on improving the interpretability of learning behavior analysis. From a cross-modal measurement perspective, Kyllonen et al. [6] designed and developed a computational model framework for assessing learning status. This framework is capable of capturing, analyzing, and measuring complex human behavior as well as analyzing noisy, unstructured, and cross-modal data. The analysis process utilizes cross-modal data, including audio, video, and activity log files, to construct an Analytic Hierarchy Process for modeling the temporal dynamics of human behavior. Riquelme et al. [7] used transponders to collect students' voice data to analyze the continuity and motivation of group collaboration. Poria et al. [8] used an ensemble feature extraction method to develop a new cross-modal information extraction agent. In particular, the developed method exploited joint tri-modal features (text, audio, and video) to infer and aggregate semantic and emotional information on user-generated cross-modal data. In terms of analysis methods, Scherer [9] used a cross-modal sequence classifier to analyze the expressions of laughter in the process of multi-party dialogue in the natural environment. At the same time, multi-dimensional data channels are used to extract the frequency and spectral features from audio streams and motion-related behavioral features from video streams. On this basis, the Hidden Markov Model and the Echo State Network are used to identify the paralinguistic behavior in the communication process, and a classification model with a certain level of accuracy is obtained. Researchers from the Norwegian University of Science and Technology collected cross-modal data of learners in adaptive learning activities and used the fuzzy set qualitative comparative analysis (fsQCA) method to describe the relationship between learner participation patterns and learning performance [10]. Vicente et al. proposed a Wearable Internet of Things in Education (WIoTED) system based on IoT technology and real-time monitoring data from wearable devices and used machine learning techniques and cross-modal learning analysis methods to build a model that could "explain" student engagement. This study selects a decision tree and rule system based on a set of correlated variables, and the obtained rules could be easily interpreted by non-professionals [11].

Overall, the research on cross-modal learning analysis has a rapid speed of development in spite of its comparatively short history. In recent years, it has received extensive attention from academic groups across different disciplines. The evolutionary process covers knowledge model and framework design, cross-modal feature extraction and sentiment analysis, cross-modal representation learning, and in-depth learning. Its research ideas and research results have further enriched the fields of learning analysis and learning science. Major research teams in China are from normal universities such as Beijing Normal University, East China Normal University, and South China Normal University. By analyzing cross-modal learning from the perspective of learning science, the research mainly focuses on theoretical discussion (Such as Mu Su [12], Wang Weifu [13]) and framework construction (Such as Zhou Jin [14], Zhang Qi [15], Mu Zhijia [16], Li Qing [17], etc.).

Previous studies indicate that cross-modal learning analysis is an effective method for analyzing learning behavior and obtaining learners' status. However, current methods rarely apply theories to practice from the perspectives of computer science and data science. Meanwhile, the accuracy, comprehensiveness, and interpretability of analytical models need to be improved. Based on deep learning and cross-modal learning, we established a multi-modal sentiment analysis model for learning behavior, and the robustness of the analytical model was improved using deep learning and pre-training, implying an important research trend.

3. Research Method

3.1. Problem Definition

This paper processes and analyzes multi-modal data (visual, text, audio, etc.) through in-depth learning algorithms such as neural networks and attention mechanisms. It also constructs an intelligent analysis model of the emotional status of the learner's learning process, which can accurately capture the emotional changes produced by learners during the learning process for collaborative analysis and evaluation.

3.2. Model Architecture

By studying cross-modal data processing, cross-modal data fusion, pre-training model construction, and collaborative analysis of learner learning status, we propose an architecture to identify the mechanism of deep inquiry learning occurrence and improve the accuracy of students' emotional status analysis during the learning process. We term our architecture 'BLBA-MODEL' (Bi-LSTM and Bi-Attention Mechanism Model). Its main framework is shown in Figure 1, which can be divided into three modules: data feature extraction, data fusion and parsing, and emotional status assessment of the learning process. These three modules are illustrated as follows.

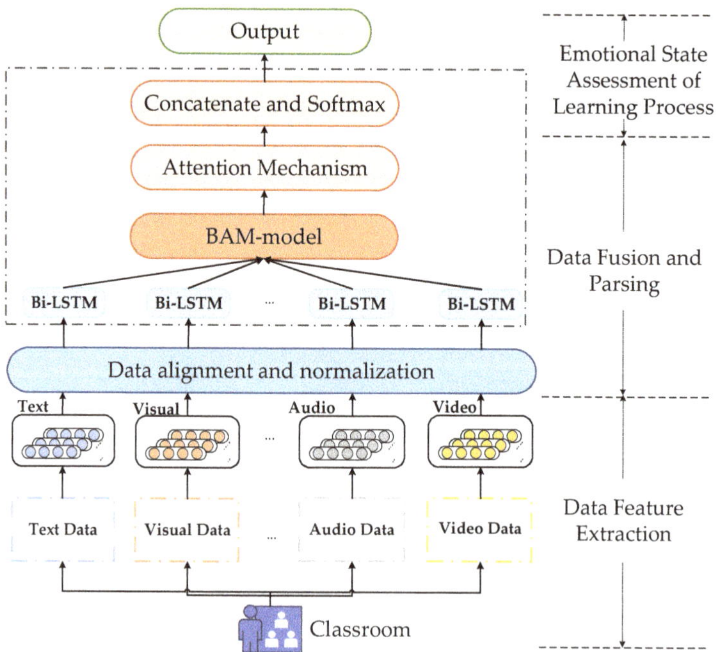

Figure 1. Overall architecture diagram of a cross-modal collaborative analysis model.

3.2.1. Representation, Normalization, and Alignment of Cross-Modal Learning Behavioral Data

The data in the learning process includes multiple modalities such as video, text, and audio. In multi-modal data, each modality provides specific information for other modalities, and there is a certain correlation between the modalities. In this paper, the text, audio, and visual in the video of the learner's learning process are used as data for processing, and the two techniques of batch normalization and layer normalization are used to remove the dimension and realize the normalization of the training data. Formula (1) is a specific representation of normalization. For input z_i^l, after calculating its mean μ and variance σ^2, normalization is performed, where γ and β represent the learnable parameter variables of scaling and translation, respectively. For each neuron, before the data enters the activation function, the mean and variance of each batch are calculated along the channel so that the data maintains a normal distribution with a mean of 0 and a variance of 1 to avoid the disappearance of the gradient.

$$\tilde{z}_i^l = \frac{z_i^l - \mu}{\sqrt{\sigma^2 + o^l}} \times \gamma + \beta \tag{1}$$

The aligned hidden vectors are spliced as the input of the subsequent module; it is the LSTM-Attention (Long Short-Term Memory and Attention Mechanism) combination module. The technical route of characterization, normalization, and alignment of cross-modal streaming media data is shown in Figure 2.

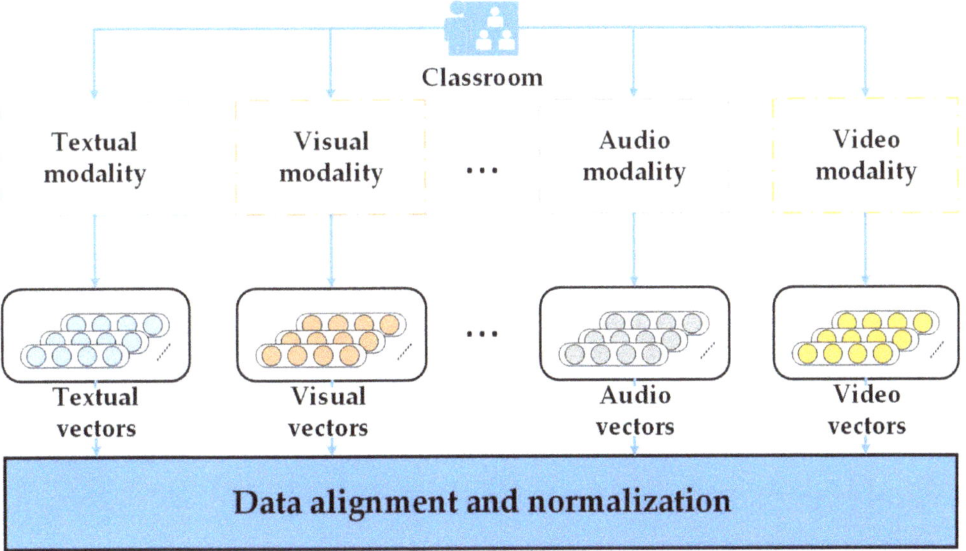

Figure 2. Representation, normalization, and alignment of cross-modal learning behavioral data.

3.2.2. Fusion and Analysis of Cross-Modal Learning Behavior Data

After normalizing and aligning the cross-modal data, the information interaction within and between modalities is identified by fusing the feature representations of various modal information. This paper proposes the LSTM-Attention combination module to realize this kind of information interaction and takes the interaction of three modalities (audio, text, and visual) in the video as an example to illustrate in detail. Since we only used text, audio, and visual as our inputs for subsequent experiments, we will no longer regard video as a separate input.

3.2.2.1. Unimodal Feature Information Interaction

Considering the video information as a set of several utterances, we use multiple independent Bi-LSTMs (Bi-Directional Long Short-Term Memory) to capture the context-related semantic information of each modality, and the output h_{ij} after the Bi-LSTM layer can be obtained.

$$h_{ij} = \overrightarrow{LSTM}\left(x_{ij}, h_{j-1}\right) \oplus \overleftarrow{LSTM}\left(x_{ij}, h_{j-1}\right) \tag{2}$$

Among them, x_{ij} represents the input feature of utterance j in video i, h_{j-1} represents the hidden layer state of utterance $j-1$, and h_{ij} represents the output of Bi-LSTM layer.

Then the matrix of video i after the Bi-LSTM layer is expressed as:

$$H_i^{S_m} = \left[h_{ij} : j \leq L_i\right], \ H_i^{S_m} \in \mathbb{R}^{q \times d} \tag{3}$$

where d represents the feature dimension.

Through this layer processing, three modal feature representations with context-related information in video i can be obtained: textual feature $H_i^{S_t}$, audio feature $H_i^{S_a}$, and visual feature $H_i^{S_v}$.

3.2.2.2. Bimodal Feature Information Interaction

After obtaining the above unimodal feature representation, three modalities are combined in pairs (text + audio, audio + visual, and text + visual) to obtain three sets of modal pairs. Next, we proposed a BAM-model (Bimodal Attention Mechanism model) to capture the interaction information between modalities and contexts. The model structure is shown in Figure 3. The specific process and formula are as follows:

The first step is to multiply the feature representation matrices obtained in Section 3.2.2.1 to obtain the cross-modal information matrix, where $u, w = \{t, a, y\}$:

$$C_i^{uw} = H_i^{S_u} \cdot H_i^{S_w T} \tag{4}$$

$$C_i^{wu} = H_i^{S_w} \cdot H_i^{S_u T}, \tag{5}$$

The SoftMax function (SoftMax logical regression) is used to model multiclass classification problems where we want the output to be a probability distribution of the different possible classes. The second step is to demonstrate a SoftMax calculation on the cross-modal information matrix to obtain the attention score:

$$\alpha_i^{uw} = softmax(C_i^{uw}), \ \alpha_i^{uw} \in \mathbb{R}^{q \times q} \tag{6}$$

$$\alpha_i^{wu} = softmax(C_i^{wu}), \ \alpha_i^{wu} \in \mathbb{R}^{q \times q}, \tag{7}$$

In the third step, the attention score is multiplied by the feature matrix, and the modal feature representations with information interaction are obtained, respectively:

$$H_i^{d1u} = \left(\alpha_{uv} H_i^u\right) \cdot H_i^u, \ H_i^{d1u} \in \mathbb{R}^{q \times d} \tag{8}$$

$$H_i^{d1w} = \left(\alpha_{wu} H_i^w\right) \cdot H_i^w, \ H_i^{d1w} \in \mathbb{R}^{q \times d}, \tag{9}$$

After the above operations, text feature representations $H_i^{d1_t}$, $H_i^{d3_t}$, audio feature representations $H_i^{d1_a}$, $H_i^{d2_a}$, and visual feature representations $H_i^{d2_v}$, $H_i^{d3_v}$ can be obtained, respectively.

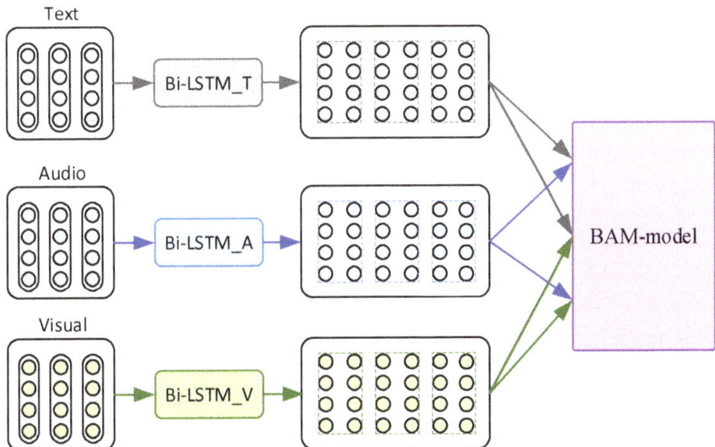

Figure 3. BAM-model.

3.2.3. Assessment of Learning Emotional Status

The feature representation of each modality with contextual interaction information can be obtained through the module processing in the first two sections. In this section, the attention mechanism is used to weigh the importance of each modality of information to filter out redundant information, classify emotions according to the final fusion information and obtain students' emotional status throughout the whole process of learning. The specific process is as follows:

First, we use the fully connected layer to splice the above modal features and obtain the feature representation to enter the attention mechanism to filter redundant information:

$$R_i^t = \tanh\left(W_i^t\left[H_i^{S_t} \oplus H_i^{d1_t} \oplus H_i^{d3_t}\right] + b_i^t\right) \tag{10}$$

$$R_i^a = \tan h\left(W_i^a\left[H_i^{S_a} \oplus H_i^{d1_a} \oplus H_i^{d2_a}\right] + b_i^a\right), \tag{11}$$

$$R_i^v = \tan h\left(W_i^v\left[H_i^{S_v} \oplus H_i^{d2_v} \oplus H_i^{d3_v}\right] + b_i^v\right), \tag{12}$$

where W_i^m is the weight and b_i^m is the bias.

Then, the feature representation R_i^m of each modality is sent to the attention mechanism model, and the multi-modal fusion feature representation is obtained by weighted summation:

$$\Upsilon_m = softmax(\tan h(W_{att}^m R_i^m + b_{att}^m)) \tag{13}$$

$$R_i^{m*} = \sum_m R_i^m \Upsilon_m^{\ T}, \tag{14}$$

Among them, W_{att}^m represents the weight, b_{att}^m represents the bias, and Υ_m represents the normalized weight.

Finally, through the fully connected layer and using the SoftMax function for emotion prediction and classification, the final original learned emotion and cognitive state of the model are obtained:

$$y_i = softmax\left(W_q\cdot\left(\tanh\left(W_p\cdot R_i^{m*} + b_p\right)\right) + b_q\right) \tag{15}$$

Among them, W_p and b_p are the weights and biases of the SoftMax layer, and W_q and b_q are the weights and biases of the fully connected layer.

4. Results and Discussion

This section systematically analyzes the effect of the proposed model on the multi-modal sentiment analysis task.

4.1. Experimental Datasets

Since the data set for students' emotional behavior in the learning process involves issues such as students' privacy, no data set has been found as an experimental data set for targeted model training and analysis. Therefore, this paper uses the universal public data set CMU-MOSEI to conduct experiments. On the one hand, it can verify the sentiment classification performance of the model. On the other hand, it can expand the scope of application of the model and facilitate subsequent targeted work.

The dataset was collected from 1000 speakers and contains a total of 3228 videos with 23,453 annotated sentences and 250 different topics. In this dataset, males and females make up about 57% and 43%. We divide the dataset into a training set and a test set; the details are shown in Table 1. The dataset has six different annotations for sentiment, namely Happiness, Sadness, Anger, Disgust, Surprise, and Fear. For classification prediction, videos with sentiment values greater than or equal to 0 are marked as positive sentiment classifications, and videos with sentiment values less than 0 are marked as negative sentiment classifications.

Table 1. Training set and test set of the CMU-MOSEI dataset.

CMU-MOSEI Dataset	Videos	Number of Utterances	Positive Sentiment	Negative Sentiment
Training set	2250	16146	7869	8277
test set	679	4634	2541	2093

4.2. Ablation Experiment

This paper divides the feature input into different combinations to verify the importance of multi-modal information and performs sentiment classification prediction, respectively. This paper uses ACC (accuracy) and $F1$ indexes to evaluate algorithm performance:

$$Accuracy = (TP + TN)/(TP + FP + TN + FN) \tag{16}$$

$$F1 = 2TP/(2TP + FP + FN), \tag{17}$$

where TP is true positive, TN is true negative, FP is false positive, and FN is false negative. The experimental results are shown in Figure 4.

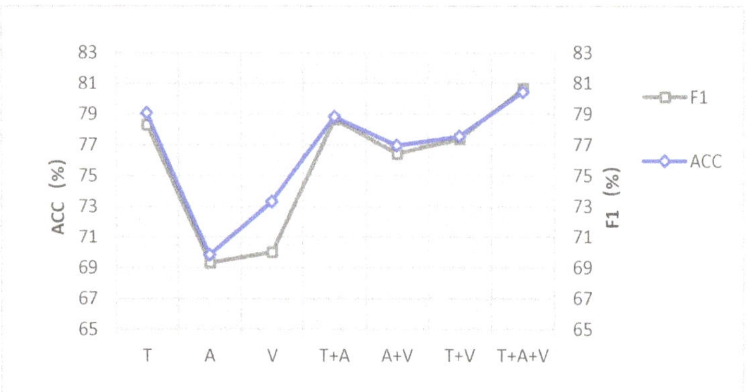

Figure 4. Ablation experiment results.

4.3. Comparative Experiment

This paper compares the proposed model with existing multi-modal sentiment analysis models, and we use the same dataset to experiment with these methods. The benchmark methods are as follows:

(1)   SVM-MD: An SVM model using multi-modal features trained by early fusion.
(2)   LF-LSTM: LSTM network with late fusion method.
(3)   MFN: A neural network structure for multi-view sequential learning based on attention networks and gated memory, which well models the interaction between modalities.
(4)   MARN: The method of combining a multi-attention module and a recurrent neural network is adopted to obtain the interaction information between three modalities through a multi-attention unit, and the recurrent network is used as a memory unit for storage.
(5)   Graph-MFN: Based on the MFN method, a method of combining the dynamic fusion graph with it is introduced; that is, the multi-modal dynamic fusion graph is combined with the context memory fusion network.
(6)   MMMU-BA: A method that utilizes the correlation of contextual information of target video segments between different modalities to assist multi-modal information fusion.

4.4. Experimental Results and Analysis

It can be seen from the ablation experiment results that the performance of text-based sentiment analysis in unimodality is the best. The sentiment analysis performance of all modal combinations in bimodality is generally better than that of the best unimodality analysis type. The trimodal sentiment analysis model performed the best among all models. Its accuracy was increased by 1.36% and 1.61% compared with the best models in the unimodal and bimodal, respectively. Its $F1$ index is improved by 2.33% and 1.98% compared to the best model in unimodal and bimodal, respectively. Therefore, the effective combination of multiple modalities can well obtain sentiment classification prediction and improve the model's performance. Moreover, during the process of students' learning, it can accurately grasp the emotional status of students by capturing the information from multiple modalities, adopting multi-modal emotional analysis, and presenting evaluation results that could provide emotional guidance to students promptly or adjust the difficulty level and teaching mode of the course to help students acquire knowledge much better and achieve the learning aims much easier.

From the experimental data in Table 2, the proposed method has achieved excellent results accuracy and $F1$ index. Compared with early fusion, the performance improvement is the most obvious. Compared with recent more advanced deep learning and other methods, the accuracy and $F1$ index are also 0.17% and 0.95% higher. Therefore, the model can extract meaningful information from multi-modalities for fusion and accurately classify and predict, which is helpful to apply to the scenarios where students learn emotional status analysis in the process of learning.

Table 2. Comparative experimental results of different models.

CMU-MOSEI	Method Model						
	SVM-MD	**LF-LSTM**	**MFN**	**MARN**	**Graph-MFN**	**MMMU-BA**	**BLBA-MODEL**
ACC (%)	68.34	80.61	77.13	76.8	76.9	80.27	80.78
F1 (%)	67.92	79.46	76.35	77.01	77.14	79.23	80.41

5. Conclusions

With the development of intelligence, the single source and simple structure of analytics data in the past have gradually become unsuitable for the era of rapid development.

Based on deep learning, this paper proposes a multi-modal emotional analysis method to obtain people's sentiment status and make accurate evaluations of people's behavior processes. All the above experiments demonstrate this study method's effectiveness and excellent performance. In future work, we will use emotional analysis in intelligent education to build a learning analysis model between different subjects and find out the correlation between them while studying the learning behavior of each subject. We will use different dataset related to the student's learning behavior process to build a complete cross-modal collaborative analysis framework.

Author Contributions: Methodology, W.W.; Formal analysis, Y.H.; Data curation, G.F.; Writing—original draft, W.W.; Supervision, Q.H. All authors have read and agreed to the published version of the manuscript.

Funding: Supported by National Natural Science Foundation of China, 2022, "Collaborative analysis and evaluation method of cross-modal knowledge elements of classroom streaming media"(Grant No. 62237001).

Institutional Review Board Statement: Not applicable.

Informed Consent Statement: Not applicable.

Data Availability Statement: Publicly available dataset was analyzed in this study. The dataset can be found here: http://multicomp.cs.cmu.edu/resources/cmu-mosei-dataset/ (accessed on 19 December 2022). The dataset is available for download through CMU Multimodal Data SDK GitHub: https://github.com/A2Zadeh/CMU-MultimodalDataSDK (accessed on 19 December 2022).

Conflicts of Interest: The authors declare no conflict of interest.

References

1. Zhong, W.; Li, R.; Ma, X.; Wu, Y. The development trend of learning and analysis technology—Research and exploration under the environment of multi-modal data. *China Distance Educ.* **2018**, *11*, 41–49.
2. Li, X.; Zuo, M.; Wang, Z. Research Status and Future Prospects of Learning Analysis—Review of the 2016 International Conference on Learning Analysis and Knowledge. *Open Educ. Res.* **2017**, *23*, 46–55.
3. Li, S.; Zhao, Q.; Zhou, Z.; Zhang, Y. Cognitive neural mechanism of image and text processing in multimedia learning. *Prog. Psychol. Sci.* **2015**, *23*, 1361–1370.
4. Wu, B.; Peng, X.; Hu, Y. The Way to Eliminate the Turnip and Preserve the Quintessence in Educational Research: From multi-modal Narration to Evidential Fairness—A Review of the American Aera 2019 Conference. *J. Distance Educ.* **2019**, *37*, 13–23.
5. Hu, Q.; Wu, W.; Feng, G.; Pang, T.; Qiu, K. Research on interpretability analysis of multi-modal learning behavior supported by in-depth learning. *Res. Audio Vis. Educ.* **2021**, *42*, 77–83.
6. Kyllonen, P.C.; Zhu, M.; Davier, A.A. *Introduction: Innovative Assessment of Collaboration*; Springer: Berlin/Heidelberg, Germany, 2017.
7. Riquelme, F.; Munoz, R.; Mac Lean, R.; Villarroel, R.; Barcelos, T.S.; de Albuquerque, V.H.C. Using multi-modal learning analytics to study collaboration on discussion groups. *Univers. Access Inf. Soc.* **2019**, *18*, 633–643. [CrossRef]
8. Poria, S.; Cambria, E.; Hussain, A.; Huang, G.B. Towards an intelligent framework for multi-modal affective data analysis. *Neural Netw.* **2015**, *63*, 104–116. [CrossRef] [PubMed]
9. Scherer, S. multi-modal behavior analytics for interactive technologies. *Künstl. Intell.* **2016**, *30*, 91–92. [CrossRef]
10. Papamitsiou, Z.; Pappas, I.O.; Sharma, K.; Giannakos, M.N. Utilizing multi-modal data through fsQCA to explain engagement in adaptive learning. *IEEE Trans. Learn. Technol.* **2020**, *13*, 689–703. [CrossRef]
11. Camacho, V.L.; Guía ED, L.; Olivares, T.; Flores, M.J.; Orozco-Barbosa, L. Data capture and multi-modal learning analytics focused on engagement with a new wearable IoT approach. *IEEE Trans. Learn. Technol.* **2020**, *13*, 704–717. [CrossRef]
12. Mu, S.; Cui, M.; Huang, X. Data Integration Method for Panoramic Perspective multi-modal Learning Analysis. *Res. Mod. Distance Educ.* **2021**, *33*, 26–37.
13. Wang, W.; Mao, M. multi-modal learning analysis: A new way to understand and evaluate real learning. *Res. Audio Vis. Educ.* **2021**, *42*, 25–32.
14. Zhou, J.; Ye, J.; Li, C. Emotional Computing in multi-modal Learning: Motivation, Framework and Suggestions. *Res. Audio Vis. Educ.* **2021**, *42*, 26–32.
15. Zhang, Q.; Li, F.; Sun, J. Multi-modal Learning Analysis: Learning Analysis Towards the Era of Computational Education. *China Audio Vis. Educ.* **2020**, *9*, 7–14.

16. Mou, Z. Multi-modal learning analysis: A new growth point of learning analysis research. *Audio Vis. Educ.* **2020**, *41*, 27–32.
17. Li, Q.; Ren, Y.; Huang, T.; Liu, S.; Qu, J. Research on application of learning analysis based on sensor data. *Res. Audio Vis. Educ.* **2019**, *40*, 64–71.

Article

Dynamic Generation of Knowledge Graph Supporting STEAM Learning Theme Design

Qingchao Ke [1],* and Jian Lin [2]

[1] School of Educational Information Technology, South China Normal University, Guangzhou 510631, China
[2] School of Computer Science and Intelligence Education, Lingnan Normal University, Zhanjiang 524048, China
* Correspondence: keqingchao@m.scnu.edu.cn

Abstract: Instructional framework based on a knowledge graph makes up for the interdisciplinary theme design ability of teachers in a single discipline, to some extent, and provides a curriculum-oriented theme generation path for STEAM instructional design. This study proposed a dynamic completion model of a knowledge graph based on the subject semantic tensor decomposition. This model can be based on the tensor calculation of multi-disciplinary curriculum standard knowledge semantics to provide more reasonable STEAM project-based learning themes for teachers of those subjects. First, the STEAM multi-disciplinary knowledge semantic dataset was generated through the course's standard text and open-source encyclopedia data. Next, based on the semantic tensor decomposition of specific STEAM topics, the dynamic generation of knowledge graphs was realized, providing interdisciplinary STEAM learning topic sequences for teachers of a single discipline. Finally, the application experiment of generating STEAM learning themes proved the effectiveness of our model.

Keywords: dynamic generation; knowledge graph; orthogonal iteration; STEAM instructional design; tensor decomposition

Citation: Ke, Q.; Lin, J. Dynamic Generation of Knowledge Graph Supporting STEAM Learning Theme Design. *Appl. Sci.* **2022**, *12*, 11001. https://doi.org/10.3390/app122111001

Academic Editors: Shan Wang, Zehui Zhan, Qintai Hu and Xuan Liu

Received: 27 September 2022
Accepted: 28 October 2022
Published: 30 October 2022

Publisher's Note: MDPI stays neutral with regard to jurisdictional claims in published maps and institutional affiliations.

1. Introduction

STEAM education involves the integration of multi-disciplinary instructional design with specific themes. However, teachers trained in a single discipline can hardly meet the requirements of STEAM multi-disciplinary theme design [1]. As a multi-disciplinary instructional framework, the subject knowledge graph, with its rich semantic relevance, is expected to solve the problem of STEAM multi-disciplinary theme design. However, the current knowledge graph of disciplines is still unable to provide teachers with appropriate STEAM learning topics according to the instructional requirements of the curriculum standards, which leads to serious disconnection of the knowledge system of various disciplines in STEAM education. Therefore, how to effectively support the generation of appropriate STEAM learning topics through the dynamic cutting and completion of the subject knowledge graph has become an important research question.

In recent years, since the tensor decomposition model has flexible computing capability for potential semantic features, it is regarded by researchers as a key algorithm model for developing the learners' knowledge graph from the basic lower cognitive stages to the higher panoramic cognitive stages [2]. Presently, the dynamic organization effect of multi-disciplinary learning resources on STEAM learning topics is not appropriate enough, and it has not been found that STEAM teachers can be configured with learning topics that meet the learning requirements of the curriculum standards. The application of the tensor decomposition model in the STEAM multi-disciplinary semantic association and its completion of the panoramic knowledge graph provides relatively ideal support for STEAM learning theme content organization. This method helps single-discipline teachers to organize, discover, configure, and recommend appropriate themes of interdisciplinary learning.

This study had three main objectives:

(1) To propose the dynamic generation technology of subject knowledge graph technology, which includes two types of technical solutions: knowledge graph generation based on basic data sets, and STEAM subject knowledge graph completion oriented to curriculum standards in order to achieve flexible STEAM subject generation and design for single-subject teachers.

(2) To integrate multiple APIs to conduct multiple trainings on data sets to meet the requirements for entity recognition and semantic modeling of the subject knowledge graph under different personalized learning scenarios. Among them, the conditional random field and tensor decomposition algorithm achieved the best semantic relationship modeling effect.

(3) To introduce the high-order singular value decomposition with orthogonal iteration (HO-SVD-OI) for the tensor decomposition algorithm of a penalty function and the high-dimensional orthogonal iteration mechanism to calculate the matches between STEAM learning resources and learners' cognitive styles in different stages. The simulation results show that the algorithm optimization strategy had good computational performance on the topic coverage of STEAM learning themes.

2. Related Works

2.1. The Interdisciplinary Instruction of STEAM Project

The research field of STEAM interdisciplinary project teaching, which originated from the design and development of interdisciplinary medical courses in colleges and universities, is very active [3]. Cuadrado and others were the first to introduce the STEAM interdisciplinary teaching field supported by ICT [4]. Many studies have confirmed that interdisciplinary teaching with ICT integration contributes to problem-solving and deep exploration in real situations [5–7]. Chen et al. studied the design and application of theme resources in STEAM interdisciplinary learning scenarios based on empirical research on the application of an interdisciplinary online learning resource recommendation system [8]. Subsequently, interdisciplinary education supported by information technology has been widely used in higher medical [9] and engineering education [10], aiming to cultivate interdisciplinary professionals with good interdisciplinary teamwork and comprehensive discipline quality. However, relevant research, such as knowledge graph and other cognitive intelligence technologies, is still relatively rare in assisting teachers to carry out STEAM theme designs.

2.2. Dynamic Generation and Completion of Knowledge Graph

With the algorithm performance improvement of machine learning and its wide application in the field of natural language processing, some scholars have conducted research on the construction methods of knowledge graphs. Based on intelligent algorithms such as machine learning or deep learning, they have endeavored to dynamically build entity data models, semantic relationship models, and graph quality control methods of the subject knowledge graphs, in order to realize large-scale entity construction and relationship organization of multi-modal unstructured data in the intelligent learning environment [11,12]. The dynamic completion and division of a knowledge graph based on representation learning and tensor decomposition is one of the frontier research fields in the current computer science community for the intelligent and dynamic construction of subject knowledge graphs. Common implementation methods include the model ConMask [13] based on entity embedding, TansH [14], TansR [15], TansE [16] based on the distance translation of neural tensor network structure, etc. However, if the number of entities or their relationship links show an extremely uneven distribution, its completion performance becomes not ideal enough. Some scholars also pay attention to semantic path sorting based on feature space similarity [17], reinforcement learning [18], and small sample learning algorithm [19], and they consider entity-embedded representation and local semantic network sparsity to optimize the completion effect of a knowledge graph. Machine learning makes this possible for large-scale entities of subject knowledge graphs and their semantic intelligent computing, thus promoting the transition of subject knowledge graphs from extensive semantics to precise semantics in subject knowledge fusion and digital education resource organization. However, due to the collaboration of learners' cognitive states, their preferences, and the process changes in multi-disciplinary learning activities, there is still a broad research space to dynamically generate and complete discipline knowledge graphs through machine learning.

2.3. Application of Knowledge Graph in Instructional Design

Presently, in the application of the international education knowledge graph in instructional design, natural language processing [20], data mining [21], visualization [22], deep learning [23], and other artificial intelligence technologies are closely related to its cutting-edge development. Chen et al. used machine learning algorithms to train heterogeneous data, such as teaching content and learning evaluation in the field of education. They proposed an automatic construction method of an education knowledge graph based on neural sequence markers and probabilistic association rule mining [24]. Jia and others implemented a personalized learning recommendation service application based on the semantic similarity calculation of a discipline knowledge graph [25]. Nayyeri and others developed the semantic link prediction by embedding the model of Trans4E in an academic resource knowledge graph based on AIDA, MAG, and other benchmark datasets [26]. Based on the knowledge graph model of science and technology resources, Chi and others built an interdisciplinary, cross-domain, and multi-application academic resource retrieval and analysis platform to help improve the intelligent organization and management of the academic resources in colleges and universities [27].

3. Methodology

This section proposes a dynamic generation method for knowledge graphs of STEAM themes, including three technical processes: the dynamic generation of cross-modal subject knowledge graphs based on natural language processing, dynamic completion of subject knowledge graphs based on the tensor semantic decomposition model, and quality evaluation of subject knowledge graphs for STEAM interdisciplinary curriculum standards.

3.1. Knowledge Graph Generation of STEAM Instructional Themes

The subject knowledge graph is the key cornerstone of interdisciplinary instructional design for realizing the multi-disciplinary knowledge collaboration of STEAM. Machine learning intelligence, with the semantic annotation of a large-scale knowledge corpus, enables teachers of a single subject to design STEAM learning themes that are appropriate for the curriculum standards. Based on our research of the key technologies of knowledge graphs at home and abroad, this study proposed a dynamic generation framework of STEAM subject knowledge graphs based on a multi-algorithm collaboration, as shown in Figure 1.

The data set for constructing a STEAM discipline knowledge graph comes from two channels: (1) unstructured data based on national discipline curriculum standards, digital textbooks, and texts; and (2) open-source semi-structured data, such as encyclopedia entries on OpenKG. The knowledge system of unstructured data, such as national discipline curriculum standards and digital textbooks in the atlas, is more precise and rigorous. The open-source encyclopedia entry data are open and collaborative, and its knowledge system in the atlas is more dynamic and extensive. The knowledge systems of the two data channels complement each other, and together they form multi-disciplinary teaching domain knowledge with strong integrity that conforms to the national discipline curriculum standards and academic quality system. At the same time, it also has the same degree of applicability for constructing the methodology of knowledge graphing for a single discipline. From the above technical work summary, it can be seen that unstructured texts on national subject curriculum standards or digital textbooks can be transformed into a structured triplet of subject knowledge graphs only through the BERT model's neural network attention-mapping mechanism. The process and method of knowledge extraction in this link are relatively mature and also applicable to semi-structured data, such as open-source encyclopedia entry data. Technically, it is still very difficult for us to map the knowledge system space in the cognitive field, which is suitable for learning characteristics in each academic stage. There are two specific possible solutions: (1) Under the metadata design mode of the current subject knowledge graph, the open-source encyclopedia entry data are manually marked, and an unstructured training set that matches the knowledge ability level described in the national subject curriculum standard is built. It provides an excellent, labeled training source for the supervised learning algorithm, but its labor cost is obviously too high. (2) Based on the standardized corpus generated by the curriculum standard or contents of digital textbooks, the remote semi-supervised learning method is employed to generate the semantic data labels of some discipline knowledges. At this time, algorithm intelligence is used to replace heavy repetitive work, reducing the cost of artificial semantic labels for large-scale datasets. However, the semi-supervised learning algorithm cannot ensure that the discipline semantic labels of the training set that was constructed fully comply with the discipline curriculum standards in terms of knowledge compliance. Therefore, strict requirements for accuracy, such as applicability, may affect the further iterative training effect of machine learning algorithms.

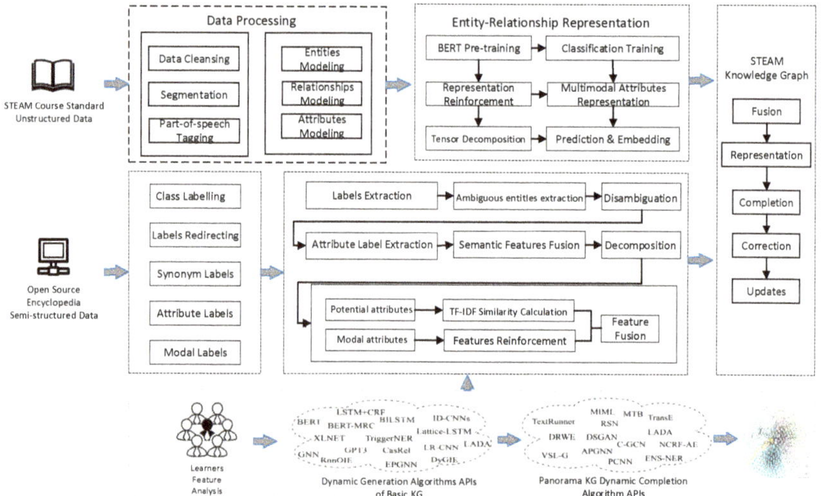

Figure 1. A dynamic generation framework of STEAM knowledge graphs.

The dynamic construction of STEAM's interdisciplinary knowledge graph is reflected not only in the integration and openness of its encapsulation algorithm library API, but also in its intelligent configurability of the knowledge graph of various disciplines, which is specifically divided into the core stages of multi-disciplinary knowledge fusion, representation, completion, error correction, and update. Multi-discipline knowledge fusion refers to entity alignment and attribute fusion of multi-group and multi-discipline knowledge data. In addition to the above basic tuple elements such as "head entity—relational predicate—tail entity", the multi-group data processed by the multi-disciplinary knowledge extraction algorithm also contain multi-group knowledge mixed with other language elements, such as time, place, degree modifier, etc. In order to realize the standardized semantic storage, linking, reasoning, etc., of the multi-group knowledge, it is necessary to regularize the tensor decomposition of the mixed elements in the multi-group; the knowledge representation is further matched with the triple representation in the current discipline knowledge graph so as to build a discipline knowledge graph configuration unit with a standardized data format, clear semantic association, and multiple reasoning paths. The representation of multi-disciplinary knowledge refers to the problem of how to map the expression forms of various multi-disciplinary knowledge in different cognitive intelligence fields. The multi-disciplinary knowledge in the learner's cognitive intelligence development field is represented as symbolic information, and the multi-disciplinary knowledge in the machine's cognitive learning intelligent field is represented as vectorized information. The disciplinary knowledge graph itself becomes the interface between the knowledge symbolic representation and vectorized representation to meet the needs of knowledge flow and integration in their respective fields. The completion, error correction, and updating of multi-disciplinary knowledge is a mechanism for the collaborative optimization of multi-disciplinary knowledge quality in the process of data fusion and the representation of graph corpus between learners and machines during multi-disciplinary learning. On the one hand, it dynamically updates the semantic relationship of associated conceptual entities among disciplines by encapsulating and integrating various panoramic knowledge graph completion algorithms' API, and it promotes the configuration and integration of discipline graphs. On the other hand, through the joint training and learning of large-scale corpus semantic data, we can detect the logical fallacy of semantic relations between different entities and update knowledge cooperatively, so as to build a cognitive intelligence dynamic architecture for the appropriate instructional design of STEAM themes.

3.2. Knowledge Graph Completion of STEAM Instructional Themes

The STEAM interdisciplinary knowledge graph has an interdisciplinary and multi-level cognitive development path. It aims to cultivate learners' multi-clue and multi-dimensional problem-solving thinking and ability. The dynamic completion of the disciplinary knowledge graph provides an intelligent cognitive framework for teachers to carry out interdisciplinary instructional design. It also enables learners to construct multi-personalized meanings for multi-disciplinary learning tasks. The completion of the STEAM knowledge graph refers to using the tensor decomposition model to convert the data of various topic maps that are related to each other or have a learning order (before and after knowledge) into a data form that is easy to be processed by the computer. Next, the semantic relationship between the subject knowledge points is mined to predict and embed the potential semantic relationship between new and old knowledge entities or their attributes. Finally, the intelligent and dynamic completion of the subject knowledge graph is realized. According to the knowledge learning path of different disciplines of STEAM, the completion path of a STEAM knowledge graph also presents multiple dynamics.

Before introducing the tensor decomposition model applied to STEAM knowledge graph embedding, the following symbol definitions were decided:

The knowledge triple is represented by (head entity, relationship type, tail entity), which is denoted as (h, r, t) and abbreviated as φ. The Greek letter $\mathcal{X}$ represents the three-dimensional knowledge tensor, and the elements in $\mathcal{X}$ are represented by $\mathcal{X}_{i,j,k}$. The capital letter A represents the matrices model, which is used to represent entity sets and relationship type sets, whereas $|A|$ represents the size of the matrices. The lowercase letter a represents a vector, where item i is represented by a_i. $\times_n$ represents the Tucker product, which is used for the product of the three-dimensional knowledge tensor and matrix. The operator "$\circ$" represents the product of a vector.

In STEAM multi-disciplinary project-based instruction, the relationships among knowledges are often numerous, and their semantics are connected; in addition, the representation of the relationships is not unique. To simplify the principle introduction, the knowledge semantic relationship is represented as a triple. Taking "Mendel's Law of Inheritance" as an example, the random combination of dominant genes and recessive genes of organisms determines various traits, and the corresponding probability can be obtained through mathematical modeling. Therefore, its knowledge semantics can be represented by a composite triple (Mendel's law, mathematical modeling/material basis, and a random combination of probability/DNA genetic information). When it is extended to the tensor model, the knowledge triplet is constructed as a three-dimensional tensor, in which two dimensions are composed of knowledge head and tail entities (such as Mendel's law or a random combination of probability/DNA genetic information) and the third dimension is the relationship between knowledges (mathematical modeling/material basis).

Take the classic algorithm of the Tucker decomposition [28] of a tensor model as an example. For the three-dimensional tensor in the subject knowledge graph, the Tucker decomposition can obtain the core tensor and three knowledge matrices, as shown in Formula (1).

$$\mathcal{X} \approx [\![G; A, B, C]\!] \equiv G \times_1 A \times_2 B \times_3 C \equiv \sum_{p=1}^{P} \sum_{q=1}^{Q} \sum_{r=1}^{R} g_{pqr} \circ a_p \circ b_q \circ c_r \tag{1}$$

Among them, the core tensor G represents the level of semantic relationship between knowledge entities and relationships, which can be used for a potential semantic analysis of knowledge points and knowledge relationships. In essence, tensor decomposition is to decompose the potential semantic relationship of new knowledge through the dimension reduction processing of complex knowledge structures and the dimension reduction algorithm of model structures, so as to generate new knowledge structures that do not exist in the original graph spectrum and to promote the STEAM project to learn multiple observation perspectives and solve multi-clue problems. Therefore, the tensor model is suitable for the dynamic completion of the subject knowledge map, and the semantic relationship cohesion of the subject knowledge is improved through the dynamic embedded model.

Combined with the tensor decomposition algorithm, this study proposed a dynamic completion model of a STEAM interdisciplinary knowledge graph as shown in Figure 2. The main idea of the model is to reflect the hidden information of knowledge through tensor decomposition into a low-dimensional matrix, which is used to estimate the incomplete part of the original discipline knowledge graph. The model extracts knowledge triplets from the knowledge graph of a single discipline to form a data set. The positive and negative samples in the data set of a single discipline are constructed into a three-dimensional knowledge tensor, which is decomposed into a low-dimensional knowledge matrix by Tucker and sent to the scoring function. The embedded entities and relationships are constantly updated by the optimizer, and the implicit relationships between knowledges are mined to supplement the STEAM interdisciplinary knowledge graph as a newly generated knowledge connection.

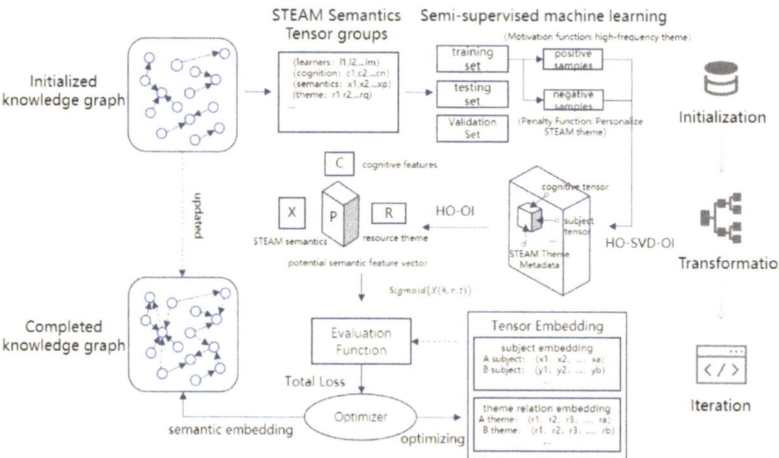

Figure 2. Dynamic completion model of STEAM knowledge graph based on tensor decomposition.

In order to establish relationships among the knowledge entities with unknown relationships, the model uses the replacement method to replace the correct knowledge triplet entities into negative triples. For prediction, the *Sigmoid* function is used as the activation function to predict the correctness of the triple through probability. For optimization, the model uses the logarithmic likelihood loss function to measure the quality of the model's prediction.

3.3. Quality Assessment of STEAM Knowledge Graph Based on Curriculum Standards

The matching degree between the subject knowledge system and the subject accomplishment evaluation index in the STEAM curriculum standard is an important indicator for measuring the quality of the generation and completion of the subject knowledge graph. Whether it is the generation of the subject knowledge graph or the dynamic completion of the subject knowledge graph based on different teaching objectives and teaching themes, the quality evaluation index should focus on the teaching connectivity and teaching appropriateness of the map. The teaching connectivity is calculated by the dynamic performance of the interdisciplinary knowledge of the atlas, while the teaching appropriateness is calculated by the cognitive and evaluation matching degree of the discipline curriculum standards related to the STEAM teaching theme of the atlas. Therefore, its quality evaluation methods include the matching degree calculation methods of the knowledge atlas completion performance based on tensor model and the academic quality evaluation indicators based on the discipline curriculum standards. We evaluated whether the constructed discipline knowledge graph had good dynamic embedding performance; that is, whether the discipline knowledge graph met the dynamic embedding of entity relations with multimodal attributes. The test data selected multi-source heterogeneous data, such as digital education resource library and public service resource metadata, through MRR, Hits@1, Hits@3, and Hits@10 to calculate the matching effect between the subject knowledge graph and the subject curriculum standard knowledge system.

The Mean Reciprocal Rank (MRR) is the average reciprocal ranking of embedded entities in the discipline knowledge graph. The larger the index is, the better the matching effect is, as shown in Formula (2).

$$\text{MRR} = \frac{1}{|S|} \sum_{i=1}^{|S|} \frac{1}{rank_i} = \frac{1}{|S|} \left(\frac{1}{rank_1} + \frac{1}{rank_2} + \ldots + \frac{1}{rank_{|S|}} \right) \tag{2}$$

Hits@N is the average proportion of triples ranking less than N in the prediction of a STEAM interdisciplinary knowledge link. The larger the index is, the better the matching effect is, as shown in Formula (3).

$$\text{Hits@N} = \frac{1}{|S|} \sum_{i=1}^{|S|} \mathbb{I}(rank_i \leq n) \tag{3}$$

Conversely, we evaluated whether the constructed subject knowledge graph matched the instructional requirements of the subject curriculum standards of each learning segment, such as whether the generated and completed subject knowledge graph met the teaching objectives, resource organization, and activity design of STEAM teachers based on interdisciplinary curriculum standards while meeting the requirement of providing students with cognitive support, resource recommendations, and learning path planning based on the subject curriculum standards. Therefore, the matching data set of the evaluation method was the text data of the curriculum standards of each discipline of STEAM. Through the knowledge extraction, knowledge annotation, knowledge matching, and other links of the curriculum standard text, the matching degree of the curriculum standard teaching requirements of each discipline can be measured according to the phased output of each link. Specifically, it includes the matching of the subject knowledge graph and the cognitive level of the subject curriculum standard, and the matching of the academic quality with literacy evaluation indicators.

4. Experiments

4.1. Data Set Construction

(1) Source of data set: The data set selected for the experiment came from the subject curriculum standard, digital textbook text, and open-source Chinese encyclopedias, mainly including the concepts and relationships of biology, chemistry, geography, and mathematics in middle schools. The resource data set came from the EdX MOOC platform, which covers mathematical science, natural science, humanities and social sciences, engineering technology, and other fields. From 2013 to 2019, 36,892 users learned and browsed 528,536 course resource files, with 372 feature status tags in total.

(2) Data set preprocessing: First, we cleaned and removed learner and course resource records with feature frequency lower than 50 to prevent the cold start of machine learning. We then removed the user data whose feature frequency was less than 1000 to prevent the interference of the crawler. Finally, we removed the curriculum resource records whose resource access frequency was less than 30 to prevent the clustering effect from discretization caused by low-frequency tags.

(3) Segmentation of training set and test set: After the above preprocessing, 68,745 effective records of resource learning with feature annotation were obtained, 85% of the data was randomly selected to construct the training set, and 15% of the data was used to construct the test set.

(4) Model training and testing: In the training set, the random algorithm used at least one marked learning situation feature tag as the learner's cognitive state feature tag. In the test set, the records of curriculum resources were fitted with tensor models to calculate relevant evaluation indicators.

4.2. Analysis of STEAM Interdisciplinary Instructional Cases and Generation of Themes

The instructional case comes from the high school STEAM interdisciplinary instructional case of the experimental school of the research group. The instructional case takes nutrition science, cultivation technology, horticultural culture, and related literary works of peas as the instructional theme and carries out project instructions from the perspectives of biological properties, geographical distribution, nutrient composition, modern farming technology, landscape culture, historical literature, and other disciplines. The knowledge points cover many basic education disciplines, such as chemistry, biology, geography, mathematics, physics, history, and literature; they include learning the principles of pea science and technology, pea gardening culture and literary works, field research practice of pea plantations, and the sharing and exchange of interdisciplinary learning achievements.

The learners are the first year students of the experimental high school. During the course of project learning, they are divided into a pea science and technology research group and a pea culture exploration group, according to the learners' cognitive preference style. Under the activity mode of group cooperation, they carry out personalized learning and research on the related scientific principles, agricultural technology, horticultural culture, literary works, and other sub-topic keywords of pea, respectively, with the support of the subject knowledge graph. Its subject learning theme keywords take "pea" as the central word, "scientific principles, agricultural technology, horticultural culture, and literary works" as the clues of knowledge and resource organization and management, and learners' cognitive style preferences and state attributes as the sequence feature tags to dynamically generate multi-disciplinary learning paths and resource allocation sequences under different cognitive style tags. The knowledge structure system based on the corresponding subject curriculum standards dynamically retraces the semantic relationship between the old knowledge and the new knowledge of learners, and it maps the hierarchical characteristics of cognitive ability between the learning sequence of learners' knowledge and the sequence of resource organization, so as to realize personalized multi-disciplinary intelligent learning. Finally, under the support of the subject knowledge graph, the learners are guided to dialectically understand the cooperative development relationship between crop science and technology and human farming culture through group communication and sharing. This section takes the STEAM instructional activity of Exploring the Scientific Principles of Peas as an example to show the dynamic generation effect of interdisciplinary knowledge graphs based on the curriculum standards of chemistry, biology, geography, and mathematics. The generated STEAM interdisciplinary thematic knowledge graph is shown in Figure 3.

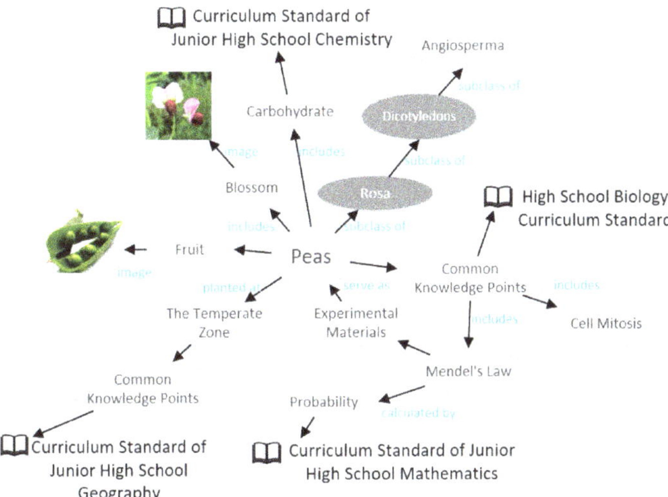

Figure 3. Interdisciplinary knowledge graph automatically generated based on keyword "peas".

5. Results and Discussion

5.1. Performance Results of Comparison Simulation Experiment

In the training set data, the corresponding TopN list of each learner was calculated by the multi-disciplinary semantic tensor decomposition algorithm HO-SVD-OI in this section. The accuracy recall curve was used to reflect the changing trend of the accuracy of the method in this paper and the mainstream classical algorithm in matching STEAM learning topics with course standards, and the performance gap of different algorithms was reflected by the harmonic mean F1 index [29]. In this section, we selected the Kmeans IDF algorithm [30] based on semantic tags and K-means clustering of user access resources, the classical tensor decomposition Co SClu algorithm [31] based on semantic tag co-occurrence spectrum clustering, and the classical TD algorithm [32] based on original tensor decomposition as the reference comparison algorithm.

The P-R curve and F1 index curve simulation results of the above four algorithms are shown in Figure 4. In the P-R curve comparison chart, as the training scale of the data set increased, the accuracy and recall rate of the algorithm showed a reverse proportion change trend. The more the values gathered to the upper right side of the image area, the better the matching effect of the digital learning resources organized by the algorithm on the learner's knowledge tag. In the F1 index curve comparison chart, the more values gathered above the image area, the higher the accuracy of the learning topics generated by the algorithm for the coverage of multi-disciplinary curriculum standards. In addition, whether it was the HO-SVD-OI algorithm applied in this chapter or the high-dimensional semantic tensor decomposition resource organization matching algorithm based on Kmeans clustering or spectral clustering proposed by our predecessors, its performance was significantly better than that of the simple tensor decomposition algorithm. By observing the F1 index curve, we found that when the scale of STEAM learning topics approached about 20, the four algorithms all achieved the best performance in the coverage accuracy of the STEAM discipline curriculum standard.

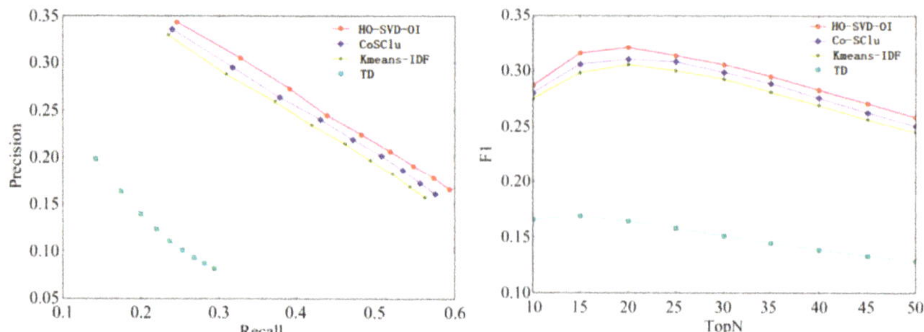

Figure 4. Simulation experiment results of the P-R curve and F1 index curve of the dataset under four algorithms.

5.2. Discussion on Learning Path Based on STEAM Interdisciplinary Knowledge Graph

When learners complete STEAM project learning, they generally have a relatively complete discipline knowledge structure in the lower learning stage. In the context of STEAM interdisciplinary project learning, the dynamic completion of a discipline knowledge graph is essentially a process of personalized meaning construction of learners for high-level learning objects, thus promoting the gradual formation of high-level knowledge and literacy of disciplines.

In the key words knowledge graph of the project theme shown in Figure 3, the teaching goal of the project is to solve the problem of "the causes of the diversity of pea genetic traits", while the original knowledge structure of the learners is the relevant scientific principle "Mendelian genetic law conducts mathematical modeling through the probability learned in junior high school, and the expression of biological genetic traits is determined by their DNA". The new knowledge that learners need to master is that "the deep reason for the establishment of Mendel's Law is that biological genetic material DNA has the random characteristics of free separation and combination, and the random combination characteristics of genetic material lead to the diversification of genetic traits, thus contributing to the diversity of biological genetics and evolution directions". The STEAM interdisciplinary knowledge graph supports single-discipline teachers to design project learning path steps as follows:

Step 1: According to the high school biology curriculum standard, the triple of the original knowledge graph is marked as the literacy evaluation index V(ij), indicating the literacy level that learners should reach when learning corresponding knowledge (h, r, t), where i represents the triple knowledge sequence, and j represents the literacy evaluation level sequence.

Step 2: According to the quality evaluation index system of the subject curriculum standards, the subject project teaching topics of different subject quality levels are presented. For example, for the learning needs of mathematical logic computing quality evaluation, teachers can build appropriate mathematical models for learners to deduce and verify the "randomness of biological genetic material combination" according to the quality evaluation index labels (junior high school, mathematics, event probability, mathematical modeling, and problem-solving) and can support learners to understand the essence of problem-solving from the evolution of mathematical variables. For the learning needs of spatial structure analysis literacy, teachers can build "the combination and change law of DNA microstructure in the genetic process" for it through spatial micro-analysis according to the quality evaluation index labels (junior high school, chemistry, structure of life macromolecules, scientific inquiry, macro-identification, and micro-analysis). Different goal orientation will lead to different paths of STEAM project learning activities.

Step 3: When learners choose different ways to organize corresponding discipline project resources, teachers will provide necessary assistance and intervention according to their learning process and learners' dynamic generation and completion direction of the discipline knowledge graph based on derived theme keywords, adjust the direction and level of learners' personalized meaning construction, and finally achieve the overall goal of STEAM project teaching that meets the discipline curriculum standards.

6. Conclusions and Future Work

Presently, there are many kinds of mature natural language processing and machine learning algorithms in the computer science field. The application of a single algorithm cannot meet the context or scene adaptability requirements of the entity recognition accuracy and relational semantic modeling of the STEAM interdisciplinary knowledge graph from different perspectives. It is necessary to integrate multiple APIs to conduct multiple trainings on data sets to meet the needs of teachers in a single discipline for the design of STEAM interdisciplinary teaching themes; among them, conditional random field and tensor decomposition algorithms are relatively ideal for semantic relationship modeling. On the one hand, the dynamic generation of a STEAM subject knowledge graph can enable teachers of a single subject to flexibly organize instructional resources and improve teaching efficiency in an intelligent learning environment. On the other hand, a STEAM subject knowledge map can effectively plan personalized learning paths by adapting to learners' cognitive characteristics, thus broadening learners' multi-dimensional thinking ability.

Although the traditional tensor decomposition algorithm has good performance for the generation of STEAM multi-disciplinary topics, it is easy to form a STEAM "topic cocoon room" due to the long-term recommendation of high-frequency single discipline themes, which are not conducive to the development of learners' multiple cognitive abilities. The HO-SVD-OI tensor decomposition algorithm, which introduces a penalty function and high-dimensional orthogonal iteration mechanism, has good computational performance for the matching and coverage of STEAM learning themes and multi-disciplinary curriculum standards; it can effectively eliminate the problem of a "theme cocoon room" in STEAM instructional design. Future research will focus on the interpretability mechanism of knowledge graphs for STEAM interdisciplinary semantic learning.

Author Contributions: Conceptualization and supervision, Q.K.; Writing—original draft, J.L. All authors have read and agreed to the published version of the manuscript.

Funding: This study was funded by the National Key R&D Program of China (2022YFC3303600) and the Guangdong Provincial Philosophy and Social Science Foundation (GD22TWCXGC10).

Conflicts of Interest: The authors declare no conflict of interest.

References

1. Elaine, P.; Jen, K.B. STEAM in practice and research: An integrative literature review. *Think. Ski. Creat.* **2019**, *31*, 31–43.
2. Zhang, Z.; Cai, J.; Wang, J. Duality-Induced regularizer for tensor factorization based knowledge graph completion. In Proceedings of the 34th Conference on Neural Information Processing Systems (NeurIPS 2020), Vancouver, BC, Canada, 6–12 December 2020.
3. Levine, A.E.; Bebermeyer, R.D.; Chen, J.W.; Davis, D.; Harty, C. Development of an Interdisciplinary Course in Information Resources and Evidence-Based Dentistry. *J. Dent. Educ.* **2008**, *72*, 1067–1076. [CrossRef] [PubMed]
4. Cuadrado, M.; Molina, M.E.R.; Coca, M. Participation and performance of university students in an interdisciplinary and bilingual e-learning project. *Rev. Educ.* **2009**, *348*, 279–280.
5. Van Til, R.P.; Tracey, M.W.; Sengupta, S.; Fliedner, G. Teaching Lean with an Interdisciplinary Problem-solving Learning Approach. *Int. J. Eng. Educ.* **2009**, *25*, 173–180.
6. Havnes, A. Talk, planning and decision-making in interdisciplinary teacher teams: A case study. *Teach. Teach.* **2009**, *15*, 155–176. [CrossRef]
7. Pharo, E.J.; Davison, A.; Warr, K.; Nursey-Bray, M.; Beswick, K.; Wapstra, E.; Jones, C. Can teacher collaboration overcome barriers to interdisciplinary learning in a disciplinary university? A case study using climate change. *Teach. High. Educ.* **2012**, *17*, 497–507. [CrossRef]
8. Chen, H.R.; Huang, J.G. Exploring Learner Attitudes toward Web-based Recommendation Learning Service System for Interdisciplinary Applications. *Educ. Technol. Soc.* **2012**, *15*, 89–100.
9. Briede-Westermeyer, J.C.; Perez-Villalobos, C.E.; Bastias Vega, N.; Bustamante-Durán, C.F.; Olivera-Morales, P.; Parra-Ponce, P.; Delgado-Rivera, M.; Cabello-Mora, M.; Campos-Cerda, I. Interdisciplinary experience for the design of health care products. *Rev. Med. Chile* **2017**, *145*, 1289–1299. [CrossRef]

10. Van den Beemt, A.; MacLeod, M.; Van der Veen, J.; van de Ven, A.; van Baalen, S.; Klaassen, R.; Boon, M. Interdisciplinary engineering education: A review of vision, teaching, and support. *J. Eng. Educ.* **2020**, *109*, 508–555. [CrossRef]
11. Lin, Y.S.; Chang, Y.C.; Liew, K.H. Effects of concept map extraction and a test-based diagnostic environment on learning achievement and learners' perceptions. *Br. J. Educ. Technol.* **2016**, *47*, 649–664. [CrossRef]
12. Wen, Y.; Lool, C.K.; Chen, W. Appropriation of a representational tool in a second-language classroom. *Int. J. Comput. Support. Collab. Learn.* **2015**, *10*, 77–108. [CrossRef]
13. Shi, B.X.; Weninger, T. Open-World knowledge graph completion. In Proceedings of the 32nd International Conference on Artificial Intelligence, New Orleans, LA, USA, 2 February 2018; pp. 1957–1964.
14. Socher, R.; Chen, D.; Manning, C.D.; Ng, A.Y. Reasoning with neural tensor networks for knowledge base completion. In Proceedings of the 26th International Conference on Neural Information Processing Systems, Lake Tahoe, NV, USA, 5–10 December 2013; pp. 926–934.
15. Wang, Z.; Zhang, J.; Feng, J.; Chen, Z. Knowledge graph embedding by translating on hyperplanes. In Proceedings of the 28th AAAI Conference on Artificial Intelligence, Québec City, QC, Canada, 27–31 July 2014; pp. 1112–1119.
16. Lin, Y.; Liu, Z.; Sun, M.; Liu, Y.; Zhu, X. Learning entity and relation embeddings for knowledge graph completion. In Proceedings of the 29th AAAI Conference on Artificial Intelligence, Austin, TX, USA, 25–30 January 2015; pp. 2181–2187.
17. Gardner, M.; Talukdar, P.; Krishnamurthy, J.; Mitchell, T. Incorporating vector space similarity in random walk inference over knowledge bases. In Proceedings of the 2014 Conference on Empirical Methods in Natural Language Processing, Doha, Qatar, 25–29 October 2014; pp. 397–406.
18. Xiong, W.; Hoang, T.; Wang, W.Y. DeepPath: A reinforcement learning method for knowledge graph reasoning. In Proceedings of the 2017 Conference on Empirical Methods in Natural Language Processing, Copenhagen, Denmark, 7–11 September 2017; pp. 564–573.
19. Xiong, W.; Yu, M.; Chang, S.; Guo, X.; Wang, W.Y. Oneshot relational learning for knowledge graphs. In Proceedings of the 2018 Conference on Empirical Methods in Natural Language Processing, Brussels, Belgium, 31 October–4 November 2018; pp. 1980–1990.
20. Hematialam, H.; Garbayo, L.; Gopalakrishnan, S.; Zadrozny, W.W. A Method for Computing Conceptual Distances between Medical Recommendations: Experiments in Modeling Medical Disagreement. *Appl. Sci.* **2021**, *11*, 2045. [CrossRef]
21. Liu, H.Y.; Zhang, T.C.; Li, F.; Gu, Y.; Yu, G. Tracking Knowledge Structures and Proficiencies of Students with Learning Transfer. *IEEE Access* **2021**, *9*, 55413–55421. [CrossRef]
22. Kim, M.C.; Feng, Y.Y.; Zhu, Y.J. Mapping scientific profile and knowledge diffusion of Library Hi Tech. *Libr. Hi Tech* **2021**, *39*, 549–573. [CrossRef]
23. Gaur, M.; Faldu, K.; Sheth, A. Semantics of the Black-Box: Can knowledge graphs help make deep learning systems more interpretable and explainable? *IEEE Internet Comput.* **2021**, *25*, 51–59. [CrossRef]
24. Chen, P.H.; Lu, Y.; Zheng, V.W.; Chen, X.; Yang, B. KnowEdu: A System to Construct Knowledge Graph for Education. *IEEE Access* **2018**, *6*, 31553–31563. [CrossRef]
25. Jia, B.X.; Huang, X.; Jiao, S. Application of Semantic Similarity Calculation Based on Knowledge Graph for Personalized Study Recommendation Service. *Educ. Sci. Theory Pract.* **2018**, *18*, 2958–2966.
26. Nayyeri, M.; Cil, G.M.; Vahdati, S.; Osborne, F.; Rahman, M.; Angioni, S.; Salatino, A.; Reforgiato Recupero, D.; Vassilyeva, N.; Motta, E.; et al. Trans4E: Link Prediction on Scholarly Knowledge Graphs. *Neurocomputing* **2021**, *461*, 530–542. [CrossRef]
27. Chi, Y.; Qin, Y.; Song, R.; Xu, H. Knowledge Graph in Smart Education: A Case Study of Entrepreneurship Scientific Publication Management. *Sustainability* **2018**, *10*, 995. [CrossRef]
28. Balažević, I.; Allen, C.; Hospedales, T.M. Tucker: Tensor factorization for knowledge graph completion. *arXiv* **2019**, arXiv:1901.09590:1-10.
29. Pazzani, M.; Billsus, D. Learning and Revising User Profiles: The Identification of Interesting Web Sites. *Mach. Learn.* **1997**, *27*, 313–331. [CrossRef]
30. Symeonidis, P. ClustHOSVD: Item Recommendation by Combining Semantically Enhanced Tag Clustering with Tensor HOSVD. *IEEE Trans. Syst. Man Cybern. Syst.* **2015**, *46*, 1240–1251. [CrossRef]
31. Li, H.; Hu, X.; Lin, Y.; He, W.; Pan, J. A Social Tag Clustering Method Based on Common Co-occurrence Group Similarity. *Front. Inf. Technol. Electron. Eng.* **2016**, *17*, 122–134. [CrossRef]
32. Kolda, T.G.; Bader, B.W. Tensor Decompositions and Applications. *Coll. Res. Libr.* **2005**, *66*, 294–310. [CrossRef]

 *applied sciences*

Communication

Digital Technology in Managing Erasmus+ Mobilities: Efficiency Gains and Impact Analysis from Spanish, Italian, and Turkish Universities

Martín López-Nores [1,*], José J. Pazos-Arias [1], Abdulkadir Gölcü [2] and Ömer Kavrar [3]

[1] atlanTTic Research Centre for Information and Communication Technologies, Department of Telematics Engineering, University of Vigo, 36310 Vigo, Spain
[2] Faculty of Communication, Department of Journalism, Selçuk University, Konya 42130, Turkey
[3] Faculty of Economics and Administrative Sciences, Department of Economics, Selçuk University, Konya 42130, Turkey
[*] Correspondence: mlnores@det.uvigo.es; Tel.: +34-986813967

Citation: López-Nores, M.; Pazos-Arias, J.J.; Gölcü, A.; Kavrar, Ö. Digital Technology in Managing Erasmus+ Mobilities: Efficiency Gains and Impact Analysis from Spanish, Italian, and Turkish Universities. *Appl. Sci.* **2022**, *12*, 9804. https://doi.org/10.3390/app12199804

Academic Editor: Mirco Peron

Received: 15 August 2022
Accepted: 26 September 2022
Published: 29 September 2022

Publisher's Note: MDPI stays neutral with regard to jurisdictional claims in published maps and institutional affiliations.

Abstract: The European Union is investing in the areas of digital skills, digital infrastructures, digitisation of businesses, and public services to speed up numerous administrative processes and to facilitate access to citizens from member countries and neighbouring ones as well. This study provides a quantitative assessment of the efficiency gains that can be attained by the ongoing digital transformation in the realm of Erasmus+, the European Commission's programme for education, training, youth, and sport for the period 2021–2027. This programme manages a sizable budget allocated to education and training opportunities abroad for millions of students, teachers, and other staff of Higher Education Institutions within the EU and beyond. The management of such experiences has significantly grown in complexity over the last decades, entailing notable expenses that the EC aims to reduce through the end-to-end digitalisation of administrative procedures. Our analysis of the savings attained by the so-called Erasmus Without Paper project (EWP) was conducted by taking a close look at the workload, resources, and money invested in Erasmus+ proceedings by four universities from Spain, Italy, and Turkey. The analysis revealed significant savings in terms of paper wastage (a reduction of more than 13.5 million prints every year for the whole Erasmus+ programme) and administrative time, which may translate into lower staff effort and increased productivity, to the point of managing up to 80% more mobilities with the same resources and staff currently available.

Keywords: digital transformation; Erasmus+; EWP; ICT; student mobility

1. Introduction

Digital transformation is defined as facilitating the work of individuals, improving processes, increasing the efficiency of operation, and creating new business models with the use of various digital technologies [1]. On the axis of this definition, it is seen that the main components of digital transformation are the individual, the process, and the technology. The change affects people, strategies, structures, and competitive dynamics in many areas such as politics, health, society, economy, agriculture, and industry in line with the needs of society. According to Pereira et al. [2], educational organisations that do not take advantage of this moment to improve and transform themselves in digital transformation are in danger of disappearing or being replaced by more agile organisations. In addition to this, Căpușneanu et al. [3] argue that to ensure the successful implementation of digital transformation, organisations and companies need to guarantee that they are fully aware of the failures or risks they face over a certain period of time, for which they must identify the key influencing factors.

Higher education institutions (HEIs) draw attention as one of the areas affected by digital transformation processes, which have been studied in many research works so far. According to Rodrigues [1], HEIs are under pressure to provide new and innovative digital experiences for their stakeholders, which can only be achieved by using a framework that enables them to manage all digital initiatives and approaches in a holistic and integrated way. Benavides et al. [4] stated that digital transformation in higher education institutions requires rethinking, restructuring, and reinventing, from its multi-purpose, multi-disciplinary, multi-state, and multifactorial character. Their statement suggests that the digital transformation inside institutions does not simply imply technological progress; rather, it is more transcendental and generates changes of meaning, affecting the culture immersed in the university, administration, formative activities and their evaluations, pedagogical approaches, teaching, research, extension, and administrative processes.

According to a literature study by Mohamed et al. [5], digitalisation can provide a competitive advantage to higher education institutions as long as it is implemented with a correct combination and an integrated approach linking impactful changes. The level of digital maturity, though, varies across countries and regions. A recent UNESCO report [6] enumerated factors needed for a healthy digital transformation in the field of education in the Asia Pacific, ranging from data security, strategy unity, equality of opportunity, structural and sustainable digital technology requirements of educational institutions, family support, blended learning strategies, etc. It was noticed that while 50% of high-income countries in the region have an operational policy on digital remote learning, only 27% of low- and lower-middle-income countries do. According to the survey of Marks et al. [7], a consolidated framework covering the aforementioned factors is missing also in the context of the United Arab Emirates, where none of the examined institutions had a stand-alone digital transformation vision or plan. In developing countries, on the other hand, the discussion embraces the choice of online learning methods. For example, the cross-sectional surveys carried out by Essel et al. in Ghana [8] and by Argüelles-Cruz et al. in Latin America [9] showed that the choice might be largely influenced by the availability of IT infrastructure and by the professors' levels of IT skills. Finally, an OECD Report on the Digital Transformation of Higher Education in Hungary [10] identified policy recommendations to address persistent gaps in the current policy framework supporting digitalisation, which is common in other EU countries as well.

Digital transformation processes call to rethink and plan international student and academic mobility in HEIs in a multifaceted way. In this line, the Erasmus+ programme—one of the largest student and academic mobility activities worldwide—faces a challenge due to the significant size it has achieved. The number of HEIs that have the Erasmus Charter for Higher Education has increased from a few hundred at the beginning of the 2000s to more than 5200 as of July 2022, from 33 countries (the 27 EU members plus Iceland, Liechtenstein, Norway, North Macedonia, Serbia, and Turkey). Accordingly, the number of students and staff benefiting from the Erasmus+ programme has been growing by the day, entailing a plethora of challenges in management and coordination. To overcome these challenges, the European Commission (EC) has decisively invested in digitisation, launching the Erasmus Without Paper (EWP) project [11] to define common interfaces, document types, and management procedures. On such foundations, it is possible to develop an ecosystem of digital applications that will facilitate management and participation. Integration of these developed digital platforms was declared among the priority areas by the EC, which made it compulsory to complete the digitalisation process for all HEIs by 2025.

This communication reports on an experiment conducted in the SUDTE project ("Supporting Universities in the Digital Transformation in Erasmus+") [12] to numerically reveal the benefits of the digital transformation process by comparing traditional methods and digital tools used in the execution of Erasmus+ mobility processes between HEIs. Knowing that many processes are affected by institutional or national regulations and their peculiarities and circumstances, the project's partner institutions have surveyed such details as paper wastage, workload, and time invested in managing mobility processes in their respective countries (namely Spain, Italy, and Turkey) using different digital applications:

three from different commercial providers and one built in-house. Thereupon, a statistical analysis was conducted to characterise the savings that may derive from digitalisation, providing the first quantitative evidence of savings in terms of paper wastage and administrative time, which may translate into lower staff effort, increased productivity, and greater diligence from the point of view of the beneficiaries of the mobilities: fewer errors, shorter or no delays, etc.

This study focuses on the savings in workload, paper, and time savings gained by the digital transformation process that higher education institutions are going through within the scope of the Erasmus+ programme for international offices. The structure of this paper is as follows. The academic literature related to the study is discussed in the background section after the introduction. Next, the methodology section presents information and justifications of the processes used in the study. In the results section, the data obtained from the intervening staff. The computed results are given for later analysis in the discussion section. In the conclusion, a roadmap is drawn for new studies to be carried out in the future.

2. Background

Since it was launched in 2016, the EWP project has made substantial progress in the implementation of a framework supporting the electronic exchange of students' data by interlinking the databases of thousands of HEIs. Those databases may be hosted by in-house systems or third-party providers that cater for different needs and resources, from large universities with thousands of outgoing students every year to small institutions that send just a few of them whenever a specific collaboration opportunity with a foreign institution appears. In the short term, EWP is seen by the EC as a strategic means to more efficiently manage a budget of 26,000 million euros allocated for international mobilities, attaining time savings, reductions in paper usage and increases in staff productivity [13]. In the longer term, it would provide the foundations to develop and implement far-reaching digital transformation plans for all types of education and training institutions, all levels and for all sectors.

Erasmus+ mobility management requires that activities that will take place over a long period of time be recorded from beginning to end. In the process, the HEIs' International Relationships Offices (IROs), the academic coordinators and the beneficiaries of the mobilities must produce and manage many documents and substantial amounts of data. In the paper-based process, even simple confirmations required several people in different institutions to print out, sign, archive and transport tens of pages. Also, HEIs and NAs required extra documents to be transferred. A poll conducted by Jahnke in 2017 [14] –getting 1050 answers from HEIs of all types and sizes from 31 countries– showed that almost 90% of the HEIs that sent out ~1000 students abroad on Erasmus mobilities per year considered the management workload "very high" or "high". Only 9.6% of the institutions answered that the workload was "average", and not even 1% answered that the workload was either "low" or "very low". Putting the workload into a historical context, the poll found that more than 67% of the HEIs perceived the workload had increased since 2014, whereas only 8.1% perceived the workload to have decreased. In one study conducted two years later, Mincer-Daszkiewicz [15] reported that the workload and the variety of documents required for the management of mobilities were still increasing steadily and that the expectation for the EWP solutions was that they would enable interoperability between more than 2000 in-house systems and more than 50 commercial systems to exchange data electronically and securely, and concerning tens of thousands of mobility activities every year.

As of August 2022, the EWP project has almost reached the foreseen gigantic dimensions, with thousands of HEIs, commercial providers and other organisations integrating into the digital ecosystem and taking up the new tools that come with it. As the implementation deadlines set by the European Commission get closer, they are also working together to detect and solve persisting integration and interoperability problems, make the solutions available in all pertinent languages, fully train the management staff, and make the processes accessible and understandable by the beneficiaries of the mobilities. At this point, it is necessary to conduct studies that numerically reveal the benefits of digitalisation, as done in the recent past for other areas such as business, industry, justice or healthcare [16–18]. As explained in [19], the agenda for HEIs has been complicated by the COVID-19 pandemic and the sudden massive adoption of online learning schemes. Numerical evidence will help strengthen the HEIs' and NAs' motivation to complete the transition from paper-based processes, notwithstanding the cost that complex software systems entail in terms of licensing, operation, maintenance and/or evolution.

This brief literature review covers almost the last six years in which the paperless Erasmus studies have been launched. This communication gives novelty and added value to the field since not much research has been conducted regarding the efficiency and gain analysis of digital transformation in the Erasmus+ programme so far.

3. Methodology

As shown in the diagram of Figure 1, the starting point for our study was the acquisition of numerical information about paper wastage, workload and time invested in managing mobility processes during the academic years before the adoption of digital processes.

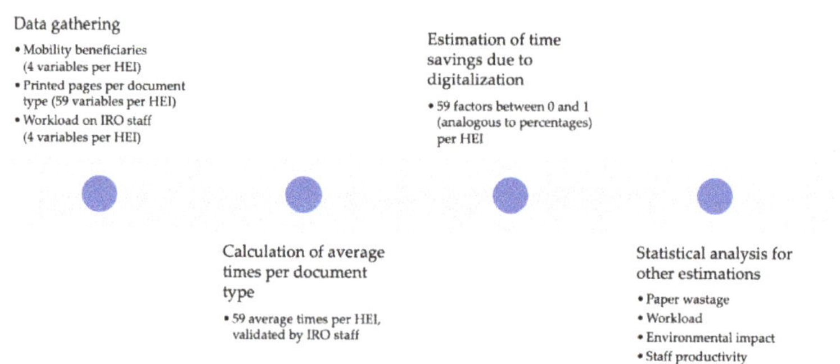

Figure 1. Methodological steps followed in the study.

Specifically, the staff of the International Relationships Offices of the four HEIs in the SUDTE consortium (SU—Selçuk University; IZTECH—Izmir Institute of Technology; UVIGO—University of Vigo; UNINA—University of Naples Federico II) were asked to provide the following data:

- The Average Numbers of Beneficiaries (ANB) of the four types of Erasmus+ mobilities: SMS (Student Mobility for Studies), SMP (Student Mobility for Placement), STA (Staff Mobility for Teaching Assignment), and STT (Staff Mobility for Training).
  - ○ This input yielded four variables for each HEI: ANB_{SMS}, ANB_{SMP}, ANB_{STA}, and ANB_{STT}.

- The Number of Printed Pages (NPP) used for each one of the documents required by each type of mobility (see Table 1), which would rely on templates provided by the EU or models created by each HEI.
  - ○ This input yielded 59 variables for each HEI, corresponding to the 59 types of documents listed in Table 1 (those in the rows of STA and STT are counted twice): NPP_{GA_SMS} for the Grant Agreements of SMS mobilities, NPP_{LA_SMS} for the Learning Agreements, etc.

- The Number of IRO Staff Members (NSM), their Yearly Hours of Work (YHW) per person, the Percentage of the yearly hours taken by SMS/SMP/STA/STT PaperWork (PMPW), the 'M' meaning 'mobilities', and the Percentage taken by Other Paper-Work (POPW), such as interinstitutional agreements, visa letters, passport letters, and internal communications.

Table 1. Documents counted for the four types of mobilities (**bold** documents are used by some institutions only).

Mobility Types	Documents
SMS (Student Mobility for Studies)	• Grant agreement • Learning agreement/work programme • Outgoing application form • Acceptance form • Certificate of arrival • Departure certificate • Transcript of records • Bank details form • Social security/tax declaration • Letter of acceptance to entrants • Copy of incoming ID/passport • Copy of incoming insurance • University nomination • Incoming application form • **Course recognition sheet** • **Document for visa** • **Course recognition confirmation** • **Transcripts** • **Student mobility survey** • **OLS results** • **Incoming learning agreement** • **Incoming final transcript**
SMP (Student Mobility for Placement)	• Grant agreement • Work programme • Outgoing application form • Acceptance form • Certificate of arrival • Departure certificate • Bank details form • Social security/tax declaration • Copy of incoming ID/passport • Copy of incoming insurance • Incoming application form • **Copy of outgoing ID/passport** • **Grant payment to outgoing students** • **Document for visa** • **Academic recognition document**
STA (Staff Mobility for Teaching Assignment) and STT (Staff Mobility for Training)	• Grant agreement • Work programme • Application form • Invitation letter • Letter of acceptance • Certificate of attendance • Report of expenses • Certificate of participation • **Assignment letter** • **Grant payment documents** • **Work programme for incoming beneficiaries**

From the initial data gathered from the IROs, the Average Processing Times (APT) taken by each type of document were calculated for the different types of mobility (second stage of Figure 1), yielding 59 figures for each HEI from the inputs gathered in the first stage: APT_{SMS_GA} for SMS Grant Agreements, APT_{SMS_LA} for SMS Learning Agreements, etc.

In the computations, a sum of the products of ANB and NPP variables determined the workload, whereas $NSM \times YHW \times (PMPW + POPW)$ determined the overall staff effort. It was assumed that, given enough experience with the management procedures, the figures to consider for paper wastage, workload, and time for each type of document were proportional to its number of pages.

It must be noted that it was more convenient to compute the APT figures this way than to ask the IRO staff directly for 59 time estimations because—due to well-known subjective biases in the perception of workload [20,21]—the overall amounts resulting from the latter approach would not match, by far, the yearly hours of work. It was much easier for the staff to check whether the calculated APT values were good approximations of what they experienced in their daily work. They did so with a level of agreement exceeding 92%, which served to validate the abovementioned assumption in the computations.

Subsequently (third stage of Figure 1), the IRO staff were asked to estimate the time savings that they would expect thanks to digitalisation, based on the type of information included in the different documents and on their current familiarity with the EWP tools used in their institutions. This yielded 59 Time Saving Factors (TSF) for each HEI, with values between 0 and 1, equivalent to percentages. For example, a factor of 0.75 means that a proceeding that has traditionally taken 4 min would be expected to take $4 \times 0.75 = 3$ min with the fully-digital approach. The staff's rationale was that fields such as personal data would require little or no verification after they had been introduced once, and university nominations and incoming application forms could be checked much more quickly than former paper-based forms and photocopies. In contrast, learning agreements and work programmes would still require careful examination because the catalogues used for recognition of the education/training attained in a foreign institution are enhanced with each new pairwise matching of sending and receiving institutions. The same goes for highly sensitive data such as transcripts of records, and bank details.

It is worth noting that the IRO staff could provide negative figures at this stage if their expectation was that some proceedings would take longer with the digital approach for whichever reason.

With the time saving estimations, a statistical analysis was conducted with SPSS as the last stage of Figure 1 and served to characterize the expectations in relation to paper wastage, workload, and time which in turn yielded findings about environmental impact and staff productivity.

4. Results

Following the methodological process explained in the preceding section, the consultations with the IRO staff of the four participating HEIs took place from September 2021 to February 2022. Table 2 shows the amounts of paper used by SU, IZTECH, UVIGO, and UNINA, both in total and per beneficiary of each type of mobility. It is noticeable that the additional documents handled by the Turkish HEIs (presented in bold in Table 1) cause a significant increase in the paper wastage per beneficiary. Most commonly, this is related to the fact that beneficiaries from Turkey need to apply for a Schengen visa in order to enter the EU countries.

Table 2. Total and per beneficiary amounts of paper.

Paper Wastage	Number of Printed Pages (NPP)			
	SU	IZTECH	UVIGO	UNINA
Per SMS beneficiary	45	79	31	39
Per SMP beneficiary	31	76	24	36
Per STA beneficiary	25	44	20	25
Per STT beneficiary	24	41	20	25
Total (including other paperwork)	23,647	5664	24,339	43,757

Table 3 shows the calculated times–in minutes–spent yearly by individual IRO staff members from SU, IZTECH, UVIGO, and UNINA with the different documents required by the four types of mobilities.

Table 3. Total Average Processing Times for the different documents.

Mobility Types and Documents		Total *Average Processing Times* (In Minutes)			
		SU	IZTECH	UVIGO	UNINA
SMS and SMP	Grant agreement	15,016	2599	18,367	23,226
	Learning agreement/Work programme	25,377	28,214	18,239	30,613
	Outgoing application form	31,497	21,279	6284	3974
	Acceptance form	1797	743	2292	2285
	Certificate of arrival	1797	743	3118	1852
	Departure certificate	1797	6480	3118	1369
	Transcript of records	2517	1772	3339	1750
	Bank details form	1502	8809	1837	2183
	Social security/tax declaration	2760	0	1837	3974
	Letter of acceptance to entrants	275	101	1214	875
	Copy of incoming ID/passport	295	1519	1281	5352
	Copy of incoming insurance	591	1316	2562	1078
	University nomination	275	4928	1214	2666
	Incoming application form	295	1772	1281	977
	Other documents	5640	13,753	0	0
STA and STT	Grant agreement	2428	1823	2793	1709
	Work programme	2890	405	1648	2466
	Application form	1465	405	410	802
	Invitation letter	425	135	410	279
	Letter of acceptance	470	34	349	279
	Certificate of attendance	470	34	549	279
	Report of expenses	1311	270	1396	773
	Certificate of participation	328	135	349	279
	Other documents	1410	1991	0	0

Table 4 lists the average time saving factors indicated by the consulted IRO staff for the different types of documents.

Table 4. Summary of the time saving factors (TSF) indicated by the consulted IRO staff for the different types of documents [1].

Mobility Types and Documents		Time Saving Factors (See Explanation In Section 3)
	Grant agreement	0.7
	Learning agreement/Work programme	0.3
	Outgoing application form	0.3
	Acceptance form	0.7
	Certificate of arrival	0.3
	Departure certificate	0.5
	Transcript of records	0.3
SMS and SMP	Bank details form	0.1
	Social security/tax declaration	0.2
	Letter of acceptance to entrants	0.7
	Copy of incoming ID/passport	0.2
	Copy of incoming insurance	0.2
	University nomination	0.8
	Incoming application form	0.8
	Other documents	0.8
	Grant agreement	0.7
	Work programme	0.7
	Application form	0.6
	Invitation letter	0.3
STA and STT	Letter of acceptance	0.7
	Certificate of attendance	0.3
	Report of expenses	0.8
	Certificate of participation	0.2
	Other documents	0.8

[1] "Other documents" represent the additional documents used by some institutions only.

From these figures and the calculations of the workload implied by each document type, Table 5 summarizes the overall savings foreseen for the four surveyed HEIs, in the following terms:

- Total working weeks saved in managing their current yearly numbers of mobilities. This was obtained by recomputing the total times invested in the management of the four types of mobilities, computing the post-digitalization values of PMPW (*Percentage of the yearly hours taken by Mobilities PaperWork*) and POPW (*Percentage taken by Other PaperWork*) and taking into account the Time Saving Factors. The difference in minutes from the original value, returned by the equation NSM x YHW x (PMPW + POPW), was turned to working weeks by making the corresponding divisions.
- Reduction of workload experienced by the IRO staff members. This is simply a percentage value computed by comparing the original amount of time devoted to mobilities paperwork and other paperwork with the updated amount that incorporated the TSF values.
- Potential increase in the number of mobilities that could be handled by the current staff under their current workload. This was computed as an updated value of the variables Average Numbers of Beneficiaries (ANB) of each HEI, not considering the

increase due to paperwork, but to additional efforts made in the preparation and management of mobilities: advice on logistics (travel, accommodation, etc.), cultural awareness, academic and linguistic support, monitoring of incidents and needs, etc. The computation was a conservative one, assuming that the staff's yearly hours not invested in mobilities paperwork or other paperwork (i.e., $YHW \times (100 - PMPW - POPW)$) would be devoted to those efforts in the Erasmus+ programme.

Table 5. Overall savings foreseen for the four surveyed HEIs as a result of the digitalisation of mobility management.

Productivity Gains Derived from Time Savings	SU	IZTECH	UVIGO	UNINA
Total working weeks saved in managing the current yearly numbers of mobilities.	185	240	128	220
Reduction of workload experienced by the IRO staff members (in % of time devoted to paperwork).	56.6%	56.9%	53.2%	54.8%
Potential increase in the number of mobilities that could be handled by the current staff under their current workload.	75.9%	75.7%	87.8%	82.5%

5. Discussion

Our analysis reveals a potential to achieve substantial gains as a result of digitalising the management of Erasmus+ mobilities. The following are the key takeaways derived from the results of the preceding section:

- With more than 300,000 higher education students participating annually in Erasmus+ mobilities, saving an average of 45 printed pages in management paperwork (see Table 2) implies a reduction of more than 13.5 million prints every year, which entails enormous environmental impact not only in relation to paper, but also transport, packaging, ink, electricity, and storage. The mobilities of teachers and administrative staff imply an extra 3.6 million prints saved yearly.
- In terms of staff productivity, the figures in Table 5 reveal an expectation of average reductions above 55% of the time spent on paperwork, which would result in a reduced workload for the IRO staff members, and contribute significantly to improving management diligence from the point of view of the beneficiaries of the mobilities: fewer errors, shorter or no delays, etc. The IRO staff would be able to use the saved time on aspects that have traditionally stood out in the satisfaction polls as needing improvements or more extensive coverage [22], such as the cultural preparations and logistic support given to beneficiaries before their mobilities, the follow-up and support offered during their stays abroad, and the dissemination of the attained results after returning.
- From a different perspective, the time savings reveal an opportunity to manage an average of 80% more mobilities with the same resources and staff currently available. Effectively, this removes one bottleneck that has prevented HEIs from offering the Erasmus+ experiences to a greater number of people, thus paving the road for more effective usage of the programme's budget and multiplication of the return from the massive public investment, whose figures are already positive as reported by D'Hombres [23].

6. Conclusions

From the aforementioned figures, it is concluded that, although the digital transformation in the Erasmus+ mobilities of higher education seems difficult and costly, it would provide a great convenience in the effective and efficient management of HEI processes, fully in line with the Digital Decade targets of the European Union [24] in the areas of digital skills, digital infrastructures, digitalisation of businesses, and public services. Nevertheless, thinking about the long term, it is necessary to assess the environmental sustainability of the paradigm shift taking into consideration the fact that, for every HEI that goes paperless, there will be an expanding data centre footprint and significant amounts of e-waste [25]. Thus, as highlighted in a World Economic Forum report from 2019 [26], data centres may lead to an unintentional but unchecked negative impact of digital technology, exacerbated by quick technology obsolescence. It will therefore be necessary to assess the question of whether it is possible to decouple digital transformation in higher education from e-waste and the negative impact of digital technology.

In line with the aims of the European Commission, with this change of paradigm, the administrative burden for the four HEIs participating in this study to manage Erasmus mobilities has been reduced drastically by using digital tools. In relation to the specific implications of these achievements on the Erasmus+ programme, the first highlight is that it is possible to carry out more mobilities with fewer staff members, thanks to the reduction in work intensity. Alternatively, in keeping the same staff, the saved extra time could be devoted to improving the quality of the mobilities or to different focal points. Beneficiaries are expected to show more interest in the programme thanks to the reduction of bureaucratic procedures and the lower probability of administrative mishaps. It is recommended to develop digital technologies like an Erasmus app to be used by academic and administrative staff in the new Erasmus Without Paper programme, which demands an investigation of expectations and possibilities.

This communication reports on one dimension of the digital transformation of HEIs, which–as noted in the introduction–is an extremely complex endeavour that also touches many different psychological, social, economic, and legal aspects. The literature on these aspects is extremely scarce as of 2022, calling for intense research in the following years. As part of our ongoing research, besides including the cases of more universities to further substantiate the findings of this communication, it is planned to conduct research on the points of view of students, professors, and other staff about the ease with which they go through the new digital procedures, with a particular interest in recruiting a sufficient sample of people who engaged in Erasmus+ mobilities both before and after digitalisation.

Author Contributions: Conceptualisation, M.L.-N., J.J.P.-A., A.G. and Ö.K.; methodology, M.L.-N., J.J.P.-A., A.G. and Ö.K.; software, M.L.-N. and J.J.P.-A.; investigation, M.L.-N., J.J.P.-A., A.G. and Ö.K.; data curation, M.L.-N. and J.J.P.-A.; writing—original draft preparation, M.L.-N. and J.J.P.-A.; writing—review and editing, A.G. and Ö.K. All authors have read and agreed to the published version of the manuscript.

Funding: This research was funded by the Erasmus+ programme of the European Union through the "Supporting Universities in the Digital Transformation in Erasmus+" (SUDTE) project, Grant Number 2020-1-TR-KA203-093849.

Institutional Review Board Statement: Not applicable.

Informed Consent Statement: Not applicable.

Data Availability Statement: Data is contained within the article.

Conflicts of Interest: The authors declare no conflict of interest.

References

1. Rodrigues, L.S. Challenges of Digital Transformation in Higher Education Institutions: A Brief Discussion. In Proceedings of the 30th IBIMA Conference, Seville, Spain, 8–9 November 2017.
2. Pereira, C.S.; Durão, N.; Fonseca, D.; Ferreira, M.J.; Moreira, F. An Educational Approach for Present and Future of Digital Transformation in Portuguese Organizations. *Appl. Sci.* **2020**, *10*, 757. [CrossRef]
3. Căpușneanu, S.; Mateș, D.; Tűrkeș, M.C.; Barbu, C.-M.; Staraș, A.-I.; Topor, D.I.; Stoenică, L.; Fűlöp, M.T. The Impact of Force Factors on the Benefits of Digital Transformation in Romania. *Appl. Sci.* **2021**, *11*, 2365. [CrossRef]
4. Benavides, L.M.C.; Tamayo Arias, J.A.; Arango Serna, M.D.; Branch Bedoya, J.W.; Burgos, D. Digital Transformation in Higher Education Institutions: A Systematic Literature Review. *Sensors* **2020**, *20*, 3291. [CrossRef]
5. Mohamed, H.M.; Tlemsani, I.; Matthews, R. Higher education strategy in digital transformation. *Educ. Inf. Technol.* **2022**, *27*, 3171–3195. [CrossRef]
6. UNESCO. Digital Transformation in Education in Asia Pacific. 2022. Available online: https://transformingeducationsummit.sdg4education2030.org/system/files/2022-06/Digital%20Transformation-min.pdf (accessed on 30 August 2022).
7. Marks, A.; Al-Ali, M.; Atassi, R.; Abualkishik, A.Z.; Rezgui, Y. Digital Transformation in Higher Education: A Framework for Maturity Assessment. *Int. J. Adv. Comput. Sci. Appl.* **2020**, *11*, 504–513. [CrossRef]
8. Essel, H.B.; Vlachopoulos, D.; Adom, D.; Tachie-Menson, A. Transforming higher education in Ghana in times of disruption: Flexible learning in rural communities with high latency internet connectivity. *J. Enterpris. Communities People Places Glob. Econ.* **2021**, *15*, 296–312. [CrossRef]
9. Argüelles-Cruz, A.J.; García-Peñalvo, F.J.; Ramírez-Montoya, M.S. Education in Latin America: Toward the Digital Transformation in Universities. In *Radical Solutions for Digital Transformation in Latin American Universities, Lecture Notes in Educational Technology*; Burgos, D., Branch, J.W., Eds.; Springer: Singapore, 2021. [CrossRef]
10. OECD. *Supporting the Digital Transformation of Higher Education in Hungary, Higher Education*; OECD Publishing: Paris, France, 2021. [CrossRef]
11. Erasmus without Paper. Available online: https://erasmuswithoutpaper.eu/ (accessed on 16 July 2022).
12. Supporting Universities in the Digital Transformation in Erasmus+. Available online: https://sudte.iyte.edu.tr/ (accessed on 16 July 2022).
13. Kavrar, Ö.; Çankaya Kurnaz, S. Digitalisation readiness of university students in Erasmus+: A case study of Turkey, Italy and Spain. *J. Selçuk Univ. Soc. Sci. Inst.* **2022**, *47*, 328–341. [CrossRef]
14. Jahnke, S. Desk Research: Erasmus without Paper. European University Foundation. 2017. Available online: https://uni-foundation.eu/uploads/2017_EWP%20desk%20research%20final%20version.pdf (accessed on 8 August 2022).
15. Mincer-Daszkiewicz, J. What Is Erasmus without Paper and Why It Matters? European University Information Systems Organisation Annual Congress, Student Mobility SIG Workshop. 2019. Available online: https://www.eunis.org/wp-content/uploads/2019/07/3-eunis2019-mobility-slides-jmdEWP.pdf (accessed on 8 August 2022).
16. Chakravarty, S.; Mishra, R. Using social norms to reduce paper waste: Results from a field experiment in the Indian Information Technology sector. *Ecol. Econ.* **2019**, *164*, 106356. [CrossRef]
17. Strohmaier, R.; Schuetz, M.; Vannuccini, S. A systemic perspective on socioeconomic transformation in the digital age. *J. Ind. Bus. Econ.* **2019**, *46*, 361–378. [CrossRef]
18. Lázaro-Alemán, W.; Manrique-Galdós, F.; Ramírez-Valdivia, C.; Raymundo-Ibáñez, C.; Moguerza, J.M. Digital transformation model for the reduction of time taken for document management with a technology adoption approach for construction SMEs. In Proceedings of the 9th International Conference on Industrial Technology and Management (ICITM), Oxford, UK, 11–13 February 2020.
19. García-Morales, V.J.; Garrido-Moreno, A.; Martín-Rojas, R. The transformation of higher education after the COVID disruption: Emerging challenges in an online learning scenario. *Front. Psychol.* **2021**, *12*, 616059. [CrossRef]
20. Moore, T.M.; Picou, E.M. A Potential Bias in Subjective Ratings of Mental Effort. *J. Speech Lang. Hear. Res.* **2019**, *61*, 2405–2421. [CrossRef]
21. Moss, S.; Wilson, S.; Davis, J. Which Cognitive Biases can Exacerbate our Workload? *Australas. J. Organ. Psychol.* **2016**, *9*, 1. [CrossRef]
22. European Commission. Combined Evaluation of Erasmus+ and Predecessor Programmes. 2017. Available online: https://ec.europa.eu/assets/eac/erasmus-plus/eval/icf-volume1-main-report.pdf (accessed on 16 July 2022).
23. D'Hombres, B. International mobility of students in Italy and the UK: Does it pay off and for whom? *High. Educ.* **2021**, *82*, 1173–1194. [CrossRef]
24. European Commission. State of the Union: Commission Proposes a Path to the Digital Decade to Deliver the EU's Digital Transformation by 2030. 2021. Available online: https://ec.europa.eu/commission/presscorner/detail/en/ip_21_4630 (accessed on 16 July 2022).
25. Feroz, A.K.; Zo, H.; Chiravuri, A. Digital transformation and environmental sustainability: A review and research agenda. *Sustainability* **2021**, *13*, 1530. [CrossRef]
26. World Economic Forum. A New Circular Vision for Electronics. Time for A Global Reboot. 2022. Available online: https://www.weforum.org/reports/a-new-circular-vision-for-electronics-time-for-a-global-reboot (accessed on 16 July 2022).

applied sciences

Article

Practice Promotes Learning: Analyzing Students' Acceptance of a Learning-by-Doing Online Programming Learning Tool

Sidra Iftikhar *, Ana-Elena Guerrero-Roldán and Enric Mor

Computer Science, Multimedia and Telecommunications Department, Universitat Oberta de Catalunya, 08018 Barcelona, Spain
* Correspondence: siftikhari@uoc.edu

Abstract: Learning-by-doing is a pedagogical approach that helps in learning skills through practice. An online learning-by-doing tool, CodeLab, has been introduced to students undertaking the digital design and creation bachelor's degree program at the Universitat Oberta de Catalunya. The tool has been used to facilitate and engage students, not well-acquainted with problem-solving techniques, in an introductory programming course. The aim of this study was to examine the factors that play vital roles in students' acceptance of learning-by-doing tools that facilitate the development of problem-solving skills. The Unified Theory of Acceptance and the Use of Technology (UTAUT) model was used for this purpose and extended by adding the factor of motivation, which is essential for educational contexts. The results highlight that there is a strong relationship between acceptance and motivation, implying that students would use online learning-by-doing tools, such as CodeLab, depending on the amount of motivation and engagement while practicing the learning activities. A positive relationship between motivation and acceptance clearly supports the primary aim of using learning-by-doing tools in problem-solving courses.

Keywords: learning-by-doing; practice-based tools; online learning tools; learning to code; technology acceptance; UTAUT model

Citation: Iftikhar, S.; Guerrero-Roldán, A.-E.; Mor, E. Practice Promotes Learning: Analyzing Students' Acceptance of a Learning-by-Doing Online Programming Learning Tool. *Appl. Sci.* **2022**, *12*, 12613. https://doi.org/10.3390/app122412613

Academic Editors: Shan Wang, Zehui Zhan, Qintai Hu and Xuan Liu

Received: 27 October 2022
Accepted: 7 December 2022
Published: 9 December 2022

Publisher's Note: MDPI stays neutral with regard to jurisdictional claims in published maps and institutional affiliations.

1. Introduction

Technology-enhanced learning (TEL) has changed the mechanism of acquiring education [1]. From primary to higher levels of education, TEL has made it possible to learn online, full-time as well as part-time [2]. Online learning has numerous benefits, ranging from self-paced learning to exceptional circumstances when going to educational institutions is not a possibility [3,4]. Despite the numerous benefits of online learning, there are still many problems associated with it. One of the main problems is unsatisfactory educational models and teaching techniques leading to a lack of motivation and engagement among students, resulting in less participation in online courses and, in extreme cases, dropout from online education [5].

Numerous models and strategies have been presented for improving education over time. An effective approach to disseminate education is learning-by-doing [6]. The learning-by doing approach is defined as learning that is a result of ones' own actions, efforts, and experiences [7].

It requires students to be actively involved in the learning process. It is not just the teacher who delivers the lectures; rather, students are expected to practice and participate in learning [8]. The learning-by-doing education approach also plays a vital role in online learning contexts. Students are expected to practice by performing learning activities as part of their learning process, which are dependent on the nature of the course [9]. Similar to face-to-face courses, memorization techniques are also usually applied in online courses that aim to disseminate knowledge. On the contrary, online courses that require the development of problem-solving skills, for example, in the cases of learning to program, learning to solve mathematical problems, and solving scientific problems, the learning-by-doing approach has proven beneficial [10,11]. In the context of online programming courses, especially for novice and non-steam learners, we believe that a virtual laboratory environment can be used. This environment should facilitate learners to develop programming skills by practicing, similar to the practice that can be performed in a face-to-face programming laboratory. It should also help the learners to achieve their intended pedagogical goals.

An important factor in the evaluation of online learning tools is students' acceptance of and willingness to use the system [12]. Several studies have been conducted to evaluate these systems for acceptance that consider various models of acceptance [13,14]. In addition to students' evaluations, teachers' acceptance evaluations of online learning management systems have been conducted [15]. A widely used and validated model is the Unified Theory of Acceptance and Use of Technology (UTAUT) model, which takes into account the essential constructs for technology acceptance and use [16]. A construct is an idea that is based on theoretical explanations and is empirically verified [17]. Additionally, with acceptance evaluation through UTAUT, the context of usage and the necessary constructs specific to the context can also be evaluated [14].

Given the significance of the learning-by-doing approach and practice in the domain of problem solving and online programming education, a learning tool has been introduced to novice students to help them to learn to code. The tool, providing the opportunity to practice, has been incorporated into an introductory programming course for the Design and Digital Creation bachelor's degree at the Universitat Oberta de Catalunya (UOC) with the aim of helping and engaging learners. The learners of this degree program are less familiar with programming and problem-solving concepts. The primary aim of this study was to identify the major constructs for acceptance of learning-by-doing practice-based tools in online education using the validated UTAUT model. The model was extended by including the construct of motivation, which is vital for online learning tools and systems. The results of the study highlight that motivation is an important construct to consider when evaluating practice-based online tools regarding acceptance. When exploring the relationships amongst the constructs included in this study, a statistically significant relationship was found between facilitating conditions and motivation. There was also a significant relationship found between effort expectancy, trust in the system, and social influence. These significant relationships highlight the fact that online learning tools based on practice, if designed by taking the significant constructs into account, can be successfully accepted by online learners.

Online education, blended or fully online, has proven its advantages and was a necessity during the COVID-19 pandemic, when going to educational institutes was not possible [18]. Students' acceptance of and willingness to use practice-based online learning tools in their courses is crucial, as it will help promote student engagement and willingness to learn. Through successful acceptance of practice-based tools, students can experience learning with great potential and achieve their learning goals associated with skill acquisitions, such as learning to code. For instructors and institutions disseminating knowledge, engaging students is important to ensure quality in online higher education.

2. Theoretical Background

Several theories have been proposed for learning and educational contexts. In online education, basic learning theories, such as behaviorism, constructivism, humanism, and connectivism, have been applied [19]. Additionally, other theories, such as social learning [20], transformative learning [21], and experiential learning, have also been used in e-learning. In terms of learning skills, the approach of modern education is a feasible one that encourages the use of creative techniques and problem-solving-based approaches for learning [22]. The educational approach of learning-by-doing is aligned with the modern educational approach. This approach is derived from experiential and active learning and encourages learners to practice and acquire hands-on experience to learn rather than relying on theory alone [23]. The approach of experiential learning, learning-by-doing, dates to John Dewey's theory of progressive education and is still applicable in this era [6]. The idea is to learn by practice, that is, learning is to be achieved through one's own hands-on experiences, which could be mental or physical [24]. Learning-by-doing has been applied in a variety of contexts and fields of education and has provided favorable results. Learning-by-doing holds significance for scientific education, especially when real-life simulations are considered [25]. In situations where learning skills are of interest, a learning-by-doing approach proves to be feasible [24]. Knowledge about a domain can be acquired by traditional means of education, such as reading theory and attending lectures, but for the domains that require concepts to be developed as skills, practice is required [26]. Learning a concept by theory first, then applying it several times helps in acquiring a skill. Problem-solving skills are also developed by practicing similar problems based on a particular concept [27]. As for skills, learning to code is a skill that can be only developed by virtue of practice [28].

In this context, Nantsou et al. have presented learning-by-doing for physics and electronics education [29]. Similarly, in the domain of marketing education, experiential learning and learning-by-doing has provided beneficial experiences [30]. George et al. explored the effects of applying learning-by-doing in the field of criminal justice. The study based on undergraduate students concluded that there are several benefits of learning-by-doing in terms of professional development [31]. Niiranen also conducted a study that assessed the use of the learning-by-doing approach in the field of craft and technology education [32].

In online education, where there is a lack of face-to-face experience, learning-by-doing has been used. Not only does it help develop skills and effectively teach concepts, it also regulates student engagement [33]. For the learning of problem-solving skills in online contexts, various tools have been discussed in the literature [34–36]. In the specific case of online programming tools and environments, which also belong to the category of problem solving and take into account the modern education perspective, several learning tools and systems have been proposed for students, for various programming languages, ranging from novice [37] to advanced levels [38]. Rossano et al. proposed a tool for the programming of digital logic and design based on learning-by-doing and practice for high-school students [39]. In another study, Hosseini et al. proposed a tool that aims to teach Python to students in higher education through practice [40]. Minecraft is also a learning-by-doing tool that helps learners develop problem-solving skills through the process of gamification [41]. Similarly, w3school is an online learning platform that provides the opportunity to learn to code by practicing [42]. Code.org [43] and CodeAcademy [44] are also based on the ideas of experiential learning. However, not all tools and systems for teaching and learning programming are built upon the idea of encouraging students to learn to code through practice; rather, there is a limited set of activities and tools focusing on assessment alone.

Acceptance of technology and systems has been a concern among the research community for many years now [45]. Several models and theories have been proposed for this purpose [46–48]. Unified Theory of Acceptance and Use of Technology (UTAUT) has been used in several domains to evaluate the acceptance of technological systems and products [16,49]. Similarly, in the domain of education in general and online learning in particular, UTAUT has been used extensively to evaluate the factors that lead to the acceptance of systems [50–54]. In the case of a programming course, Morais et al. assessed the introduction of programming practices in a degree program of arts and design [55]. However, most studies on UTAUT in the domain of e-learning and online education evaluate acceptance in terms of a system having the scope to be implemented in a particular university or institution. The educational approach that the system uses is usually not evaluated for acceptance by students. In this study, the focus was on identifying and evaluating the factors that lead to acceptance of online learning tools that are based on the educational approach of learning-by-doing and practicing—an approach that is essential to learning to code and developing problem-solving skills.

3. Research Methodology

3.1. The Study Context

This research centered on an online mandatory course taught as part of the online bachelor's degree program of "Design and Digital Creation" at the Universitat Oberta de Catalunya (UOC)—a fully online university. UOC has over 70,000 students and more than 4500 teaching and research staff and aims to provide online learners with quality personalized learning for online higher education. To address these aims of UOC, the CodeLab programming tool has been designed, developed, and incorporated into the course [56]. The learning tool aims to facilitate and engage novice learners in their first programming course of the degree, who typically have little to no programming experience. The students are provided with the opportunity to practice the concepts continuously, as learning to code requires practice and reflection. CodeLab considers the cognitive processes and social interactions of the students as they experiment and test the ideas and concepts and communicate with their peers.

CodeLab is based on a practice approach and contains a detailed set of learning itineraries related to coding concepts using the P5.js language. Not only is it possible to write and execute a piece of code, students can also visualize the executed code. First, the students are introduced to programming concepts that establish a basic understanding, and later they are expected to practice these concepts for the acquisition of programming skills related to these concepts. Once they feel confident, they are encouraged to move to advanced and complex concepts and to practice to continue learning. In the online course considered in this study, five modules in the tool were defined based on specific concepts, including several learning activities for assessment, recommended and complementary in nature, the latter two encouraging the practice of concepts. In this study, the CodeLab tool was used for one semester spanning 16 weeks, and each module was 3 weeks in duration. The students could ask questions, discuss, and enhance their knowledge and experience by interacting with teachers and peers through a communication channel provided within the tool.

Taking into account the unique experience of an online course that mainly involves interacting with an online programming tool, students' acceptance evaluations of the tool seemed vital. This paper discusses the acceptance evaluation of a pilot study of the CodeLab tool and is part of the on-going research project on the learning-by-doing practice-based e-learning systems. The research methodology that is applied in this context is the "design and creation" methodology [57]. This research methodology helps in providing the opportunity to recognize, articulate, reflect, and design artifacts to address the problems at issue (Figure 1). The pilot study aimed to improve the educational and interaction design of the tool and to elicit qualitative and quantitative data from the students and teachers. The data generated through the usage of the CodeLab tool were also used for the evaluation of student behavior and engagement. Several e-learning tools and systems were evaluated for acceptance to help determine their continued usage by the students. The acceptance evaluations in this paper were not only specific to the tools but would also help to determine the factors that would need to be considered when developing and evaluating similar e-learning systems that encourage practice for learning.

3.2. Research Model and Hypothesis Development

The UTAUT model takes into account the behavioral intention (BI) to use a product or system for acceptance evaluation. Calder discusses that a construct is an idea that is based on theoretical explanations; however, in most cases constructs are not limited to theory alone; rather, they are empirically verified [17]. In this study, we considered this point of view, similar to the original UTAUT model, i.e., theoretical explanations were provided first and later we verified the constructs through statistical testing. To evaluate BI, Venkatesh et al. took into account the constructs of performance expectancy (PE), effort expectancy (EE), facilitating conditions (FCs), and social influence (SI) (Figure 2).

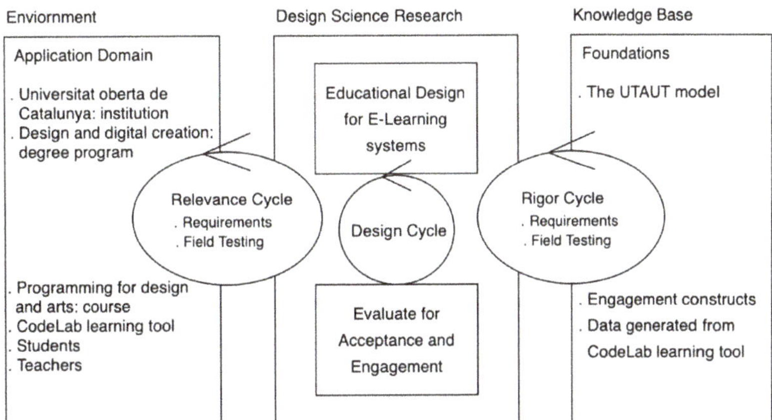

Figure 1. Design and creation adapted for CodeLab.

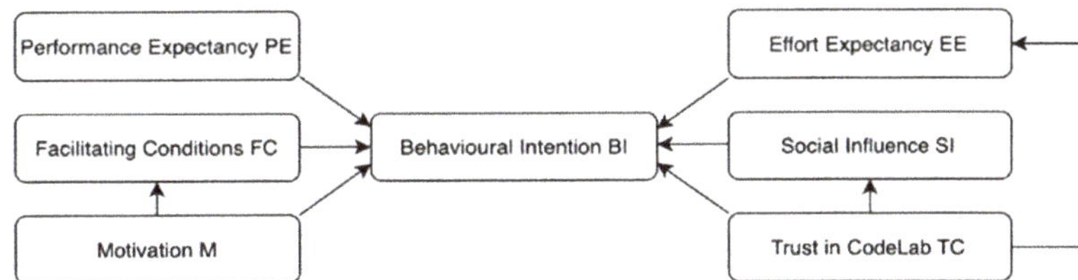

Figure 2. Model adapted from [45].

Performance expectancy (PE) refers to the degree of belief of an individual that by using a system they will attain benefits in their job or benefits associated with the intended purpose of using the system. In the case of CodeLab, the intended purpose is to learn effectively while being engaged in the course. As the students are encouraged to learn to code by practicing the programming-based learning activities, it was to be inquired into whether the tool helps them to learn. Based on the above discussion, the following hypothesis was proposed:

H1. *Performance expectancy has a positive effect on the behavioral intention to use CodeLab.*

Effort expectancy (EE) is related to the degree of ease of use of the system. For e-learning systems, ease of use is an important factor in acceptance and effectiveness [58]. In the specific case of CodeLab, the students were to be asked whether they thought and whether they found that the CodeLab tool was easy to use and whether it was easy for them to be skillful in learning to use CodeLab.

H2. *Effort expectancy has a positive effect on the behavioral intention to use CodeLab.*

Facilitating conditions (FCs) are the user's perceptions of the factors in the environment of the system that help to use the system. In addition to general e-learning systems, FCs have also been used in studies related to tools for teaching programming [59]. The students were to be asked if they were provided with the necessary resources and knowledge and whether someone was available to help them in case they encounter difficulties when using CodeLab.

H3. *Facilitating conditions have a positive effect on the behavioral intention to use CodeLab.*

Social influence (SI) is the perceived effect of the opinion of other people and affects the intention to use. The construct of SI was evaluated by inquiring of the students whether the people who influence their behavior and are important, for example, their friends and family, believe that they should use CodeLab.

H4. *Social influence has a positive effect on the behavioral intention to use CodeLab.*

Trust is an important construct that has been used and validated in the context of e-learning systems [60], even though the original UTAUT model does not emphasize trust as a factor that influences the intention to use. For the acceptance of mobile learning systems, trust is an important construct [61]. The construct of trust has also been considered for CodeLab, not in terms of security but in terms of trust in the system (TC) that it will help them to learn to code. Trust is not a core construct of UTAUT; it has been used and validated in e-learning [62]. In this study, we considered a mediating effect between TC and EE and SI as the degree of ease of use of the system and that the people who influence a person's behavior can inculcate TC.

H5. *Trust has a positive effect on the behavioral intention to use CodeLab.*

H5.a. *The positive effect of trust in CodeLab on behavioral intention to use will be mediated through effort expectancy and social influence.*

Motivation and engagement are considered essential constructs when it comes to education [59] and e-learning tools [63]. Reports in the literature have strongly emphasized the fact that the motivation and engagement of students is essential for learning and is a primary concern of researchers and teachers [64]. As CodeLab is an online educational tool, motivation is to be considered; motivated students will be able to learn productively by being engaged. As this is one of the first studies on the evaluation of acceptance of the tool, the construct of motivation has been taken into account by asking students if they feel motivated to learn to program because they are using the learning tool. In addition to the direct relationship between motivation (M) and BI, a mediating relationship with FCs was also evaluated.

H6. *Motivation has a positive effect on the behavioral intention to use CodeLab.*

H6.a. *The positive effect of motivation with respect to CodeLab on the behavioral intention to use it will be mediated through facilitating conditions.*

The behavioral intention to use (BI) a construct is the acceptance of the technological system in the core UTAUT model. However, for various situations, depending on the context of usage and the study, acceptance can differ. In the specific case of the CodeLab tool, the meaning of acceptance is also related to the usage context. The students who use the tool to study for the course will be using the CodeLab tool for one semester. The acceptance in this scenario is twofold: firstly, the acceptance to use CodeLab during the semester, and, secondly, to accept that they are willing to use similar tools that are based on an approach of learning-by-doing.

3.3. Procedure for Data Collection

The questionnaire used in the study was based on UTAUT as this is a validated and widely used model for examining users' perceptions regarding the acceptance of a system and one that has been used in several domains, including education [49], as mentioned in Section 2. Other acceptance models could have been used, for example, TAM; however, it was observed that UTAUT is validated in the field of education, and the constructs it contains are relevant to this study. The UTAUT questionnaire uses the constructs of PE, EE, SI, FCs, and TC and any additional ones that are vital for the context, and in this case motivation was also used. The basic structure of the questionnaire was maintained; however, where necessary, it was adapted to the CodeLab context. The questionnaire consisted of 33 questions divided into 3 sections. The first section contained questions based on the UTAUT constructs, the second was about the usability of the system, and the third section captured demographic information. For the first section, a 7-point Likert scale ranging from strongly agree to strongly disagree was used to capture the responses. The questionnaire was developed using Qualtrics software [65] and was sent to the students via anonymous links on the classroom board, so that it was voluntary for them to answer. Qualtrics has been used in this context as it provides a variety of features that facilitate the development, distribution, and preparation of data for analysis as compared with other survey creation tools, such as Google Forms and SurveyMonkey. Additionally, it supports various languages and could easily be integrated with the learning environment of the university, facilitating the questionnaire-based research. As for the timing of sending the questionnaire out, it was sent in the final weeks of the semester, when the students were finishing the last module of activities, such that a post-usage evaluation of UTAUT was conducted.

3.4. Sample

The classrooms consisted of 116 students in total. There were 39 students in the Catalan classroom and 77 in the Spanish classroom. Answering the questionnaire was completely voluntary for the students. A consent form was first given to the students, and those who consented took part in the survey. A total number of 87 responses were captured, making for a 75% response rate. Later, the valid responses were evaluated that did not contain any missing data. Of the valid responses, 38% were from male students, with the majority of 62% being from female students. The students were also asked about their course type, i.e., full-time or part-time: 31% of the students were full-time; 69% of the students were part-time. As regards age, 90% of the students were within the age group of 21 to 45 years.

3.5. Data Filtering Software and Techniques

To test the model and the stated hypotheses, partial least squares (PLS) was used [66]. As has been mentioned, our overall sample space comprised 116 enrolled students and the total number of valid responses was 72, such that the sample size with respect to the context of the study was significant, though it can be generally considered low in magnitude. For this purpose, PLS was considered for the analysis as it has the potential to provide robust results even with smaller sample sizes [67]. The most complex construct in the study had 3 items; therefore, the minimum sample size required was 30, i.e., 10 times the maximum number of items in a construct [66]. The PLS analysis was performed using the smartPLS tool, version 3.3.3. Data collected from the survey were filtered for valid responses, and the partially completed responses were removed from the valid set of responses. The descriptive statistics and data related to gender, age, and student type were evaluated in SPSS.

4. Results

As the process for analysis with PLS-SME suggests, first, the measurement model was created and evaluated. In the second step, the structural model was created to verify the relationships between the constructs for statistical significance [66]. The model used in this research was reflective in nature; taking the nature of the model into account, the consistent version of the PLS algorithm was used [68].

4.1. Variance of the Endogenous Variable: R2 and Indicator Loadings

The coefficient of determination of variance, that is, the in-sample explanatory power, for the target dependent variable, R2, for BI was 0.417, i.e., 41%. This implies that the six latent variables PE, EE, SI, FCs, T, and M explained 42% of the variance in BI. For the indirect effects, the coefficient of determination, R2, for FCs, the latent variable, was 0.461; this implies that M explained 46% of the variance in FCs. Similarly, for SI, 44% of the variance of TC was explained by SI. However, for EE, 69% of the variance of TC was explained by EE. The R2 values in the case of this study moderately explain the variance for the respective dependent variables [69].

The standardized path coefficients for the outer loadings, M has the strongest effect on BI, 0.466. FCs has a path coefficient of 0.155, EE one of 0.319, PE one of -0.192, SI one of 0.011, and TC one of -0.078. The path coefficients for FCs, EE, and M were statistically significant; however, for SI, the path coefficient was insignificant, since it was lower than 0.1, and the path coefficients for PE and TC highlighted a negative relationship with BI [69]. For the mediated effects, the path coefficient of TC to EE was 0.843, and from TC to SI was 0.664; both in this case were statistically significant. The path coefficient from M to FCs was 0.679, and was significant too. Regarding the total indirect effects, the path TC-EE-BI was 0.0228, and was significant. The path TC-SI-BI was -0.004, showing insignificance and a negative correlation. The path M-FCs-BI was 0.128, which was close to a significant value.

4.2. Indicator Reliability

For indicator reliability, the squares of all the outer loadings of the items were considered; a value ≥ 0.7 was preferred, while 0.4 was acceptable [69]. Table 1 highlights the indicator reliabilities for all the constructs in the UTAUT study for CodeLab within the range. Internal consistency reliability depicted in smartPLS as composite reliability and evaluated by Cronbach's alpha should be equal to or higher than 0.5 [70]. In the case of this study, the composite reliability was greater than 0.7 and reasonably high for all the latent variables. Another view about the reliability of PLS constructs was proposed in [68]—that the rho_A value should be ≥ 0.7 and ≤ 1. In the case of this study, the rho_A value for all the latent variables was also within this range (Table 1).

Table 1. Indicator and composite reliability: AVE values.

Variables	Indicators	Loadings	Indicator Reliability	Composite Reliability	AVE	rho_A
	PE_1	0.889	0.790			
PE	PE_2	0.930	0.865	0.964	0.899	0.964
	PE_3	1.021	1.042			
	EE_1	0.677	0.458			
EE	EE_2	0.671	0.450	0.759	0.517	0.777
	EE_3	0.843	0.711			
	SI_1	1.041	1.083			
SI	SI_2	0.910	0.828	0.935	0.829	0.954
	SI_3	0.795	0.632			
	FC_1	0.791	0.625			
FCs	FC_2	0.869	0.755	0.872	0.694	0.873
	FC_3	0.836	0.698			
	T_1	0.867	0.751			
TC	T_2	0.884	0.781	0.889	0.728	0.891
	T_3	0.807	0.651			
M	M_1	1.067	1.138	0.893	0.813	0.966
	M_2	0.698	0.487			
BI	BI_1	1.000	1.000	1.000	1.000	1.000

4.3. Convergent and Discriminant Validity

As for convergent validity, it depends on the average variance extracted (AVE) values. The AVE values should be greater than 0.5 [71]. In the case of this study, the AVE of values of all the constructs were greater than 0.5; however, for EE the AVE was equal to 0.5 (Table 1), highlighting convergent validity for all the constructs except EE. Discriminant validity was shown by the AVE values and the correlation of latent variables. F is used to evaluate discriminant validity, which suggests that the square root of AVE for each latent variable should be greater than its correlation with other latent variables. For this study, the criterion of [72] was satisfied for all the latent variables except EE, i.e., the square root of each latent variable was greater than its correlation (Table 2).

Table 2. Discriminant validity: the bolded values are AVE squares.

	EE	FC	BI	M	PE	SI	TC
EE	**0.179**						
FC	0.788	**0.833**					
BI	0.592	0.528	**1.000**				
M	0.738	0.641	0.586	**0.901**			
PE	0.695	0.633	0.453	0.812	**0.948**		
SI	0.582	0.491	0.390	0.627	0.648	**0.911**	
TC	0.840	0.776	0.558	0.836	0.771	0.648	**0.853**

4.4. Hypothesis Testing

The measurement model was evaluated for statistical significance through the process of bootstrapping [67]. *t*-statistics and *p*-values were generated for the paths between the items of each latent construct and amongst the constructs. When performing bootstrapping in smartPLS, the sub-samples used were 5000, and the one-tailed test with a 0.05 significance level was used. As for the interactions between the latent constructs, four paths that were significant are highlighted in Table 3.

Table 3. Structural model results (bold rows highlight significant paths).

Path	*t*-Statistics	*p*-Values
EE-BI	1.243	0.107
FC-BI	1.000	0.159
M-FC	**6.965**	**0.000**
M-BI	**1.715**	**0.043**
PE-BI	0.593	0.277
SI-BI	0.121	0.452
TC-EE	12.616	0.000
T-BI	0.176	0.430
TC-SI	8.247	0.000

The path from motivation to behavioral intention was significant. It can be concluded that motivation is an important factor in the intention to use the CodeLab tool (Table 4). For the indirect effects, the measurement model was also bootstrapped to identify the statistical significance and as per the results the total indirect paths were not significant. The relationship between motivation (M) and the intention to use (BI) was positive and significant (Table 3), verifying H6. It can be concluded that motivation is an important construct when evaluating the intention to use e-learning systems that encourage learning-by-doing. As this was the first UTAUT evaluation of CodeLab, in the future, studies on other online learning systems that encourage learning-by-doing, not just CodeLab, can consider educational factors, such as student engagement, which is a positive outcome of student motivation.

Table 4. Hypothesis testing summary.

	Statement	Measurement Model (Preliminary Significance)	Structural Model (Statistical Significance)
H1	Performance expectancy (PE)	Reject (negative relationship)	Reject
H2	Effort expectancy (EE)	Accept	Reject
H3	Facilitating conditions (FCs)	Accept	Reject
H4	Social influence (SI)	Reject	Reject
H5	Trust in CodeLab (TC)	Reject (negative relationship)	Reject
H5.a	TC, EE, and SI	Accept	Accept
H6	Motivation	Accept	Accept
H6.a	M and FCs	Accept	Accept

Additionally, there was a positive and significant relationship between motivation (M) and facilitating conditions (FCs) (Table 3), partially supporting H6.a, implying that the presence of facilitating conditions contributes to the motivation of students when using the system. Facilitating conditions, as previously mentioned, are personal qualities or factors that contribute positively to the learning process. For future evaluations of CodeLab and similar e-learning tools, in-context facilitating conditions can also be considered, for example, course forums, teachers' feedback, and channels for peer communication that facilitate the learning process.

Similarly, the relationships between trust (TC) and effort expectancy (EE) and between TC and social influence (SI) were positive and significant, partially supporting H5.a, since the total indirect effect was not significant. Regarding the degree of ease of use, EE does inculcate trust in CodeLab. e-learning systems that are easy to use can be relied upon and trusted in several respects, including those specific to education.

Taking into account the hypothesis for the PE construct, for H1 there was a negative and a non-significant relationship. As for the degree of ease of use, effort expectancy (EE) was positive but insignificant for CodeLab (H2). Similarly, in the case of the facilitating conditions, the effect was positive yet insignificant, highlighted by H3. For the construct of social influence represented in H4, the relationship between SI and acceptance was non-significant.

5. Discussion

Learning-by-doing promotes students' motivation and engagement [32]. In this study, similar to the study conducted by [47], the relationship between motivation and acceptance was found to be prominently positive [52]. The CodeLab tool motivated the students to learn and actively participate in the course. This implies that if e-learning systems are designed with innovative educational strategies that keep students motivated and encourage their involvement, students are likely to willingly accept the e-learning tools and learn. As this was the first UTAUT evaluation of CodeLab, in the future, studies not just on CodeLab but on other online learning systems that encourage learning-by-doing can consider educational factors, such as student engagement, which is a positive outcome of student motivation.

Regarding the non-significant relationships determined in this study, an interesting study by Williams concluded that 23 out of a total of 116 studies on UTAUT did not find significant impacts of PE on students' acceptance [45]. As has been mentioned, skills such as programming are acquired by learning-by-doing [28], and even though CodeLab helped the students to acquire programming skills, the non-significant relationship between PE and acceptance can be explained by an observation reported by the course teachers. In the classrooms using CodeLab, the students were only performing learning activities of a mandatory nature and not of the recommended and complementary nature, signifying that there was less practice being performed by the students to learn to program. For the next semester, it has been considered that the students should be made aware of the fact that the purpose of the CodeLab tool is to help them to learn to code by practicing. The more they do, i.e., practice the programming-based activities provided in CodeLab, the more they will learn.

Similarly, several studies have found an insignificant relationship between the intention to use and effort expectancy [45]. Additionally, as suggested in a UTAUT for computer programming, effort expectancy can improve overtime and could be low for novice learners [55]. In the case of the CodeLab tool, it can be claimed that there were some technical issues of a minor nature that were encountered by some students which could possibly have led them to believe that the EE for CodeLab was lower than it actually was. For example, the option to save the code did not work as expected for some students.

For the non-significant relationship with FCs, in the case of CodeLab, as it is an e-learning programming tool, a person, more specifically, a teacher, can be considered a facilitator who is available to help the students. Additionally, as the development of the

CodeLab tool progresses, other ways and measures can be incorporated that facilitate the students in terms of learning to code using the tool. For example, a communication channel that encourages asynchronous communication not just between teachers and students but also encourages peer-to-peer interaction could be helpful and facilitate the students in using CodeLab in the course and enhance their learning.

In the case of online learning tools, the social aspect is vital, as highlighted in [20]. The relationship between social influence and acceptance in this study having been found to be insignificant, it could be argued that the items of SI used in this study are the core items of SI that are considered by the core UTAUT model. These items did not coincide well with the context of CodeLab. The SI items suggest that friends and family could influence the intention to use (BI) among the students. However, reflecting on the results for SI, we may consider context-specific people involved. In the case of CodeLab, the context-specific people could be the teachers teaching the course, friends, or colleagues in related or similar domains, including those who have studied online courses that encourage the learning-by-doing model and alumni of the course in which CodeLab was used. In future studies related to CodeLab and similar tools that encourage learning by virtue of practice, context-specific people and persons who have social influence on students should be considered.

6. Conclusions and Future Work

Learning-by doing is an educational approach that helps in the acquisition of skills in face-to-face, blended, and fully online contexts. The approach encourages students to learn by practice. The purpose of this study was to identify the factors that play an important role in successful acceptance of the practice-based online learning tools that are based on the pedagogical approach of learning-by-doing. As learning to program is a skill that is acquired by practice, the CodeLab tool, based on the learning-by-doing approach, has been incorporated into a programming course for a bachelor's degree program. The learners are new to programming concepts and the tool is likely to help them acquire problem-solving skills. The UTAUT model was used for the evaluation of the CodeLab tool to identify the factors that lead to students' acceptance of it and similar e-learning tools and systems that encourage learning-by-doing.

The core constructs of UTAUT—performance expectancy (PE), effort expectancy (EE), social influence (SI), and facilitating conditions (FCs)—were not significant according to the results of our study. The factors that possibly could have led to insignificant results for the mentioned constructs have been discussed and recommendations for the design of e-learning tools that encourage practice have been provided. It cannot be claimed that these constructs are not important in the context of e-learning systems; however, the design of the e-learning system should be supportive enough for these constructs to play a significant role in the acceptance of the systems. The results highlight that motivation is an important construct for practice-based tools that should be taken into account for the acceptance of these systems. In the specific case of learning-by-doing and practice-based tools, such as the case of CodeLab tool, these tools and systems are designed and incorporated into online courses with the aim of enhancing student motivation and engagement. The aim of incorporating CodeLab into an online programming course was to indeed help the students become involved and engaged actively in the course, which is supported by the results of this study.

In future, additional constructs that determine the successful acceptance of practice-based tools can be used. As this was the first pilot study on the CodeLab tool, the results can be acknowledged as important, yet additional pilot studies with more than one course could help determine the factors leading to acceptance more precisely. The relationship between student motivation and acceptance and willingness to use is interesting, as signified by this study; however, more insightful relationships, for example, between student engagement and its various dimensions, student performance and grades in online classrooms, and satisfaction, can be examined as determinants of successful acceptance of online learning-by-doing tools and systems. Additionally, apart from determining the factors that lead to successful adoption and acceptance, other important factors, for example, student engagement and student awareness of progress in the course and learners' experiences can also be considered for the evaluation of learning-by-doing tools and systems. Another important aspect concerning such tools and systems could be to take into account teachers' perspectives on the practice-based tools and to enhance the teaching experiences that would consequently positively impact the entire educational process.

Author Contributions: Conceptualization, S.I., A.-E.G.-R. and E.M.; Methodology, S.I., A.-E.G.-R. and E.M.; Validation, A.-E.G.-R.; Formal analysis, S.I.; Investigation, S.I., A.-E.G.-R. and E.M.; Writing— original draft, S.I.; Writing—review & editing, S.I., A.-E.G.-R. and E.M. All authors have read and agreed to the published version of the manuscript.

Funding: This work is part of the R + D + i project PID2021-128875NA-I00, funded by MCIN/AEI/ 10.13039/501100011033/ERDF A way of making Europe.

Institutional Review Board Statement: The study was conducted in accordance with the Ethics Committee of UOC (15 October 2019).

Informed Consent Statement: Informed consent was obtained from all subjects involved in the study.

Data Availability Statement: The data used in this project are confidential and came from the learning management system of UOC, the CodeLab tool, and the questionnaires that included personal identifying information about the participants which cannot be disclosed.

Conflicts of Interest: The authors declare no conflict of interest.

References

1. Goodyear, P.; Retalis, S. *Technology-Enhanced Learning*; Sense Publishers: Rotterdam, The Netherlands, 2010; Volume 6.
2. Appana, S. A review of benefits and limitations of online learning in the context of the student, the instructor and the tenured faculty. *Int. J. E-Learn.* **2008**, *7*, 5–22.
3. Rapanta, C.; Botturi, L.; Goodyear, P.; Guàrdia, L.; Koole, M. Online university teaching during and after the Covid-19 crisis: Refocusing teacher presence and learning activity. *Postdigit. Sci. Educ.* **2020**, *2*, 923–945. [CrossRef]
4. Verawardina, U.; Asnur, L.; Lubis, A.L.; Hendriyani, Y.; Ramadhani, D.; Dewi, I.P.; Darni, R.; Betri, T.J.; Susanti, W.; Sriwahyuni, T. Reviewing online learning facing the COVID-19 outbreak. *Talent. Dev. Excell.* **2020**, *12*, 385–392.
5. Rostaminezhad, M.A.; Mozayani, N.; Norozi, D.; Iziy, M. Factors Related to E-learner Dropout: Case Study of IUST Elearning Center. *Procedia Soc. Behav. Sci.* **2013**, *83*, 522–527. [CrossRef]
6. Williams, M. John Dewey in the 21st Century. *J. Inq. Action Educ.* **2017**, *9*, 91–102.
7. Reese, H.W. The learning-by-doing principle. *Behav. Dev. Bull.* **2011**, *17*, 1. [CrossRef]
8. Niiranen, S.; Rissanen, T. Learning by Doing and Creating Things with Hands: Supporting Students in Craft and Technology Education. *PATT Proc.* **2017**, *34*, 150–155. Available online: https://www.iteea.org/File.aspx?id=115739&v=21dfd7a (accessed on 4 December 2022).
9. Kanakana-Katumba, M.G.; Maladzhi, R. Online Learning Approaches for Science, Engineering and Technology in Distance Education. In Proceedings of the 2019 IEEE International Conference on Industrial Engineering and Engineering Management (IEEM), Macao, China, 15–18 December 2019; pp. 930–934.
10. Frache, G. A Constructively aligned Learning-by-Doing Pedagogical Model for 21st Century Education. *Adv. Sci. Technol. Eng. Syst. J.* **2017**, *3*, 1119–1124.
11. Hettiarachchi, E.; Huertas, M.A. Introducing a Formative E-Assessment System to Improve Online Learning Experience and Performance. *J. Univers. Comput. Sci.* **2015**, *21*, 1001–1021.
12. Olasina, G. Human and social factors affecting the decision of students to accept e-learning. *Interact. Learn. Environ.* **2019**, *27*, 363–376. [CrossRef]
13. Fathema, N.; Shannon, D.; Ross, M. Expanding the Technology Acceptance Model (TAM) to examine faculty use of Learning Management Systems (LMSs) in higher education institutions. *J. Online Learn. Teach.* **2015**, *11*. Available online: https://jolt.merlot.org/Vol11no2/Fathema_0615.pdf (accessed on 4 December 2022).
14. Persada, S.F.; Miraja, B.A.; Nadlifatin, R. Understanding the Generation Z Behavior on D-Learning: A Unified Theory of Acceptance and Use of Technology (UTAUT) Approach. *Int. J. Emerg. Technol. Learn.* **2019**, *14*, 20–33. [CrossRef]
15. Dindar, M.; Suorsa, A.; Hermes, J.; Karppinen, P.; Näykki, P. Comparing technology acceptance of K-12 teachers with and without prior experience of learning management systems: A Covid-19 pandemic study. *J. Comput. Assist. Learn.* **2021**, *37*, 1553–1565. [CrossRef] [PubMed]
16. Venkatesh, V.; Morris, M.G.; Davis, G.B.; Davis, F.D. User acceptance of information technology: Toward a unified view. *MIS Q.* **2003**, *27*, 425–478. [CrossRef]

17. Calder, B.J.; Brendl, C.M.; Tybout, A.M.; Sternthal, B. Distinguishing constructs from variables in designing research. *J. Consum. Psychol.* **2021**, *31*, 188–208. [CrossRef]
18. Xie, X.; Siau, K.; Nah, F.F.-H. COVID-19 pandemic–online education in the new normal and the next normal. *J. Inf. Technol. Case Appl. Res.* **2020**, *22*, 175–187. [CrossRef]
19. Arghode, V.; Brieger, E.W.; McLean, G.N. Adult learning theories: Implications for online instruction. *Eur. J. Train. Dev.* **2017**, *41*, 593–609. [CrossRef]
20. Hill, J.R.; Song, L.; West, R.E. Social learning theory and web-based learning environments: A review of research and discussion of implications. *Am. J. Distance Educ.* **2009**, *23*, 88–103. [CrossRef]
21. Boyer, N.R.; Maher, P.A.; Kirkman, S. Transformative Learning in Online Settings: The Use of Self-Direction, Metacognition, and Collaborative Learning. *J. Transform. Educ.* **2006**, *4*, 335–361. [CrossRef]
22. Pllana, D. Creativity in Modern Education. *World J. Educ.* **2019**, *9*, 136–140. [CrossRef]
23. Zhang, G.X.; Sheese, R. 100 years of John Dewey and education in China. *J. Gilded Age Progress. Era* **2017**, *16*, 400. [CrossRef]
24. Gutiérrez-Carreón, G.; Daradoumis, T.; Jorba, J. Integrating learning services in the cloud: An approach that benefits both systems and learning. *Educ. Technol. Soc.* **2015**, *18*, 145–157.
25. Bot, L.; Gossiaux, P.-B.; Rauch, C.-P.; Tabiou, S. 'Learning by doing': A teaching method for active learning in scientific graduate education. *Eur. J. Eng. Educ.* **2005**, *30*, 105–119. [CrossRef]
26. Ekici, D.I. The Use of Edmodo in Creating an Online Learning Community of Practice for Learning to Teach Science. *Malays. Online J. Educ. Sci.* **2017**, *5*, 91–106.
27. Samani, M.; Putra, B.A.W.; Rahmadian, R.; Rohman, J.N. Learning Strategy to Develop Critical Thinking, Creativity, and Problem-Solving Skills for Vocational School Students. *J. Pendidik. Teknol. Dan Kejuru.* **2019**, *25*, 36–42. [CrossRef]
28. Bareiss, R.; Griss, M. A story-centered, learn-by-doing approach to software engineering education. In Proceedings of the 39th SIGCSE Technical Symposium on Computer Science Education, Portland, OR, USA, 12–15 March 2008; pp. 221–225.
29. Nantsou, T.; Frache, G.; Kapotis, E.C.; Nistazakis, H.E.; Tombras, G.S. Learning-by-Doing as an Educational Method of Conducting Experiments in Electronic Physics. In Proceedings of the 2020 IEEE Global Engineering Education Conference (EDUCON), Porto, Portugal, 27–30 April 2020; pp. 236–241. [CrossRef]
30. Sangpikul, A. Challenging graduate students through experiential learning projects: The case of a marketing course in Thailand. *J. Teach. Travel Tour.* **2020**, *20*, 59–73. [CrossRef]
31. George, M.; Lim, H.; Lucas, S.; Meadows, R. Learning by doing: Experiential learning in criminal justice. *J. Crim. Justice Educ.* **2015**, *26*, 471–492. [CrossRef]
32. Niiranen, S. Supporting the development of students' technological understanding in craft and technology education via the learning-by-doing approach. *Int. J. Technol. Des. Educ.* **2021**, *31*, 81–93. [CrossRef]
33. Adkins, D.; Bossaller, J.; Brendler, B.; Buchanan, S.; Sandy, H.M. Learning by Doing: Using Field Experience to Promote Online Students' Diversity Engagement and Professional Development. *Expand. LIS Educ. Universe* **2018**, *17*, 123–127. Available online: http://hdl.handle.net/2142/99018 (accessed on 4 December 2022).
34. Gunawan, G.; Harjono, A.; Sahidu, H.; Herayanti, L. Virtual laboratory to improve students' problem-solving skills on electricity concept. *J. Pendidik. IPA Indones.* **2017**, *6*, 257–264. [CrossRef]
35. Hasibuan, A.M.; Saragih, S.; Amry, Z. Development of Learning Materials Based on Realistic Mathematics Education to Improve Problem Solving Ability and Student Learning Independence. *Int. Electron. J. Math. Educ.* **2019**, *14*, 243–252. [CrossRef] [PubMed]
36. Hettiarachchi, E. *Technology-Enhanced Assessment for Skill and Knowledge Acquisition in Online Education*; Universitat Oberta de Catalunya: Barcelona, Spain, 2014.
37. Sim, T.Y.; Lau, S.L. Online Tools to Support Novice Programming: A Systematic Review. In Proceedings of the 2018 IEEE Conference on e-Learning, e-Management and e-Services (IC3e), Langkawi, Malaysia, 21–22 November 2018; pp. 91–96.
38. Salleh, S.M.; Shukur, Z.; Judi, H.M. Analysis of Research in Programming Teaching Tools: An Initial Review. *Procedia-Soc. Behav. Sci.* **2013**, *103*, 127–135. [CrossRef]
39. Rossano, V.; Roselli, T.; Quercia, G. Coding and Computational Thinking: Using Arduino to Acquire Problem-Solving Skills. In *Technology Supported Innovations in School Education*; Springer: Cham, Switzerland, 2020; pp. 91–114. [CrossRef]
40. Hosseini, R.; Akhuseyinoglu, K.; Brusilovsky, P.; Malmi, L.; Pollari-Malmi, K.; Schunn, C.; Sirkiä, T. Improving Engagement in Program Construction Examples for Learning Python Programming. *Int. J. Artif. Intell. Educ.* **2020**, *30*, 299–336. [CrossRef]
41. de Andrade, B.; Poplin, A.; de Sena, Í.S. Minecraft as a Tool for Engaging Children in Urban Planning: A Case Study in Tirol Town, Brazil. *ISPRS Int. J. Geo-Inf.* **2020**, *9*, 170. [CrossRef]
42. Syntax, J. 2012. Available online: https://www.w3schools.com/js/js_json_syntax.asp (accessed on 4 December 2022).
43. Code.org. Available online: https://code.org/ (accessed on 4 December 2022).
44. Code Academy. Available online: https://www.codecademy.com/catalog (accessed on 4 December 2022).
45. Williams, M.D.; Rana, N.P.; Dwivedi, Y.K. The unified theory of acceptance and use of technology (UTAUT): A literature review. *J. Enterp. Inf. Manag.* **2015**, *28*, 443–488. [CrossRef]
46. Davis, F.D. Perceived usefulness, perceived ease of use, and user acceptance of information technology. *MIS Q.* **1989**, *13*, 319–340. [CrossRef]
47. Hale, J.L.; Householder, B.J.; Greene, K.L. The theory of reasoned action. *Persuas. Handb. Dev. Theory Pract.* **2002**, *14*, 259–286.
48. Tolman, E.C. A cognition motivation model. *Psychol. Rev.* **1952**, *59*, 389. [CrossRef]

49. Venkatesh, V.; Thong, J.Y.L.; Xu, X. Unified theory of acceptance and use of technology: A synthesis and the road ahead. *J. Assoc. Inf. Syst.* **2016**, *17*, 328–376. [CrossRef]
50. Altalhi, M.M. Towards understanding the students' acceptance of MOOCs: A unified theory of acceptance and use of technology (UTAUT). *Int. J. Emerg. Technol. Learn. (IJET)* **2021**, *16*, 237–253. [CrossRef]
51. Guggemos, J.; Seufert, S.; Sonderegger, S. Humanoid robots in higher education: Evaluating the acceptance of Pepper in the context of an academic writing course using the UTAUT. *Br. J. Educ. Technol.* **2020**, *51*, 1864–1883. [CrossRef]
52. Khechine, H.; Raymond, B.; Augier, M. The adoption of a social learning system: Intrinsic value in the UTAUT model. *Br. J. Educ. Technol.* **2020**, *51*, 2306–2325. [CrossRef]
53. Ma, M.; Chen, J.; Zheng, P.; Wu, Y. Factors affecting EFL teachers' affordance transfer of ICT resources in China. *Interact. Learn. Environ.* **2019**, *27*, 1–16. [CrossRef]
54. Veiga, F.J.M.; de Andrade, A.M.V. Critical success factors in accepting technology in the classroom. *Int. J. Emerg. Technol. Learn.* **2021**, *16*, 4–22. [CrossRef]
55. Morais, E.; Morais, C.; Paiva, J.C. Computer Programming Acceptance by Students in Higher Arts and Design Education. In Proceedings of the ICERI2018, Seville, Spain, 12–14 November 2018; pp. 2058–2065.
56. Garcia-Lopez, C.; Mor, E.; Tesconi, S. Code. In *International Conference on Human-Computer Interaction*; Springer: Cham, Switzerland, 2021; pp. 437–455.
57. Hevner, A.; Chatterjee, S. Design science research in information systems. In *Design Research in Information Systems*; Springer: Berlin/Heidelberg, Germany, 2010; pp. 9–22.
58. Rahmi, B.; Birgoren, B.; Aktepe, A. A meta analysis of factors affecting perceived usefulness and perceived ease of use in the adoption of e-learning systems. *Turk. Online J. Distance Educ.* **2018**, *19*, 4–42.
59. Wrycza, S.; Marcinkowski, B.; Gajda, D. The enriched UTAUT model for the acceptance of software engineering tools in academic education. *Inf. Syst. Manag.* **2017**, *34*, 38–49. [CrossRef]
60. Casey, T.; Wilson-Evered, E. Predicting uptake of technology innovations in online family dispute resolution services: An application and extension of the UTAUT. *Comput. Hum. Behav.* **2012**, *28*, 2034–2045. [CrossRef]
61. Almaiah, M.A.; Alamri, M.M.; Al-Rahmi, W. Applying the UTAUT model to explain the students' acceptance of mobile learning system in higher education. *IEEE Access* **2019**, *7*, 174673–174686. [CrossRef]
62. Han, J.; Conti, D. The use of UTAUT and post acceptance models to investigate the attitude towards a telepresence robot in an educational setting. *Robotics* **2020**, *9*, 34. [CrossRef]
63. Kahn, P.; Everington, L.; Kelm, K.; Reid, L.; Watkins, F. Understanding student engagement in online learning environments: The role of reflexivity. *Educ. Technol. Res. Dev.* **2017**, *65*, 203–218. [CrossRef]
64. Mandernach, B.J. Assessment of student engagement in higher education: A synthesis of literature and assessment tools. *Int. J. Learn. Teach. Educ. Res.* **2015**, *12*, 1–14.
65. Qualtrics. Available online: https://www.qualtrics.com/ (accessed on 4 December 2022).
66. Latan, H.; Noonan, R.; Matthews, L. Partial least squares path modeling. In *Partial Least Squares Path Modeling: Basic Concepts, Methodological Issues and Applications*; Springer: Cham, Switzerland, 2017.
67. Haenlein, M.; Kaplan, A.M. A Beginner's Guide to Partial Least Squares Analysis. *Underst. Stat.* **2004**, *3*, 283–297. [CrossRef]
68. Dijkstra, T.K.; Henseler, J. Consistent partial least squares path modeling. *MIS Q.* **2015**, *39*, 297–316. [CrossRef]
69. Hock, M.; Ringle, C.M. Local strategic networks in the software industry: An empirical analysis of the value continuum. *Int. J. Knowl. Manag. Stud.* **2010**, *4*, 132–151. [CrossRef]
70. Hair, J.F.; Black, W.C.; Babin, B.J.; Anderson, R.E.; Tatham, R. *Multivariate Data Analysis Uppersaddle River*; Pearson Prentice Hall: Hoboken, NJ, USA, 2006.
71. Anderson, J.C.; Gerbing, D.W. Structural equation modeling in practice: A review and recommended two-step approach. *Psychol. Bull.* **1988**, *103*, 411. [CrossRef]
72. Fornell, C.; Larcker, D.F. *Structural Equation Models with Unobservable Variables and Measurement Error: Algebra and Statistics*; Sage Publications Sage CA: Los Angeles, CA, USA, 1981.

Article

Gamification in Engineering Education: The Use of Classcraft Platform to Improve Motivation and Academic Performance

Luisa Parody [1],*, Jenifer Santos [2], Luis Alfonso Trujillo-Cayado [3],* and Manuel Ceballos [4]

[1] Departamento de Métodos Cuantitativos, Universidad Loyola Andalucía, Av. de las Universidades, s/n, 41704 Dos Hermanas, Spain
[2] Facultad de Ciencias de la Salud, Universidad Loyola Andalucía, Av. de las Universidades, s/n, 41704 Dos Hermanas, Spain
[3] Departamento de Ingeniería Química, Escuela Politécnica Superior, Universidad de Sevilla, c/ P. García González, 1, 41012 Sevilla, Spain
[4] Departamento de Ingeniería, Universidad Loyola Andalucía, Av. de las Universidades, s/n, 41704 Dos Hermanas, Spain
* Correspondence: mlparody@uloyola.es (L.P.); ltrujillo@us.es (L.A.T.-C.); Tel.: +34-954-556447 (L.A.T.-C.); Fax: +34-954-556441 (L.A.T.-C.)

Abstract: Pedagogical innovation involving information and communications technology (ICT) may offer teachers the opportunity to create engaging learning environments in engineering courses. In this paper, we present a gamification teaching experience whose primary objective is to improve motivation, and we obtained results for students of a mathematics course during their first year at university. For this case study, we used Classcraft®, which is a role-playing game supported by a digital platform and a mobile application that has been developed to answer teachers' classroom management needs. We hypothesized that using this application as ICT could enhance learning and promote the development of the four "super skills" (or the Four C's): critical thinking, communication, collaboration, and creativity. In order to explore the educational effectiveness of the methodology, a comparison between a gamification group of students and a control group was carried out. Our results showed that the mean mark obtained by the control group students was lower than that obtained by the gamification group students. In addition, the Nemenyi test showed that the Four C's were improved thanks to the Classcraft® activities and group project. Overall, course participants positively evaluated the use of the gamification platform.

Keywords: 21st century skills; gamification; motivation; digital platform; mathematics

Citation: Parody, L.; Santos, J.; Trujillo-Cayado, L.A.; Ceballos, M. Gamification in Engineering Education: The Use of Classcraft Platform to Improve Motivation and Academic Performance. *Appl. Sci.* **2022**, *12*, 11832. https://doi.org/10.3390/app122211832

Academic Editor: Vicente Julian

Received: 31 October 2022
Accepted: 18 November 2022
Published: 21 November 2022

Publisher's Note: MDPI stays neutral with regard to jurisdictional claims in published maps and institutional affiliations.

1. Introduction

Today, a review of education quality is strongly required so that we can provide instructions that enable effective learning in the future [1]. In this sense, governments should encourage innovation in educational techniques, methodologies, and strategies in order to promote the competencies adapted to the emerging knowledge-based societies of the 21st century [2]. Critical thinking, communication, collaboration, and creativity were established as the most important skills for education in the 21st century by the Partnership for 21st Century Skills (also called P21) in 2002. Therefore, curricula should be changed to ensure that what students learn is important for them as individuals and members of modern society [3]. This matter is already being considered in mathematics [4].

As previously mentioned, critical thinking, communication, collaboration, and creativity (the Four C's) are important aspects of modern teaching and education. The National Council for Excellence in Critical Thinking defined critical thinking as an intellectually disciplined process of actively and skillfully conceptualizing, applying, analyzing, synthesizing, and/or evaluating information gathered from or generated by observation, experience, reflection, reasoning, or communication as a guide to belief and action [5]. In this sense, critical thinking involves clarity, accuracy, precision, consistency, relevance, sound evidence, reasoning, depth, breadth, and fairness. Another key skill is communication, which has not attracted the same amount of attention as critical thinking, creativity, and collaboration. This skill can be divided into mediated, digital, written, and oral communication. In order to assess whether students gain communication competences, educators should effectively teach how to communicate. In this sense, educators should focus on building a stronger and more empirically grounded framework for teaching by learning these skills. Collaboration is attracting more and more attention as an important educational outcome and a key educational skill; it is important for not only school but also for career and life success. This ability is critically important to effectively work with others and to facilitate team building and team-based work. Finally, creativity is well-known for being a key 21st century skill. Recently, it has been shown to be integral to a wide range of skills, including scientific thinking, designed thinking, and mathematics [6]. In 2010, more than 15,000 CEOs from 33 industries and 60 countries were interviewed for a study about leadership qualities. The results showed that the most important quality needed to meet the challenge of increasing complexity and uncertainty in the world was creativity [7]. According to Mihaly Csikszent-mihalyi, "Most of the things that are interesting, important, and human are the results of creativity" [8].

The implementation of information and communications technology (ICT) in classrooms through mobile applications is intended to improve students' learning, motivation, and participation [9–11]. To support the aforementioned skills through gamification, researchers have proposed the use of Classcraft®, which is a web and mobile application that allows an educator to manage a role-playing game with students [12,13]. This application enables the creation of student teams, with an avatar assigned to every student. In this way, students can obtain experience points as rewards for positive achievements and can even be penalized with negative points in the case of faults. It is important to mention the ease of use of Classcraft®, including for non-technical educators. The relevance of this use is supported by its success in some studies [5] and the importance of using new technologies to teach 21st century skills [7].

The main goal of this paper was to determine how the use of Classcraft® could motivate students in a mathematics course during their first-year engineering degrees at Universidad Loyola Andalucía. For this purpose, we conducted a comparison of marks obtained by students who followed a traditional teaching methodology and by students who used the Classcraft® application. Furthermore, the reliability of this method was assessed using different statistical methods.

2. Materials and Methods

2.1. Methodology

2.1.1. Participants

The participants of this study were students from Universidad Loyola Andalucía (Spain), which is a private university with sites at Seville and Córdoba. This study was carried out in the 2021/2022 academic year. Undergraduate and postgraduate programs can be studied in both Spanish and English at this university. Different degrees related to engineering, education, social sciences, law and economics, and business can be obtained.

The participants were the 38 students enrolled in the courses of Mathematics I and Mathematics II during their first-year engineering degrees. These degrees are degree in electromechanical engineering, engineering degree on industrial organization, degree in mechatronic and robotic engineering, and degree in computer engineering and virtual technology. From a descriptive point of view, 18.42% of the participants were women and 81.58% were men. The age range of the participants was from 19 to 21. The mean age was 20.12, and the coefficient of variation of the age was 0.0543.

2.1.2. Procedure

We intended to compare the marks obtained by the students who followed a traditional learning model (control group) with those obtained by students who used the innovative methodology (gamification group). To conduct this comparison, we designed five different tests to be solved by students at the end of each chapter of their study. In addition, students had to form groups and deal with some projects. All students (control group and gamification group) completed the same activities but in different forms. Control groups worked in class using paper and pen, while the gamification group used the Classcraft® platform.

Firstly, students in the gamification group had to choose a character: warrior, mage, or healer.

To build motivation, students were able to obtain experience points if they showed different class behaviors. For example:

- Asking an interesting question in class;
- Correctly answering a question in class;
- Correcting a mistake in class;
- Showing positive attitudes in one of the Four C's;
- Helping another student with a class task;
- Passing a class test;
- Finishing a project before its deadline.

Students could also lose experience points if they demonstrated the following behaviors:

- Arriving late to class;
- Using a mobile phone or laptop during class;
- Using inappropriate language;
- Showing weakness in one of the Four C's;
- Failing a class test;
- Finishing a project after its deadline.

As a consequence of these points, students were able to learn in-game powers, level up, and fall down in battle [14]. Their final marks were calculated by adding the score obtained in an exam (70%) and the points obtained by their behavior in class (30%).

Furthermore, the promotion of Four C-related behaviors in students was analyzed as follows. A real-world problem in the field of engineering regarding a vibration model was explained to the students, who then had to complete a research study and solve some problems related to this matter. They had to search for information about the problem, collect data of previous knowledge on the problem, and apply the techniques explained during the course in order to solve the problem. Moreover, they participated in the evaluation process of their classmates' projects. Each group had to present its work, and another group acted as reviewers, describing the strengths and weaknesses of the project in order to develop their critical skills. Finally, each group also evaluated their own participation in the process. Related to communication skills, students had to write a technical report and explain their research and obtain results from their classmates. For the gamification group of students, all of these steps were conducted with the Classcraft® platform. Concerning the communication skill, students had to collaboratively choose the name and the banner of their groups on the Classcraft® platform such that the students of each group had different characters. Moreover, students had to distribute all potentially learnable powers and abilities to each member of each group. Consequently, students were encouraged to have a collaborative attitude. Regarding creativity, students were asked to design new real-life problems similar to the one that they had to solve. We also asked students to propose new characters and powers for the Classcraft® platform. We submitted all their ideas to the official forum so that our students could receive some feedback on their initiatives. For the control group, the research study used to evaluate the students' aptitude for the Four C's was carried out as a traditional writing work.

2.1.3. Likert Scale

Different statistical tests were applied in order to study the influence of the Classcraft®
platform on the marks and motivation of our students. The Wilcoxon signed-rank test was
carried out to evaluate the differences in the median mark, and the Friedman test and the
Nemenyi test were conducted to assess the differences in the Four C's scores. A level of
significance of $\alpha = 0.05$ was considered in all performed tests.

A satisfaction questionnaire was provided to the students in order to understand
what they thought about this innovative methodology and the use of the Classcraft platform. They had to complete the questions by assigning a number to each of the ten
statements where

- 1 means strongly disagree;
- 2 means somewhat disagree;
- 3 means neither agree nor disagree;
- 4 means somewhat agree;
- 5 means strongly agree.

3. Results and Discussion

Table 1 summarizes the basic descriptive statistics of the marks obtained by the students who followed the traditional learning methodology (control group) and those who
learned using Classcraft® (gamification group). It is important to note that the mean mark
obtained by the gamification group was higher than that obtained by the control group,
suggesting that the use of gamification enhanced the learning ability of the students. In
addition, according to the Shapiro–Wilk test, the distribution of the marks of the gamification group followed a normal model but those obtained by the control group did not.
These results were validated the Shapiro–Wilk p-values and the level of significance of
$\alpha = 0.05$. Moreover, considering the values of the standard deviation and the coefficient of
variation, it seems that the data were more heterogeneous in the gamification group than
in the control group. Considering these results, students who used Classcraft® showed
generally better learning capabilities than those who followed a traditional methodology.

Table 1. Descriptive statistics about the mark obtained by the students using a traditional methodology (control group) and using Classcraft® (gamification group).

Statistic	Control Group	Gamification Group
Number of cases	38	38
Minimum	0.6	1.5
Maximum	9	10
Arithmetic Mean	4.69	5.63
Standard Deviation	2.60	2.06
Coefficient of Variation	0.56	0.37
Shapiro–Wilk Statistic	0.92	0.96
Shapiro–Wilk p-value	0.0098	0.2687

After this descriptive study, the Wilcoxon signed-rank test was applied to analyze the
median difference of both samples. Considering that the sample size was 38, which was
large enough for us to use normal approximation (Z-statistic), the results of this test are
shown in Table 2.

Table 2. Results of the Wilcoxon signed-rank test.

Sum of positive ranks (W^+)	133
Sum of negative ranks (W^-)	608
Test statistic	103
Z-statistic	-3.4444

Since the significance level was $\alpha = 0.05$, the critical value was 1.96, and the null hypothesis assuming that the medians of the two samples would be the same was rejected.

The average marks regarding the Four C's obtained by students of the gamification and control groups are shown in Figure 1. All marks were higher than 2.5 for all students except for those regarding communication for the students of the control group; the maximum score was five. There was a substantial improvement in the results of the gamification group in comparison with those of the control group, proving that the use of gamification in a classroom can enhance behavior related to the Four C's. The predominant improvement following the use of Classcraft® was in creativity (from 2.45 to 3.42), followed by communication (from 2.96 to 3.90). It should be also highlighted that critical thinking seemed to be the best skill in both groups.

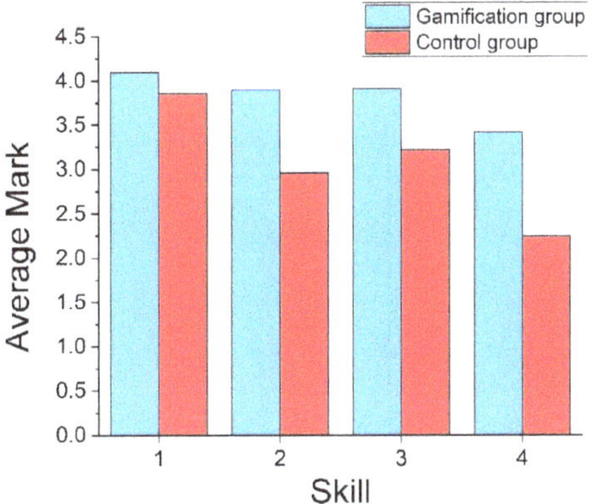

Figure 1. Average Four C's marks: (1) critical thinking, (2) communication, (3) collaboration, and (4) creativity.

The significance of the results was analyzed by using the Friedman test and the Nemenyi statistical test. The Friedman test assesses whether the differences in a group of results are significant, while the Nemenyi test evaluates which comparable pairs have significant differences. A significant level of $\alpha = 0.05$ was used. The results of the Friedman test proved that there were statistical differences among the obtained Four C-related marks. The results of the Nemenyi test showed that the students' Four C-related skills were improved thanks to the activities and group project. The Nemenyi test also revealed that critical thinking was better promoted than other skills.

A satisfaction questionnaire was filled in by the students at the end of the course. Table 3 shows the average marks for every question in the survey. Generally, the Classcraft® platform was very positively evaluated by students. They also highlighted that the use of this platform provoked an increase in their motivation. In addition, they liked working in groups, especially in comparison with the control students. Finally, students thought that Classcraft® could be used in other courses. These results show that it is a good idea to implement this information and communications technology (ICT) in classrooms, regardless the subject.

Finally, some answers to question 10 (Do you have any additional comments?) are shown.

- "I have enjoyed very much using the Classcraft platform, and it has helped to improve my marks in Mathematics. However, I think that the powers should be revised and configured in a better way".
- "It was great to deal with Classcraft. It allowed me to improve my interest and motivation in Mathematics".
- "Our teachers, by using Classcraft, have been able to get my attention and recover my interest in Mathematics".

- "I have had some problems dealing with my groupmates. I think that teachers should design the groups in the future".
- "I think that the game of Classcraft has been useful, and I got a better mark after using it".
- "I wish that other teachers used Classcraft".
- "We have had some disagreements working in groups to prepare the oral presentation".

Students pointed out two areas that should be revised: some aspects of the configuration of the game and some communication issues when working in groups. In a future study, all recommendations provided by students in order to improve their motivation and results will be considered.

Table 3. Average marks of the quantitative questions of the opinion questionnaire.

Question	Average Mark	
	Control Group	Gamification Group
The use of Classcraft platform has increased my interest and motivation for the course.	-	4.5
My implication in the use of this tool has been appropriated.	-	4.12
I am satisfied with the results I have obtained after using Classcraft.	-	3.96
I think that this platform is attractive and advisable for teaching purposes.	-	4.02
The behaviors designed has been useful for my study and progress in the course.	-	3.18
The battles and sentences dealt with have helped me during the course.	-	3.97
The experience points have been configured properly.	-	4.21
I have enjoyed working in teams with my classmates.	3.18	4.38
I would like that other courses during my degree use this platform.	-	4.62

4. Conclusions

In this study, the use of Classcraft® as an ICT to improve student education in two courses during first-year engineering degrees resulted in a measurable increase in knowledge. An analysis of the results illustrated that the gamification group's students' critical thinking, communication, collaboration, and creativity (the Four C's) skills were improved following the use of the aforementioned platform in comparison with a control group. In addition, the platform was well-received by students. Therefore, Classcraft® can be considered to be a promising new tool for the classroom implementation of active learning strategies, as well as an evaluation system in engineering degrees.

Author Contributions: Conceptualization, L.P. and M.C.; methodology, M.C.; software, L.A.T.-C. and M.C.; validation, L.P., J.S., L.A.T.-C. and M.C.; formal analysis, M.C.; investigation, M.C.; resources, L.P. and M.C.; data curation, J.S.; writing—original draft preparation, L.P. and M.C.; writing—review and editing, J.S. and L.A.T.-C.; visualization, L.A.T.-C.; supervision, J.S.; project administration, J.S.; funding acquisition, M.C. All authors have read and agreed to the published version of the manuscript.

Funding: This research received no external funding.

Informed Consent Statement: Informed consent was obtained from all subjects involved in the study.

Conflicts of Interest: The authors declare no conflict of interest.

References

1. Felder, R.M.; Brent, R. How to improve teaching quality. *Qual. Manag. Manag. J.* **1999**, *6*, 9–21. [CrossRef]
2. Crow, G.; Bardsley, N.; Wiles, R. Methodological innovation and developing understandings of 21st century society. *Twenty-First Century Soc.* **2009**, *4*, 115–118. [CrossRef]
3. Jones, B.F. Thinking and learning: New curricula for the 21st century. *Educ. Psychol.* **1991**, *26*, 129–143. [CrossRef]
4. Suh, J.; Seshaiyer, P. Informing Practice: Mathematical Practices That Promote Twenty-First Century Skills. *MatheMatics Teach. Middle Sch.* **2013**, *19*, 132–137. [CrossRef]
5. Kohn, L.Y. Engaging students to use their minds well: Exploring the relationship between critical thinking and formative assessment. *Inq. Crit. Think. Across Discip.* **2013**, *28*, 36–45. [CrossRef]
6. Sheffield, L.J. Creativity and school mathematics: Some modest observations. *ZDM* **2013**, *45*, 325–332. [CrossRef]
7. Capitalizing on Complexity: Insights from the Global Chief Executive Officer Study. 2010. Available online: http://public.dhe.ibm.com/common/ssi/ecm/gb/en/gbe03297usen/GBE03297USEN.PDF (accessed on 15 October 2022).
8. Csikszentmihalyi, M. *Creativity: Flow and the Psychology of Discovery and Invention*; HarperCollinns: New York, NY, USA, 1997.
9. Santos, J.; Parody, L.; Ceballos, M.; Alfaro, M.C.; Trujillo-Cayado, L.A. Effectiveness of mobile devices as audience response systems in the chemistry laboratory classroom. *Comput. Appl. Eng. Educ.* **2019**, *27*, 572–579. [CrossRef]
10. Saxena, M.; Mishra, D.K. Gamification and Gen Z in Higher Education: A Systematic Review of Literature. *Int. J. Inf. Commun. Technol. Educ. (IJICTE)* **2021**, *17*, 1–22. [CrossRef]
11. Corral Abad, E.; Gomez Garcia, M.J.; Diez-Jimenez, E.; Moreno-Marcos, P.M.; Castejon Sisamon, C. Improving the learning of engineering students with interactive teaching applications. *Comput. Appl. Eng. Educ.* **2021**, *29*, 1665–1674. [CrossRef]
12. Krishnan, S.D.; Norman, H.; Md Yunus, M. Online Gamified Learning to Enhance Teachers' Competencies Using Classcraft. *Sustainability* **2021**, *13*, 10817. [CrossRef]
13. Zhang, Q.; Yu, L.; Yu, Z. A Content Analysis and Meta-Analysis on the Effects of Classcraft on Gamification Learning Experiences in terms of Learning Achievement and Motivation. *Educ. Res. Int.* **2021**, *2021*, 9429112. [CrossRef]
14. Sanchez, E.; Young, S.; Jouneau-Sion, C. Classcraft: From gamification to ludicization of classroom management. *Educ. Inf. Technol.* **2017**, *22*, 497–513. [CrossRef]

 *applied sciences*

Article

Developing Computational Thinking: Design-Based Learning and Interdisciplinary Activity Design

Dongqing Wang [1,†], Liqiang Luo [2,†], Jing Luo [1,†], Sihong Lin [3] and Guangjie Ren [1,*]

1 School of Information Technology in Education, South China Normal University, 55 Zhongshan Dadao West, Tianhe District, Guangzhou 510631, China
2 Primary School Affiliated to South China Normal University, 55 Zhongshan Dadao West, Tianhe District, Guangzhou 510631, China
3 Dongguan Science Technology School, No.2, Zhijiao Road, Hengli Town, Dongguan 523000, China
* Correspondence: renguangjie@m.scnu.edu.cn
† These authors contributed equally to this work.

Abstract: As research progresses, integrating computational thinking (CT) and designing interdisciplinary activities to teach various disciplines have gradually emerged as new ideas and important ways to develop the CT of students. This paper introduces the concept of design-based learning (DBL) and analyzes the internal connections between DBL and CT teaching. In this study, an interdisciplinary activity design model was constructed based on an analysis of existing design-based scientific cycle models and research into STEAM education, which is an approach to learning that uses science, technology, engineering, the arts, and mathematics as access points for guiding student inquiry, dialogue, and critical thinking. Next, specific activities with a focus on CT were designed to teach graphical programming to fifth grade students using Scratch. This quasi-experimental research was carried out to test the promotion effects of interdisciplinary activity design and traditional programming activities on the CT of students. Finally, the results showed that the proposed interdisciplinary activity design could develop the CT levels of students more effectively than traditional programming activities.

Keywords: computational thinking; design-based learning; interdisciplinary activity design; graphical programs

Citation: Wang, D.; Luo, L.; Luo, J.; Lin, S.; Ren, G. Developing Computational Thinking: Design-Based Learning and Interdisciplinary Activity Design. *Appl. Sci.* **2022**, *12*, 11033. https://doi.org/10.3390/app122111033

Academic Editor: Antonio Visioli

Received: 20 September 2022
Accepted: 25 October 2022
Published: 31 October 2022

Publisher's Note: MDPI stays neutral with regard to jurisdictional claims in published maps and institutional affiliations.

1. Introduction

With the development of artificial intelligence, big data, and other emerging technologies, exploring practical paths for the development of the high-order thinking abilities of students has become a research hot spot in the study of school education. Computational thinking is regarded as a necessary skill for the intelligent era, which has been studied and defined by many researchers. CT was first introduced by Papert in 1996. Wing outlined the basic definition of CT as a way of "solving problems, designing systems and understanding human behavior by drawing on the concepts of computer science" [1]. Since then, CT has been considered by a growing number of researchers, especially within elementary education. The International Society for Technology in Education (ISTE) and the Computer Science Teacher Association (CSTA) defined the vocabulary for CT as: data collection, data analysis, data representation, problem decomposition, abstraction, and four other words [2]. CT has become a key digital literacy skill that digital citizens must master in the 21st century. Its popularization in K–12 education represents the inevitable trend of globalization [3].

As an element of the core information technology literacy discipline, CT has been included in the curriculum standards of senior high schools in China since 2017. Through the teaching reform of information technology courses in primary and secondary schools, researchers from China have gradually explored effective ways to develop CT in practice. With reference to the example of CT training in British school computing curricula,

Xie Zhong-Xin and Cao Yang-Lu analyzed how to realize CT training in Chinese school information technology curricula [4]. Combined with the use of App Inventor, Guo Shou-Chao et al. proposed learning models for the processes of problem posing, abstract description, reduction, decomposition, modular methods, and module splicing programming to help students to solve problems in class [5]. Liang Yun-zhen et al. took Brennan's framework of CT [6] as a theoretical reference for sixth grade information technology classes in primary school, and proposed a human–computer collaborative precision teaching mode that was oriented toward the cultivation of CT. They used a computational thinking test, Dr. Scratch, evaluation scales, and other tools to test the levels CT among students [7].

With the in-depth study of STEAM education in recent years, how to organically embed CT into interdisciplinary teaching practice has gradually emerged as a new idea and an important way to develop CT. For example, in language learning, Burke used Scratch to develop the writing skills of students [8]. Zhou Ping-hong et al. developed a STEM engineering design teaching model that integrated CT. The application of this model to teaching the STEM course "Plant Factory" showed that the integration and practice of CT in each step of STEM engineering design could significantly improve the attitudes of students toward STEM and their CT ability [9].

In CT education, learners can create games to help them to learn more effectively. Specifically, since CT education is based on algorithms and programming education, learners acquire skills and knowledge through the process of making programs or games themselves [10]. SooJin Jun et al. considered how CT education could create learning environments by actively using DBL. They performed an experiment and the results indicated that DBL improved CT in their sample of elementary school students [11].

Therefore, on the basis of our systematic analysis of the internal connections between DBL and CT, we constructed an interdisciplinary activity design to develop the CT of students. We then verified the effectiveness of our design through quasi-experimental research. The research questions of this study were as follows:

1.   How can interdisciplinary activities be designed to develop the CT of students?
2.   How do interdisciplinary activities promote the CT of students?

2. Background

Different from traditional linear and centralized learning, interdisciplinary STEAM learning emphasizes the integration and application of knowledge from different subjects. STEAM focuses on problem-oriented cooperation and communication, as well as forming solutions to develop innovative talents [12]. The development of CT has offered the dual attributes of computational science and education, which not only reflects the characteristics of interdisciplinary learning, but also highlights the development modes of innovative talents.

2.1. The Evolution of the Concept of CT

Since Papert first put forward the concept of "procedural thinking" in 1996, CT has developed over more than 20 years. During this period, many scholars and research institutions have defined this concept. In addition to the definitions from Wing, the ISTE, and the CSTA, Brennan and Resnick also proposed three key dimensions of CT, which are widely accepted. In summary, Brennan, Resnick, and other researchers expounded CT using an operational definition: the framework of CT includes concepts (sequences, loops, parallelism, events, conditionals, operators, and data), practices (being incremental and iterative, testing and debugging, and reusing and remixing, as well as abstracting and modularizing), and perspectives (expressing, connecting, and questioning) [13]. However, Fraillon, Ainley, and Schulz expounded CT using a more general definition: CT refers to an individual's ability to recognize aspects of real-world problems that are appropriate for computational formulation, and to evaluate and develop algorithmic solutions to those problems so that the solutions can be operationalized using a computer. This definition involves concepts such as conceptualization, data collection, data representation, planning evaluation, and implementation [14].

CT originated from the discipline of computational science, and its definition has mainly evolved from programming education. With the rise of STEAM education, interdisciplinary concepts have been introduced to the definition of CT. CT parallels the core practices of science, technology, engineering, and mathematics (STEM) education, and is believed to effectively support the learning of scientific and mathematical concepts [15]. In a system of Computational Thinking in STEM, CT includes modeling and simulation practices, computational problem-solving practices, and data practices and systems thinking practices, which are used to evaluate students' mastery of CT within the STEM discipline [16]. Until now, the concept of CT has been integrated into various disciplines.

2.2. The Strategy and Mode of Developing CT

How to effectively develop CT has become the focus of current research. The authors of [17] conducted a meta-analysis of CT research in SSCI/SCI (Social Sciences Citation Index/Science Citation Index) journals from 2006 to 2017 and identified 13 kinds of CT training strategies. These strategies included problem-based learning, collaborative learning, project-based learning, game-based learning, scaffolding, storytelling, systematic computational learning, aesthetic experiences, concept-based learning, human–computer interaction teaching, embodied learning, universal design for learning, and design-based learning.

Researchers have carried out in-depth studies on these strategies. Farris [18] introduced music and ViMAP, combined with experiential teaching, into the teaching of fifth grade physics to train students' physical concepts and aesthetic experiences, and to train students to master CT methods and steps. Soleimania [19] used CyberPLAYce to organically integrate digital stories and physics, and verify its role in promoting CT. In addition, Mustafa [20] compared the effects of DBL activities and programming tasks on CT skills through similar experimental studies, and found that DBL activities were similar to programming tasks in terms of playing a positive role in promoting CT. In recent years, DBL has been considered as a critical approach to improving CT [21].

3. Methodology

3.1. DBL and CT Development

DBL and CT development represents the relationship between means and objectives. They both emphasize the diversification of the teaching process: firstly, practical teaching content is helpful for the deep understanding of abstract concepts in CT as practical and interdisciplinary practices can help to make CT more concrete and visual; secondly, comprehensive interdisciplinary problems help students to deeply understand the process of problem-solving, since DBL focuses on "design" as the intermediary of inquiry learning, with the goal of changing existing situations into learning situations, thereby helping learners to carry out interdisciplinary problem-based learning and not only helping with the deeper understanding of abstract concepts, but also with experiences of problem-solving situations; thirdly, iterative learning processes are helpful for mastering CT methods; fourthly, product learning results are helpful for the visual expression of CT; finally, diversified teaching evaluations contribute to the comprehensiveness of CT evaluation, and the multiple teaching evaluations advocated by DBL provide a variety of ways to evaluate and improve CT.

3.2. Interdisciplinary Activities and CT

Interdisciplinary activity design is a research trend within CT training. CT has broken through the boundaries of computational science and has started to be developed more extensively within other disciplines. Computational thinking should not be limited to computer programming, but should instead become one of the necessary skills for daily life, as with reading and writing [22,23]. CT can be transferred to fields other than programming (e.g., science, social sciences, humanities, etc.) [22,24], and can be combined with various disciplines [25], such as English [8,26], art [27], biology [28,29], and mathematics [30]. Snodgrass [31] introduced Scratch into primary school mathematics to promote complex mathematical problem-solving among disabled students and to develop their CT ability. Matsumoto and Cao applied Excel to high school chemistry teaching, and trained students' CT ability through simulations and modeling, experimental data analyses, and coding/programming and algorithm reasoning, as well as statistics and probability activities [32]. In addition, the development of the CT of students through the STEM teaching process of "determine the problem, plan, propose solutions, modify and communicate" has been proposed [33]. In terms of interdisciplinary curriculum design, different countries have explored new forms of CT curricula through interdisciplinary ideas. For example, the Finnish National Education Council turned programming into a whole discipline and formed a "computing" course to develop the CT of students [34], while Malaysia [35] incorporates CT, problem-solving, and information technology into all subjects in primary schools, and encourages students to use tools to solve complex practical problems so as to develop their CT and problem-solving abilities.

3.3. Design-Based Scientific Cycle Model and CT

The design-based scientific cycle model was proposed by Kolodner, Paul, and David [36] according to the iterative cycle characteristics of DBL. In this model, there are two cyclic and iterative processes, namely design/redesign and investigate/explore. The process of design/redesign includes the understanding of the challenge, the planning process, and the design (using science). The process of investigate/explore mainly involves conducting investigations, including design investigations, and more. It should be noted that "need to do" and "need to know" are the links between the iterative processes of design/redesign and investigate/explore.

These iterative learning processes are the core features of the design-based scientific cycle model. Helping students to identify, analyze, and solve problems through the design/redesign and investigate/explore processes is the key element of the model. Different from traditional inquiry learning, DBL starts with projects and tasks and then promotes students' mastery of problem-solving processes through the iterative cycle of planning, designing, understanding the challenge, etc.

3.4. CT-Oriented Interdisciplinary Activity Design

Combined with the training goal of CT and under the guidance of STEAM interdisciplinary teaching, we constructed an interdisciplinary activity design based on DBL [37], which took active learning as the goal and the cycle of discovery, design, and expression practices as the main teaching approach. In this model, the teacher's activities included proposing tasks and offering support and guidance, while the students' activities included clarifying, modifying, and optimizing those tasks. See Figure 1 for more detail.

Discovery practices are involved in the beginning of learning activities, and are key to determining benign problems. To facilitate discovery practices, teachers need to design and issue learning task sheets according to the teaching objectives. Different from traditional learning task sheets, these not only include the information technology subject knowledge required to solve the preset problem but also other subject knowledge related to the problem. Then, according to the task sheets, students complete previews before class. The teachers determine the students' original levels using these previews, answer questions in class to set up a gradient and real problem situation, and then help the students to solve the real problem. The purpose of this process is to help teachers to understand learning situations through self-study before class and to help students to develop the habit of autonomous learning and problem awareness.

The main purpose of design practices is to develop the problem-solving abilities of students. Firstly, teachers are required to guide students to decompose the problems that need to be solved and use multidisciplinary knowledge to design solutions. Secondly, teachers are required to organize groups to carry out research, employ creative learning, and share and exchange solutions; then, they need to guide students to optimize their solutions. If the solutions cannot solve the specific problems, then the students must return to the task decomposition stage, carry out the design practice again, and form new solutions through exploration and practice.

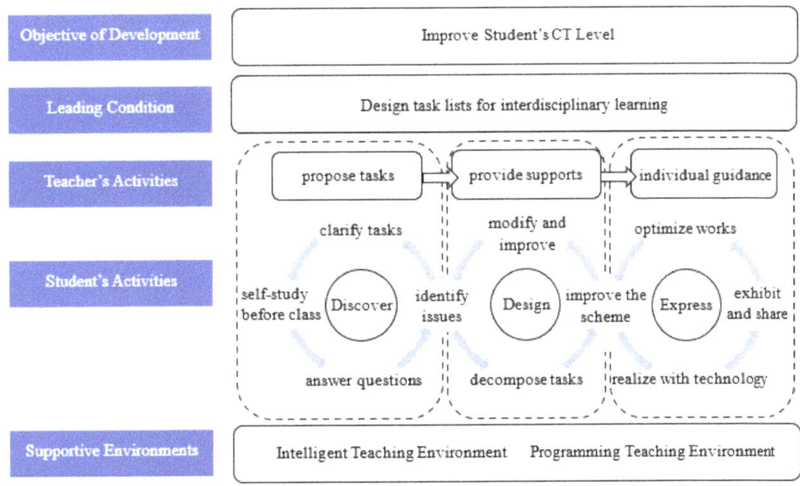

Figure 1. Our interdisciplinary activity design for the development of CT.

Expression practices require teachers to guide students to choose the appropriate technology implementation tools (e.g., Scratch, APP Inventor, etc.) for independent programming creations, and provide technical guidance to realize visual expressions of creativity. After developing independent programming creations, teachers must organize students to carry out group research and creative learning to allow them to share ideas with each other and put forward points to be improved. After that, the students should revise their creations according to the feedback. If students cannot complete their programming work according to solutions based on multidisciplinary knowledge, then they need to return to the design, exhibition, and sharing stages, and complete many design practices until they can complete their programming work. It should be noted that the knowledge of non-information technology disciplines that is required by students to solve these problems should not prevent them from carrying out scheme designs and work implementations. The main purpose of expression practices is to help students to combine information technology with other disciplines and develop their CT on the basis of enhancing a deep understanding of the application of subject knowledge by solving real and comprehensive problems using their subject knowledge and information technology literacy.

4. Implementation and Practice

4.1. Practice Plan

Taking a Scratch course from an information technology textbook from Guangdong Education Press as an example, we were able to design and develop a teaching resource package for this study. The package served as an interdisciplinary activity to develop CT. A quasi-experimental study of 10 class hours was carried out to compare the effects of this interdisciplinary activity design and traditional programming activities on the CT of students. In the design of the teaching content, the experimental class adopted the interdisciplinary activity design that was oriented to the development of CT by reorganizing and integrating the teaching content of Scratch into five interdisciplinary themed activities: "I'm the story king"; "I'm the little test king"; "I'm the little operator"; "I'm the explorer"; and "I can get the average". The same knowledge points were used for the control class. See

Table 1 for more detail. The teaching for the control class was based on Scratch curriculum content for traditional programming design and teaching. The CTt and Dr. Scratch were then selected as the evaluation tools for the practical effects.

Table 1. Teaching content of experimental class and control class.

Class Type	Teaching Content	Required Disciplines	Cycle of Instruction
Experimental Class	1. I'm Story King: use Scratch to complete Chinese composition writing 2. I am quiz king: use scratch to complete the English self test 3. I am an arithmetic expert: using Scratch to realize the design of addition, subtraction, multiplication, and division 4. I am an explorer: use Scratch to draw regular polygons 5. I can calculate the average: use Scratch to complete the program of average of a group of data	Information Technology Chinese English Mathematics	2 class hours, 10 class hours in total
Control Class	1. Little magician: role and appearance 2. Happy Bouncing: Background, Rotation, and Moving 3. Funny shooting game: positioning and translation 4. Cool Cat Playing Football: Repetitive Execution and Control 5. Happy Pig Race: keyboard control and condition detection 6. IQ challenge: keyboard information acquisition and detection 7. I'll pick apples: role control and variables 8. Remote control helicopter: role orientation and stop execution 9. Interesting graphics: brush command 10. Making promotional films: broadcasting and receiving broadcasting	Information Technology	1 class hour, 10 class hours in total

Among them, the CTt (computational thinking test) is a set of CT test questions designed by Román-González et al., based on basic psychological and problem-solving abilities, which mainly involves the concepts of sequences, cycles, events, conditions, parallelism, operators, data, etc. [38]. It contains 28 items and takes 40 min to complete. Before the experiment, a CTt pre-test was carried out and the Cronbach's alpha reliability of the CTt was calculated as 0.815. The KMO validity (Kaiser–Meyer–Olkin) value was found to be 0.829, which indicated that the overall reliability and validity of the CTt was acceptable. The reliability and validity were good, so the test data were reliable.

4.2. Design and Practice of Teaching Activities

Our design highlighted "interdisciplinary teaching" and "DBL processes". It aimed to improve the programming and logical thinking of students as they solved complex problems, by guiding the students to use computer programming and multidisciplinary knowledge to develop their CT abilities. Combined with the interdisciplinary activity design for CT and according to the characteristics of the physical and mental development and learning levels of the students, specific activities on specific topics were designed to carry out the teaching activities for CT development. The following example is the specific activity for the teaching theme "I am an explorer: Discovering the mystery of regular polygons" (Figure 2).

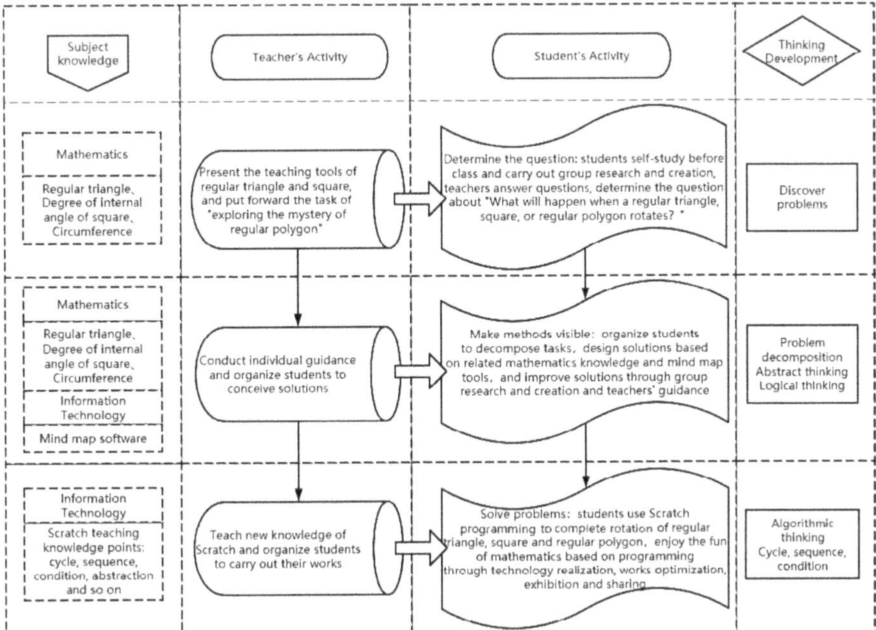

Figure 2. The specific activities for the teaching theme, "I'm an explorer: Discovering the mystery of regular polygons".

1. *Students determine the theme, investigate its background, explore independently, clarify the task, and initially form an awareness of the identified problems:* The core of our interdisciplinary activity design was to create real interdisciplinary application situations. Under the guidance of the teachers, the students independently found and proposed situations related to regular polygons. Through task decomposition, they used relevant mathematical knowledge, such as finding the degrees of internal angles and the perimeters of regular triangles and squares. The students also shared and communicated in group research and creation, and the teachers answered questions about the problems to be solved.

2. *Students draw up a plan, brainstorm ideas about the task, experience problem-solving processes, and develop their logical thinking:* After clarifying the task, the teachers needed to encourage the students to carry out self-exploration around how to solve the problems and use Scratch learning content to decompose problems, and design and improve solutions. Under the guidance of the teachers, the students learned certain knowledge points, such as cycles and sequence control, and used mind mapping tools to design solutions that promoted the visualization of their thinking process. After designing their solutions, the students were organized into groups to share their ideas and communicate with each other to optimize the solutions, thus stimulating the students' awareness of the importance of sharing and communication in developing abstract and logical thinking.

3. *Students creatively express, share, display, and perfect their works in combination with peer communication and teacher guidance, so as to promote their programming thinking and creativity:* After determining the optimal solution, the next task was to operate. The teachers helped the students to complete the tasks by providing technical guidance and solutions for students who were having difficulties. Next, the results were reported, with the students organized in groups to discuss the decomposition and form problem-solving ideas, which could not only provide guidance for students

with learning difficulties, but also help students to find the shortcomings and points to be improved in other works. Then, according to the feedback from the student groups, they revised and improved their works until the learning task was completed successfully. Finally, the teachers summarized the class content. After learning each topic, the teachers needed to summarize what the students had learnt, paying attention to student participation and the completion of tasks to help the students to understand their learning situations. Figure 3 shows the effects of some completed works, including modules built using Scratch and a schematic diagram of regular polygon rotation.

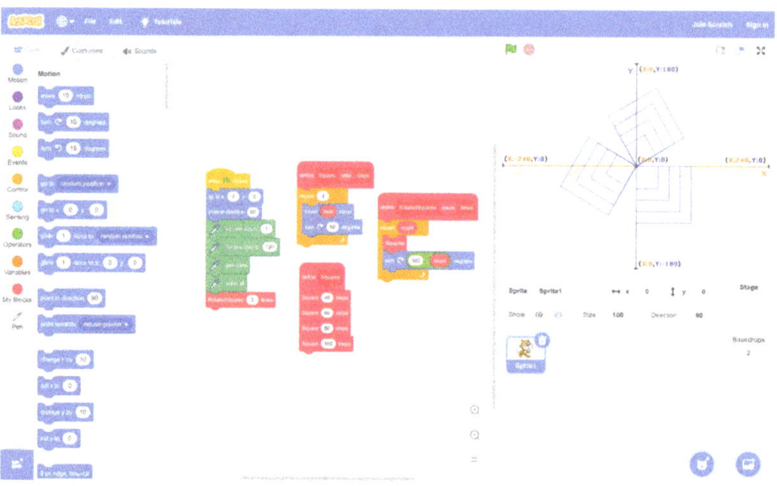

Figure 3. The effects of the work of students in the experimental class.

4.3. Research Results

4.3.1. CT Performance

To explore the differences between the experimental group and the control group, the CT scores of the two groups were analyzed before and after the experiment, as shown in Table 2. It can be seen from the results that the CT levels of the two groups before the experiment were basically the same, and that there was no significant difference between the students' CT scores (Sig. = 0.899). After the experiment, there was a significant difference in CT score between the experimental group and the control group (Sig. = 0.018), with the CT score of the experimental group being significantly higher than that of the control group.

Table 2. Descriptive statistics and independent sample *t*-test of students' CT performances in experimental class and control class.

	Group	N	Mean	Standard Deviation	*t*	Sig (Bilateral)
CT (pre-test)	Experimental Class	44	5.924	3.326	−0.128	0.899
	Control Class	40	6.008	2.675		
CT (post-test)	Experimental Class	44	9.23	3.75	−2.404	0.018
	Control Class	40	7.50	2.69		

To further explore the changes in the CT scores of the students in the experimental and control groups before and after the experiment, the pre-test and post-test CT scores of the two groups were analyzed. The results are shown in Table 3. A paired sample *t*-test was carried out on the pre- and post-test data from the experimental group and it was found that the CT performance of the experimental group significantly improved after the experiment (Sig. = 0.000). At the same time, a paired sample *t*-test was carried out on the pre- and post-test data from the control group, and it was found that the CT performance of the control group also significantly improved after the experiment (Sig. = 0.027).

Table 3. Paired sample *t*-test of students' CT performances in experimental class and control class.

		N	Mean	Standard Deviation	df	*t*	Sig (Bilateral)
CT (Control Class)	pre-test	40	5.92	3.33	39	−2.294	0.027
	post-test	40	7.50	2.69			
CT (Experimental Class)	pre-test	44	6.01	2.68	40	−4.938	0.000
	post-test	44	9.23	3.75			

A comparison of the changes in CT knowledge levels between the experimental class and the control class was performed using the post-test data (Figure 4). Scores closer to the edges indicated that the students had higher levels of mastery of CT, and scores that were closer to the middle indicate that the students had lower levels of mastery of CT. Through this analysis, we found that after the teaching experiment, the students in the experimental class had a better grasp of the orders, cycles, events, conditions, and other dimensions from the same level of pre-test CT.

Figure 4. The results of post-test data on CT scores.

4.3.2. Practical CT Skills

Through the Dr. Scratch analysis of the students' completed works, we could measure the students' influence on their practical CT skills. The results showed that the parallelism (multiple scripts), loops (repeated execution), abstraction (defining new blocks), data representation (no operation on variables and tables). and synchronization (no simultaneous use of background and broadcast blocks) scores were very low. In the case of the same level of pre-test CT knowledge, after the teaching experiment was completed, the students in the experimental class had a better grasp of the abstract, problem-solving, logical thinking, synchronization, parallelism, and sequence control dimensions, among others. However, on the whole, these students achieved low scores in data representation, user interaction, and other dimensions. The reason for this might depend on the difficulty and emphasis of the interdisciplinary activity design. The themed design of the example teaching content focused on the preliminary integration of interdisciplinary activities, and the combination of subject knowledge and Scratch was not emphasized enough (Figure 5).

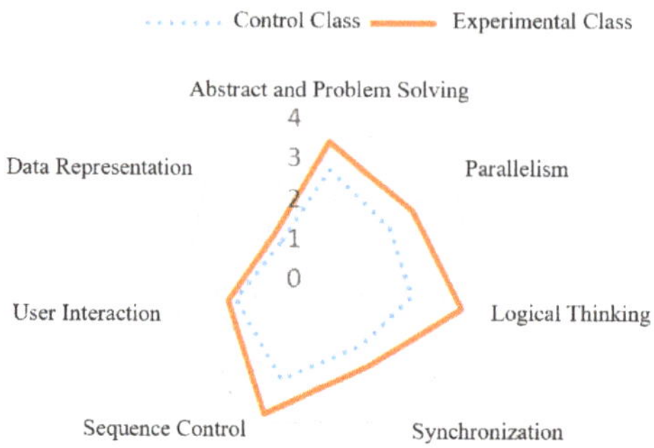

Figure 5. The results of post-test data on CT practices.

5. Conclusions and Discussion

5.1. Teaching Suggestions

5.1.1. Enhance the Practices in Existing Models or Strategies, and Enrich Research on the Development of CT

In-depth research into CT can be carried out for different audience groups, occupations, and classroom environments, as well as to explore the impacts of CT on daily learning and life [20]. Deepening the horizontal and vertical comparative study of existing models and strategies would be conducive to the further development of CT education. In terms of CT training activities, teachers need to choose the appropriate training modes according to the needs of their students, determine the teaching content and the teaching organization form, scientifically design teaching and learning activities, and then deliver CT education.

5.1.2. Improve the Educational Resources for CT and Promote the Collaborative Development of Disciplines

According to collaborative theory, as a universal basic skill that can develop students' problem-solving and creativity, CT is the key to promoting the collaborative development of disciplines [39]. STEAM, interdisciplinary design, digital stories, and visual programming have been proven to be important strategies for developing CT, but CT resource systems are not perfect. In the future, CT education teachers should seize opportunities from the development of STEAM, use programming courses to build interdisciplinary teaching resources for CT, construct programming courses, gradually explore the "block-based programming to text-based programming + subject teaching" teaching resource system, and promote the application and popularization of CT in K–12 education, and even in higher education.

5.1.3. Build a Multimodal Evaluation Method to Support Personalized Learning and Accurate Teaching

Domestic research into CT assessment has mainly focused on two aspects: the development of CT assessment scales, such as the CT scale based on programming [40] and the self-efficacy scale based on programming [41]; and the reference and application of the existing evaluation tools. Taking 1015 Chinese middle school students as a research sample, Bai Xuemei et al. [42] conducted a localization study on the CT scale (CTS) developed by Korkmaz et al. Through their data analysis, they verified that the scale could also be used to measure the CT levels of K–12 students in China. It is not difficult to see that although there are scientific measurement tools and scales for the evaluation of CT, how to automatically record and collect dynamic process data in CT education is still a dilemma. In the future, we should explore the functions of technology, such as the coding of CT learning behavior, the development of CT teaching data acquisition and analysis systems, and the automatic collection and analysis of the CT behaviors of students. Examining these areas could help to improve the accuracy of teaching and lead to the development of personalized recommendations.

5.2. Conclusions

The experimental group was taught using an interdisciplinary activity design that was oriented to CT, and the control group was taught using traditional programming teaching activities. From our statistical analysis of the independent sample *t*-test and the paired sample *t*-test, we found that there were no significant differences in CT knowledge between the experimental group and the control group before the experiment. After the experiment, both the interdisciplinary activity design and the traditional programming teaching activities had significant effects on promoting the students' CT performances, although the former had a greater impact than the latter. By further analyzing the post-test knowledge levels of these two groups, we found that the knowledge of the experimental group was significantly higher than that of the control group regarding the dimensions of orders, cycles, events, and conditions. This showed that the interdisciplinary activity design could better promote student learning in those knowledge dimensions.

Following a Dr. Scratch analysis of the students' completed works, the results showed that after the experiment, knowledge of the dimensions of abstract thinking, problem-solving, logical thinking, synchronization, parallelism, and sequence control was significantly higher in the experimental group than in the control group. This showed that the interdisciplinary activity design could better promote student learning in those knowledge dimensions; however, neither strategy significantly promoted knowledge of the dimensions of data representation or user interaction.

5.3. Discussion

The research found that an interdisciplinary activity design based on the concept of DBL could significantly promote the CT of students, especially in the dimensions of abstract thinking, problem-solving, and logical thinking. A study by Mustafa Saritepeci in 2020 also proved that DBL activities had an "intermediate" effect size on the creativity and problem-solving dimensions [20]. At this point, our findings seemed to show that DBL could be of utmost importance in students' CT skills when they solve problems creatively. This correlated fairly well with the findings of Wing, who suggested that two of the key issues in the acquisition and development of CT skills are formulating solutions for technological problems or contexts and achieving creative processes, including presenting concrete products. Zhang Yi et al. [43] also performed a similar experiment and their results showed that DBL teaching in STEM courses could significantly improve the calculation thinking and problem-solving abilities of primary school students, and play a certain role in improving their logical thinking. Therefore, DBL teaching could significantly improve the creativity of pupils.

6. Limitations

This research was an exploration and an attempt to apply interdisciplinary activities to the development of CT. However, some limitations should be noted. First, only two classes of students participated in this research because of the limitations of the implementation conditions, so the sample size was small, which could have affected the final results. Secondly, the integration of subject knowledge and Scratch into the interdisciplinary activity design was not emphasized enough. The specific teaching content for the experimental and control groups needed to be improved. Thirdly, the reasons for why our interdisciplinary activity design based on the concept of DBL could significantly promote students' abstract thinking, problem-solving, and logical thinking need further investigation and research.

Author Contributions: Conceptualization, D.W.; Formal analysis, L.L.; Methodology, D.W. and G.R.; Software, L.L. and J.L.; Writing—original draft, L.L. and J.L.; Writing—review & editing, D.W., J.L., S.L. and G.R. L.L. carried out the experiment, and analyzed and interpreted the student data regarding the CT performance and the practical skills of CT. All authors were contributors in writing the manuscript. All authors have read and agreed to the published version of the manuscript.

Funding: This research received no external funding.

Institutional Review Board Statement: Not applicable.

Informed Consent Statement: Informed consent was obtained from all subjects involved in the study.

Data Availability Statement: Not applicable.

Conflicts of Interest: The authors declare no competing interests.

References

1. Wing, J. Computational Thinking. *Commun. ACM* **2006**, *49*, 33–35. [CrossRef]
2. CSTA and ISTE. Computational Thinking in K–12 Education Leadership Toolkit. Available online: http://csta.acm.org/Curriculum/sub/CurrFiles/471.11CTLeadershiptToolkit-SP-vF.pdf (accessed on 10 April 2020).
3. Bocconi, S.; Chioccariello, A.; Dettori, G. *Developing Computational Thinking in Compulsory Education*; JRC Working Papers; Publications Office of the European Union: Luxembourg, 2016.
4. Xie, Z.-X.; Cao, Y.-L. Strategies and methods for developing Computational Thinking of information technology students in primary and secondary schools. *China Audio Vis. Educ.* **2015**, *11*, 116–120.
5. Guo, S.-C.; Zhou, R.; Deng, C.-M.; Di, C.-Y.; Zhou, Q.-G. Research on it classroom teaching based on App Inventor and Computational Thinking. *China Audio Vis. Educ.* **2014**, *3*, 91–96.
6. Brennan, K.; Resnick, M. New frameworks for studying and assessing the development of computational thinking. In Proceedings of the 2012 Annual Meeting of the American Educational Research Association, Vancouver, Canada, 13–17 April 2012; pp. 1–25.
7. Liang, Y.Z.; Liu, R.X.;Ren, L.L. Research on the precise teaching mode of human-computer cooperation oriented to the cultivation of computational thinking—Take the sixth grade information technology course "The Silk Road" as an example. *Mod. Educ. Technol.* **2022**, *3*, 51–60.
8. Burke, Q. The Markings of a New Pencil: Introducing Programming as-Writing in the Middle School Classroom. *J. Media Lit. Educ.* **2012**, *2*, 121–135. [CrossRef]
9. Zhou, P.H.; Niu, Y.K.; Wang, K.; Zhang, Y.; Li, X.; Shang, V.W. STEM engineering design teaching mode and application oriented to computational thinking training. *Mod. Distance Educ. Res.* **2022**, *1*, 104–112.
10. Kim, S.H.; Han, S.G. Design-Based Learning for Computational Thinking. *J. Korean Assoc. Inf. Educ.* **2012**, *16*, 319–326.
11. Jun, S.; Han, S.; Kim, S. Effect of design-based learning on improving computational thinking. *Behav. Inf. Technol.* **2016**, *36*, 43–53. [CrossRef]
12. Zhao, H.-C.; Zhang, N.-Y.; Ma, J.-W. STEM education interdisciplinary learning community: Promoting the change of learning style. *Open Educ. Res.* **2020**, *3*, 91–98.
13. Wang, X.-Q. Research on the development and evaluation of Computational Thinking in foreign primary and secondary schools for three-dimensional goals. *Res. Audio Vis. Educ.* **2014**, *7*, 48–53.
14. Fraillon, J.; Ainley, J.; Schulz, W. *IEA International Computer and Information Literacy Study 2018 ASSESSMENT FRAMEWORK*; Springer: Berlin/Heidelberg, Germany, 2019; pp. 1–77.
15. Basu, S.; Biswas, G.; Sengupta, P.; Dickes, A.; Kinnebrew, J.S.; Clark, D. Identifying middle school students' challenges in computational thinking-based science learning. *Res. Pract. Technol. Enhanc. Learn.* **2016**, *11*, 1–35. [CrossRef] [PubMed]
16. Weintrop, D.; Beheshti, E.; Horn, M. Defining Computational Thinking for Mathematics and Science Classrooms. *J. Sci. Educ. Technol.* **2015**, *10*, 1–21.
17. Hsu, T.C.; Chang, S.C.; Huang, Y.M. How to learn and how to teach computational thinking: Suggestions based on a review of the literature. *Comput. Educ.* **2018**, *126*, 296–310. [CrossRef]
18. Farris, A.V.; Sengupta, P. Democratizing Children's Computation: Learning Computational Science as Aesthetic Experience. *Educ. Theory* **2016**, *66*, 279–296. [CrossRef]
19. Soleimania, A.; Herrob, D.; Greenc, K.E. Cyber PLAYce—A tangible, interactive learning tool fostering children's Computational Thinking through storytelling. *Int. J. Child-Comput. Interact.* **2019**, *20*, 9–23. [CrossRef]
20. Mustafa, S. Developing Computational Thinking Skills of High School Students: Design-Based Learning Activities and Programming Tasks. *Asia-Pac. Educ. Res.* **2019**, *1*, 35–54.
21. Zhang, Y.; Wang, J.; Zhang, J.; Zhu, Y.-H.; Zhou, W.-Q.; Wang, Y.-Y. Research on the development of primary school students' Computational Thinking by DBL teaching in STEM curriculum. *Audio Vis. Educ. Res.* **2020**, *5*, 81–88.
22. Grover, S.; Pea, R. Computational Thinking in K-12 A Review of the State of the Field. *Educ. Res.* **2013**, *1*, 38–43. [CrossRef]
23. Wing, J. Research Notebook: Computational Thinking—What and Why? *Link Mag.* **2011**, *6*, 20–23.

24. Lye, S.Y.; Koh, J.H.L. Review on Teaching and Learning of Computa tional Thinking through Programming: What is Next for K-12? *Comput. Hum. Behav.* **2014**, *41*, 51–61. [CrossRef]
25. Berl, M.; Wilensky, U. Comparing Virtual and Physical Robotics En vironments for Supporting Complex Systems and Computational Think ing. *J. Sci. Educ. Technol.* **2015**, *5*, 628–647.
26. Miller, P. Learning with a Missing Sense: What Can We Learn from the Interaction of a Deaf Child with a Turtle? *Am. Nals Deaf.* **2009**, *1*, 71–82. [CrossRef] [PubMed]
27. Lee, Y.J. Developing Computer Programming Concepts and Skills via Technology-enriched Language-art Projects: A Case Study. *J. Educ. Multimed. Hypermedia* **2010**, *3*, 307–326.
28. Sengupta, P.; Kinnebrew, J.S.; Satabdi, B. Integrating Computational Thinking with K-12 Science Education using Agent-based Compu tation: A Theoretical Framework. *Educ. Inf. Tech Nologies* **2013**, *2*, 351–380. [CrossRef]
29. Yang, Y.T.C.; Chang, C.H. Empowering Students through Digital Game Authorship:Enhancing Concentration, Critical Thinking, and Academic Achievement. *Comput. Educ.* **2013**, *68*, 334–344. [CrossRef]
30. Fessakis, G.; Gouli, E.; Mavroudi, E. Problem Solving by 5-6 Years Old Kindergarten Children in a Computer Programming Environment: A Case Study. *Comput. Educ.* **2013**, *63*, 87–97. [CrossRef]
31. Snodgrass, M.R.; Israel, M.; Reese, G.C. Instructional supports for students with disabilities in K-5 computing: Findings from a cross-case analysis. *Comput. Educ.* **2016**, *9*, 1–17. [CrossRef]
32. Matsumoto, P.S.; Cao, J. The Development of Computational Thinking in a High School Chemistry Course. *J. Chem. Educ.* **2017**, *9*, 1217–1224. [CrossRef]
33. Zhu, K.; Jia, X.-X. Research on the development strategy of Computational Thinking Ability from the perspective of STEM. *Mod. Educ. Technol.* **2018**, *12*, 115–121.
34. Heintz, F.; Mannila, L.; Farnqvist, T. A Review of Models for Introducing Computational Thinking, Computer Science and Computing in K-12 Education. In Proceedings of the 2016 IEEE Frontiers in Education Conference (FIE), Erie, PA, USA, 12–15 October 2016; pp. 1–9.
35. Ismail, A.R. Computational Thinking to be Integrated into Primary and Secondary Curriculum. Available online: http://www w.hardwarezone.com.my/tech-news-computationalthinking-be-integratedprimary-and-secondary-curriculum (accessed on 5 August 2019).
36. Kolodner, J.L.; Paul, J.C.; David, C. Problem-based learning meets case-based reasoning in the middle-school science classroom. *J. Learn. Sci.* **2003**, *4*, 495–547.
37. Doppelt, Y.; Mehalik, M.M.; Schunn, C.D. Engagement and Achievements: A Case Study of Design-Based Learning in a Science Context. *J. Technol. Educ.* **2008**, *2*, 18.
38. Marcos, R.G.; Juan-carlos, P.G.; Carmen, J.F Which cognitive abilities underlie Computational Thinking? Criterion validity of the Computational Thinking Test. *Comput. Hum. Behav.* **2016**, *9*, 678–691.
39. Yu, Y.; Xie, S.-X. Enabling computing thinking for education in the age of intelligence: Interpretation and Enlightenment of ISTE Computational Thinking Ability Standard (educators). *J. Distance Educ.* **2020**, *3*, 38–46.
40. Chu, Y.K.; Liang, J.C.; Tsai, M.J. Development of a Computational Thinking Scale for Programming. In *Proceedings of the International Conference on Computational Thinking Education 2019*; The Education University of Hong Kong: Hong Kong, China, 2020.
41. Kong, S.C. Development and validation of a programming self-efficacy scale for senior primary school learners. In *Conference Proceedings of the International Conference on Computational Thinking Education 2017*; The Education University of Hong Kong: Hong Kong, China, 2017; pp. 97–102.
42. Bai, X.-M.; Gu, X.-Q. Construction and application of evaluation tools for K12 students' Computational Thinking. *China Audio Vis. Educ.* **2019**, *10*, 83–90.
43. Zhang, Y.; Wang, J.; Zhang, L.; Zhu, Y.; Zhou, W.; Wang, Y.-Y. Research on DBL Teaching in STEM Curriculum to Cultivate Pupils' Computational Thinking. *Res. Audio Vis. Educ.* **2020**, *5*, 81–88.

 *applied sciences*

MDPI

Article

EXPLORIA, STEAM Education at University Level as a New Way to Teach Engineering Mechanics in an Integrated Learning Process

Nicolás Montés [1,*], Paula Aloy [2], Teresa Ferrer [2], Pantaleon D. Romero [1], Sara Barquero [2] and Alfonso Martinez Carbonell [3]

[1] Department of Mathematics, Physics and Technological Sciences, University CEU Cardenal Herrera, C/San Bartolome 55, Alfara del Patriarca, 46115 Valencia, Spain; pantaleon.romero@uchceu.es
[2] Department of Design and Architecture, University CEU Cardenal Herrera, C/San Bartolome 55, Alfara del Patriarca, 46115 Valencia, Spain; paula.aloy@uchceu.es (P.A.); teresa.ferrer@uchceu.es (T.F.); sara@uchceu.es (S.B.)
[3] Department of Humanities, University CEU Cardenal Herrera, C/Lluis Vives 1, Alfara del Patriarca, 46115 Valencia, Spain; alfonsomc@uchceu.es
* Correspondence: nicolas.montes@uchceu.es

Citation: Montés, N.; Aloy, P.; Ferrer, T.; Romero, P.D.; Barquero, S.; Martinez Carbonell, A. EXPLORIA, STEAM Education at University Level as a New Way to Teach Engineering Mechanics in an Integrated Learning Process. *Appl. Sci.* **2022**, *12*, 5105. https://doi.org/10.3390/app12105105

Academic Editors: Shan Wang, Zehui Zhan, Qintai Hu and Xuan Liu

Received: 16 April 2022
Accepted: 13 May 2022
Published: 19 May 2022

Abstract: The objective of our research is the implementation of STEAM (Science Technology Engineering Art Mathematics) learning in the bachelor of Engineering in industrial design and product development at CEU Cardenal Herrera University through the EXPLORIA project. This article implements and develops the proposal for the first year of this bachelor, which includes 24 students aged 18–20. This article focuses on how to integrate STEAM learning within the EXPLORIA project for the improvement in the learning of the physics subject, and in particular, regarding the part of the syllabus related to mechanical engineering through different projects, challenges and milestones that allow the student to see the use in the design and development of products. The EXPLORIA project connects the competencies of the different STEAM subjects included in the bachelor, designing a learning process as a logical, sequential and incremental itinerary. Through concepts on which the fundamentals of design are based: shape, volume, color, space and structure. In particular, this article shows the adaptation made in the physical part to be able to teach the integrated mechanics part in this learning process. The complete learning was carried out through several challenges and two milestones the students had to overcome through the application of the physical knowledge learned in class. To validate the effectiveness of the proposed methodology, at the end of the paper, an ad hoc questionnaire is carried out showing the students' assessment regarding the new teaching methodology.

Keywords: STEAM; active methodologies; project based learning; challenge-based learning

1. Introduction

STEM learning (Science, Technology, Engineering and Mathematics) was developed at the beginning of the 90s, focused on non-university studies, mainly for middle and high school. STEM is a curriculum based on the idea of educating students in four specific disciplines: science, technology, engineering and mathematics, by using an interdisciplinary and applied approach. Rather than teaching the four disciplines as separate and discrete subjects, STEM integrates them into a cohesive learning paradigm based on real-world applications [1].

The United States has historically been a leader in these fields in trying to motivate high school students to choose STEM careers. This decline in student interest regarding this type of bachelor is not unique to the United States, it is a global problem, as recognized in [2]. In this context, the STEM approach becomes important since this term refers, in a generic way, to the development of initiatives and projects that promote and develop scientific-technological skills and competencies and involve the participation of STEM disciplines [3]. From an educational perspective, we seek the intentional integration of the four disciplines used to solve real-world problems [3]. In this comprehensive and interdisciplinary approach, sciences provide the scientific method. Technology and engineering provide the techniques and tools to build objects and solve technological problems, and mathematics provides a way of expression and representation that allows interpreting the environment and applying strategies to solve problems as well as promoting logical and critical thinking [4].

The implementation of STEM learning generated a deep debate on how the four disciplines should be integrated, whether independently or according to an integrated approach [5]. Of the two approaches, the integrative approach is currently the most widely accepted, in which the four disciplines constitute a single teaching–learning practice [3]. Still, there are researchers who believe that a fair interaction is the right thing to do [6], while others place one discipline above the other [7]. In [3], they observed that, although the disciplines were treated jointly, there was no true connection among them, and [8] considered that educational institutions did not agree on how the four disciplines should be established or connected. To solve this problem, in [5] a proposal has been made to include Art as a new discipline in the STEM context, which was renamed STEAM. In STEAM learning, Art, in addition to promoting interdisciplinarity, facilitates communication and understanding of reality and provides creative strategies and solutions [9]. The concept of Art proposed by [5] is a very broad concept that encompasses, in addition to the so-called fine arts, other fields such as language and social sciences. The combination of scientific and artistic disciplines, apparently opposed, provides "the variety and diversity necessary for innovative product design", and they complement each other because "science provides a methodological tool in art and art provides a creative model in the development of science" [10]. The European Parliament [11] considers the inclusion of art essential, as it leads to the acquisition of key competencies. They consider that art in STEAM is primarily concerned with creativity, and creativity includes divergent thinking [12], which leads to multiple solutions for a single problem.

1.1. Active Methodologies

In general, STEAM projects promote the use of so-called active methodologies, encouraging the student's active participation, who becomes the protagonist of the teaching–learning process and develops his/her own knowledge [13]. In [14], the importance of the activities carried out in STEM projects is demonstrated as a fundamental part to enhance attitudes, scientific creativity and motivation, generating positive emotions in learning. Active methodologies place students at the center of this process and make them protagonists of the discovery, rather than passive recipients of information [15]. There are different teaching strategies for creating an active learning environment and engaging students in it. The most common ones are project-based learning, problem-based learning, collaborative learning, etc. [15]. These methodologies allow the development of practical knowledge and critical thinking through formal analysis and creative thinking through empirical analysis and complete active learning.

1.1.1. Project-Based Learning

Project-based learning starts from an initial question or challenge and raises the objective of generating a final product, generating learning through the tasks that are carried out to develop it [13]. If any of these tasks, in addition to being part of the project, pose a new challenge or problem to solve, we will need to overcome these by using techniques taken from another methodology: problem-based learning. Both methodologies, project-based learning and problem-based learning, use the large methodological umbrella of cooperative learning, and therefore for their implementation we need a new organizational structure of the classroom, a different way of managing times and evaluation systems as well as changing the role of teachers and their training. Project-based learning is being used in many educational programs at university level, also including physics subjects, see [16,17].

1.1.2. Challenge-Based Learning

One of these active methodologies is challenge-based learning, which, based on an initial and global question or challenge, sets out the objective of guiding the students' learning to focus them on an achievable and upcoming challenge, which allows them to get personally involved in the search for effective and plausible solutions [13]. Learning is based on a complete process of research, ideation, documentation and communication, also enhancing personal skills such as teamwork, consensus, negotiation and leadership, as key elements of emotional intelligence. Challenge-based learning allows the process to be approached in a creative and innovative way, so that the process allows the detection of other challenges or problems to be solved. It therefore implies a broader vision than project-based learning. Challenge-based learning is being widely researched for its application in a multitude of Bachelors, and also as an integrating argument between mathematics and physics, see for example [18].

2. The EXPLORIA Project, a New Learning Approach for University Students

The implementation of this type of learning techniques, STEAM projects, in which active methodologies such as project-based learning and challenge-based learning are promoted, is a subject of debate among researchers in education, since for it to be successful there must be cooperation between and integration of, at least, professors of engineering, mathematics and sciences and the Institutions' management [19]. It also implies a change in the teachers' attitude since they not only have to make an effort to interconnect subjects, but also become facilitators of knowledge [20].

2.1. Integrating STEAM Projects in Bachelor Degrees

The integration of STEAM projects in bachelor degrees is not easy since there are a wide variety of Bachelor degrees in which this integration is complex or almost difficult. In Bachelors such as law, literature, history, etc., it is very hard to develop these projects. The Bachelors most likely to adapt to this methodology are the engineering ones since scientific principles are used to design and build machines, structures and other entities. We can take advantage of the accumulation of technological knowledge for innovation, invention, development and improvement of techniques and tools to satisfy the needs and solve technical problems of both people and society [21].

The engineer relies on the basic sciences (fundamentally mathematics and physics) to later apply them to his/her field of study (electronics, mechanics, product design, etc.) in which all STEAM areas and active methodologies can be used. However, currently, STEAM areas are treated separately, generating serious consequences in student performance.

2.2. University Students' Attitudes towards Maths

In [22], an in-depth study is carried out on the rejection of mathematics and negative attitudes towards it from primary education to the first year of university studies. The results show a high taste for mathematics in the initial levels, 87%, however, the taste for mathematics decreases as students go up in level, showing a 57% when they reach the first year of university. The results obtained in [22] were later corroborated in [2], which showed that 67% of the students disliked mathematics and they did not fully understand it. On the contrary, only 38% of them showed an interest and liking for this discipline.

Recently, in [23] a study has been carried out on attitudes towards mathematics in university students. We tested 1293 students in the study (830 women and 453 men) from different bachelors, Agri-food Engineering, Biology, Food Science and Technology, Preschool and Primary Education, IT and Tourism. As a result, the average percentage in attitude obtained was 54% which shows that, in general, men have a more positive attitude towards mathematics, agreeing with other existing studies in this regard, such as in [24,25]. Additionally, in [23], it was found that students in engineering Bachelors showed a better attitude towards mathematics than the rest, agreeing with other studies such as in [26]. These bachelors tend to have a greater number of men than women.

2.3. University Students' Attitudes towards Maths and Their Effect in Physics

The rejection towards mathematics has direct consequences in the rest of the subjects and especially in the other basic subject, physics, since, as indicated in [27], the role of mathematics is to be the language of physics. Thus, the success or failure of students in physics subjects can be predicted taking into account their math skills [28–33]. The research results show that having insufficient skills in mathematics, such as analytical skills, algebraic processing skills, geometry, calculation skills, tables and graph interpretation skills, etc., required to solve physics problems, is the cause of low student performance in physics subjects.

2.4. Basic Subjects vs. Engineering Subjects

In the same way, Math and Physics results predetermine the results of the engineering subjects. Poor performance in mathematics may be the cause of a downward trend in student performance in engineering subjects [34]. In [35], a study is carried out on how it affects the performance of basic subjects (mathematics and physics) related to engineering subjects (Machines, Electrical Engineering, Topography and Building). In this study we did not find a direct relationship between the grades obtained in the basic and specific subjects, but we found a relationship between the years it takes students to pass the basic subjects and the number of years it takes to complete the bachelor. Another important conclusion drawn from the study presented in [35] shows that students do not usually perceive a solid relationship between the basic and applied subjects analyzed in the study.

In this context, STEAM projects applied to the university environment, together with active methodologies such as challenge and project-based learning and collaborative learning can be the way to generate positive emotions in the students who will be able to change their perception and improve their academic performance.

2.5. The EXPLORIA Project

The EXPLORIA project was born from the need to update university learning methodologies to the new trends and requirements of the labor and professional market.

In this sense, the CEU Universities (CEU San Pablo, CEU Cardenal Herrera and CEU Abat Oliva), are developing different pilot projects for bachelor degrees such as Advertising, Political Science, Business Administration, Journalism and Engineering in Industrial Design and Product Development to rethink the learning processes of university students for a context as the current one. Among the pilot projects for bachelors, we have the bachelor of Engineering in Industrial Design and Product Development which integrates subjects that coincide with the STEAM classification.

The EXPLORIA project aims to develop an integrated competency map of the learning process in which the subjects are no longer considered as isolated contents, by elaborating an integrated learning process in which the competencies and learning outcomes of the subjects are considered as a whole, that is, as global and comprehensive learning.

The pilot project makes use of integrated learning and temporal sequences focused on different learning objectives linked to Bloom's taxonomy, see [36]: understanding, applying, experimenting and developing. In this way, through active methodologies, the students address all levels of learning and they "learn by doing". They develop critical and creative thinking, through formal and empirical analysis, and they also develop creativity and innovation, and the capacity for global and multidisciplinary analysis, essential in the current context.

The teacher assumes the role of a learning guide who accompanies students in their personal and professional development process. The teacher leaves the instructor role, encouraging students to discover, motivating them to learn and making them aware of the need to learn from each challenge, stage or new situation that may arise. In this way, the student is prepared to deal with difficult problems in changing, unstable and equally complex contexts.

In the previous work [37], we explain the effect that this methodology may have on the perception and predisposition of students towards mathematics from previous cycles. In this case, mathematics is linked with the rest of the subjects through active methodologies that generate positive stimuli and emotions in the students.

3. Research Objectives

Our research objective is the implementation of STEAM learning in the bachelor of Engineering in industrial design and product development through the EXPLORIA project. In our previous work [37], we showed the efficiency of STEAM learning within the EXPLORIA project to improve learning in mathematics and change students' perceptions. This paper focuses on how STEAM learning can be integrated within the EXPLORIA project to improve the learning of physics, in particular, the part of the syllabus related to mechanical engineering. The proposal is to do it through different projects, challenges and milestones that will allow the student to see the use in the design and development of products.

4. Materials and Methods

4.1. Participants

The participants in the study were the students of the bachelor of Engineering in Industrial Design and Product Development from the academic year 2020/2021 at the University CEU Cardenal Herrera. The number of students included was 24, which was the total number of students registered in the first year of the bachelor, so we did not need to select any participants for the study. Most of the students were from Spain, except for three who came from South America (El Salvador, Colombia and Honduras). The participants' age ranged between 18 and 20 (similarly distributed), except for one of them aged 24.

4.2. Scope of Application

STEAM learning has been planned and applied to the first year in which the following subjects are included, see Table 1.

This article focuses on the second semester when engineering mechanics is taught. The syllabus of the physics extension course is, see Table 2.

Table 1. First-year subjects of bachelor of engineering in design and their classification into STEAM categories (Science, Technology, Engineering, Art, Math).

Semester 1	STEAM Classification	Semester 2	STEAM Classification
Physics	S, T, M	Physics Extension	S, T, M
Maths	M	Maths Extension	M
Art History	A	Anthropology	S
Basic design	A, S, T	Design Extension	A, S, T
Shape representation	A	Descriptive geometry	A, S, T

Table 2. Syllabus of the physics extension subject.

Item	Contents
1	Newton's laws
2	Moment of forces.
3	Kinematics
4	Friction
5	Centroid and center of gravity
6	Free body diagram
7	Equilibrium of a particle
8	Equilibrium of a rigid body

4.3. Tools

An experimental design was carried out through a qualitative analysis following experts in this field [38] in which multiple-question Likert scale questionnaires are reliable [39]. In our experiments, data collection was obtained through an ad hoc questionnaire, following other validated methods found in the scientific literature, such as [38]. There are 21 items in the questionnaire. The first 20 questions follow a Likert-type scale within a range of five points (from 1 = Strongly disagree to 5 = Strongly agree). The last question, item 21, is an open-ended question. The questions are shown below in Table 3.

Table 3. Questions asked in the students' questionnaire.

ID	Question
1	I know what the center of gravity is and how to apply it in product design
2	I know Newton's laws and how to apply them in product design
3	I know what the moment of a force is and how it affects product design
4	I know what friction is and how it affects product design.
5	I know what the free-body diagram is and its usefulness in product design
6	I know what equilibrium equations are and their usefulness in product design
7	I know the usefulness of integral calculus in product design
8	I know the types of traditional wood joints applied in the design of products and their physical restrictions.
9	I understand that the decision to cut a product determines it, both from the physical/functional point of view, as from the aesthetic one.
10	I recognize the aesthetic impact and the added value, in detail, that can determine the choice of any traditional wood joint applied in the product design.
11	The Pringles ring exercise helped me understand how the gravity center of friction works
12	The "Equilibrium challenge" exercise helped me understand how the center of gravity works and is calculated
13	The letter design exercise helped me understand how joints work
14	The letter design exercise helped me understand the usefulness of equilibrium equations
15	The letter design exercise helped me understand the importance of integral calculusl of several variables applied to design.
16	The letter design exercise helped me model mathematically and visualize the triple volume concept
17	The letter design exercise helped me relate the concepts of transversal section and volume
18	The letter design exercise helped me understand the importance of the center of gravity.
19	The letter design exercise helped me understand the importance of the value of detail incorporated into the design.
20	The letter design exercise helped me understand which graphic system/s to use to render them according to the required objective.
21	What do you think about the physics, Pringles ring, Equilibrium challenge and letter design exercises?

5. Design and Implementation of the EXPLORIA Pilot Project

The EXPLORIA project was born from the need to improve university methodologies to solve the deficiencies found in the use of the traditional system. In this sense, CEU University has started different pilot projects in different bachelors, including the bachelor of engineering in industrial design and product development. This bachelor in particular includes subjects easily classifiable according to the STEAM model, see Table 1.

The EXPLORIA project connects the competencies of the different STEAM subjects, in which the standard subjects disappear, designing a learning process as a logical, sequential and incremental itinerary. In this learning process, teachers do not have a fixed weekly schedule, and therefore it is designed based on the learning sequence planned at each moment.

The EXPLORIA project is designed from the concretion and synthesis of the specific and general competencies of each subject included in the curriculum for the bachelor of Engineering in Industrial Design and Product Development. It was considered appropriate to group these skills following a learning process based on Bloom's Taxonomy according to the verbs understand, apply, experiment and develop.

The EXPLORIA project for the 1st year of the bachelor of Engineering in industrial design and product development was designed by the teachers of the subjects included in the first year of the bachelor, and by part of the Faculty's management team, including multidisciplinary profiles of mathematics, engineering, fine arts and designers. The design was based on five concepts derived from the fundamentals of basic design used for the itinerary of this course and these are: shape, volume, color, space and structure. In order to adjust to the academic year of two semesters, we divided the learning itinerary of design fundamentals into two modules. These in turn are divided into three acts as shown:

MODULE I

- Act I: Shape
- Act II: Volume
- Act III: Color

MODULE II

- Act IV: Space
- Act V: Structure
- Act VI: Project

In addition, we decided to introduce a milestone at the end of each Act to strengthen the objective of each of the fundamentals worked on and obtain a global vision of the related competencies. This milestone is a challenge-based methodology in which students, actively and autonomously, and based on a general topic raised by teachers, respond to their own concerns through a challenge. This is formalized and sustained through the application in a project of the skills and learning acquired by the student during the weeks for each act. In this activity, the teacher role is to accompany and guide the student according to the needs required by each phase of the project, being flexible when intervening and adapting to the team requirements depending on their specialization.

Since one of the pillars that sustain the EXPLORIA program is the creation and consolidation of the learning community, it is therefore appropriate to develop the milestone within a team. That is how transversal competencies such as decision-making, communication, critical thinking, etc., are integrated. In addition, the group is changed for each Act, which allows the students to vary their role depending on the idiosyncrasy of the team and obtain different experiences. The project developed based on the challenge is presented by each team to the community (other teams and teachers) and evaluated on the one hand by the teaching staff, who will determine the cohesion of the acquired competencies and the learning results established for the Act through a rubric designed for this activity. The other teams, by using the Post Motorola tool (Questionnaire with 4 questions: What has gone well? What has gone bad? What have I learned? What could improve?), will qualitatively evaluate what items worked, what can be improved and what has been learned, determining a quantitative score based on the responses. Finally, the team itself, based on an attitudinal and aptitude rubric, carries out a self-evaluation and co-evaluation. The weighting of all these results will be the final grade of each student.

6. Physics (Mechanical Engineering) in EXPLORIA

Physics, as a basic subject for the first year of engineering, is part of the EXPLORIA project, which required analyzing the subject role and its connection with the students' learning. Physics is a core element that not only provides the necessary knowledge for the learning of other subjects, but also guarantees the functionality of the students' designs. This article focuses on mechanical engineering, which is particularly important in product design.

The physics sessions can be of two types, sessions of theoretical concepts (master class) and work sessions in the classroom. In any of these, there is a challenge the student must overcome. Four challenges are proposed, two for act IV and two for act V where, through the resolution of the challenge, the student must understand, apply, experiment and develop the concepts of physics, linking them to the contents from other subjects such as mathematics, geometry and design. The milestone makes it possible to evaluate the acquired physics competencies, applied to real problems and designs proposed by the students, who must also establish a link with the other competencies developed in the other subjects.

6.1. Description of Sessions and Timing

The sessions carried out and the subjects included in Act IV and V are shown below (EXPLORIA project module II).

6.1.1. The Transportation Challenge

The challenge involves transporting an object in a LEGO EV3 robot, from an initial to a final position without falling, by using a Bézier curve for the robot's path, see previous work [40]. Students select an object and calculate, through the force moment, the acceleration limit so that the object does not fall. Subsequently, the students design a Bézier curve using an application developed in Geogebra where, by moving the control points, they design the trajectory so that it does not exceed the calculated acceleration limit. Finally, the students test the trajectory generated in the LEGO EV3 robot through an application programmed in Matlab, see [40]. Figure 1 shows the Geogebra application used for the design of the Bézier curve, see Figure 1 (left) as well as photos of the test with the LEGO EV3 robot, Figure 1 (right).

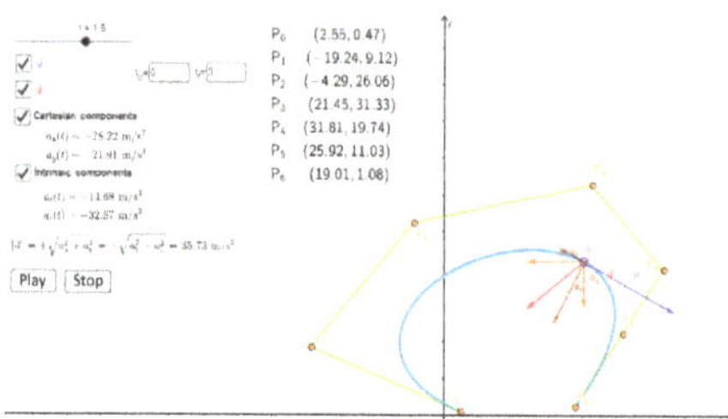

Figure 1. The transport challenge.

The development of each of the 7 challenge sessions is specified below:

- Session 1. The project begins with a motivational session explaining the problem to be solved from the point of view of mobile robotics, AGVs (Automated Guided Vehicle), AMR (Autonomous Mobile Robot), autonomous robots used to transport materials in industry and society.
- Session 2. A theory session is held in which Newton's 3 laws are explained. Inertial frames of reference are defined and the limitations of Newton's laws are explained.
- Session 3. A math session is held in which students work on the vector and scalar product.
- Session 4. A Physics session is carried out in which the force moment ($M = D \times F$) is explained and a practical exercise is carried out on how to calculate it on a bottle of water.
- A mathematics session is held in which parametric curves, Bézier's curves, Frenet's Diedro, etc., are explained.
- Session 5. A physics session is carried out in which kinematics is explained, the accelerations suffered by a moving object when it follows a curved path and centripetal/centrifugal acceleration. After, all this is linked to the reference frames in which the centrifugal force is a fictitious force found in the non-inertial frames. Here, we explain how to calculate the limit centrifugal acceleration when an object is transported. The ability of parametric curves to approximate any curve, circle or clothoid is also explained, the latter being explained in depth given its mathematical properties and therefore its application in road design and in the generation of mobile robot trajectories [41]. Students select an object and calculate the centripetal acceleration limit. The condition of moment equilibrium and the centroid are indirectly introduced for their calculation.
- Session 6. A mathematics session is carried out in which the developed Geogebra applet is explained and the students design the path the robot will follow, with the centripetal acceleration limit calculated in the previous session, moving the control points of the Bézier curve and simulating it to verify that the restriction of centripetal acceleration limit is met.

- Session 7. A test session of the trajectories designed by the students is carried out in which they verify whether the designed trajectory manages to transport the object to the destination. Otherwise, the students will redesign the trajectory until the selected object is transported to the destination.

6.1.2. The Equilibrium Challenge

The objective of this challenge, inspired by similar tests carried out in architecture [42], is to make a sculpture by using everyday objects we can find at home, such as forks, spoons, knives, shoes, etc., and also diverse rigid objects. By completing this challenge, students will begin to learn about static balance in space and its importance in the design process. Through the exploration of the tangible materiality and the physicality of the objects, the students were implicitly advancing until reaching a series of hierarchical rules for the assignment of the objects in the space in order to achieve the global static balance of the compositions under the effect of the gravity. To finish the test, students must calculate the gravity center of each object, either by coincidence with the centroid or by alignment with the balance point. Once each object has been calculated and weighed, the student must calculate the sculpture's gravity center. Figure 2 shows one of the tests in which Figure 2 (left) shows the composition and position of the gravity centers of each object and Figure 2 (right) shows a table with the measurements of distances and masses, as well as the calculation of the gravity center of the proposed composition.

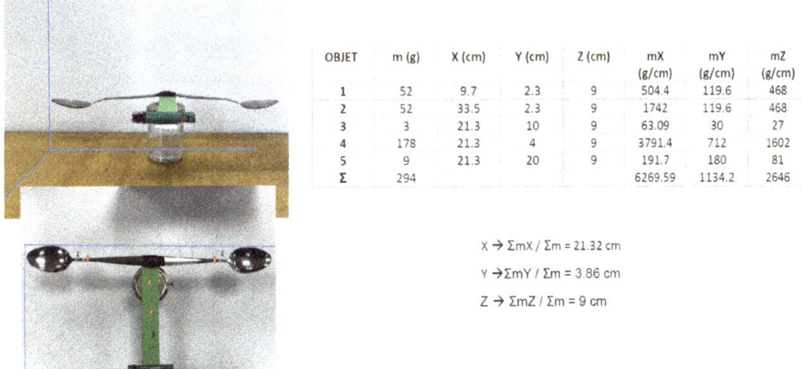

OBJET	m (g)	X (cm)	Y (cm)	Z (cm)	mX (g/cm)	mY (g/cm)	mZ (g/cm)
1	52	9.7	2.3	9	504.4	119.6	468
2	52	33.5	2.3	9	1742	119.6	468
3	3	21.3	10	9	63.09	30	27
4	178	21.3	4	9	3791.4	712	1602
5	9	21.3	20	9	191.7	180	81
Σ	294				6269.59	1134.2	2646

$$X \rightarrow \Sigma mX / \Sigma m = 21.32 \text{ cm}$$

$$Y \rightarrow \Sigma mY / \Sigma m = 3.86 \text{ cm}$$

$$Z \rightarrow \Sigma mZ / \Sigma m = 9 \text{ cm}$$

Figure 2. The equilibrium challenge.

The development of each of the 2 challenge sessions is specified below:

- Session 1. A physics session is held in which the centroid and gravity center are explained. We explain how to calculate the centroid and gravity center both analytically and experimentally in irregular objects, by finding the equilibrium point. Finally, we explain how to calculate the gravity center of composite objects and tests are carried out.
- Session 2. The students bring objects from home to make the sculpture and firstly they calculate the gravity center of each object and their weight. Later, they make their sculpture in equilibrium, measure it and calculate the composite center of gravity of the composition.

6.1.3. The Pringles Ring Challenge

The objective of this challenge, inspired by the viral challenge with the same name found on the internet, is to let students experiment with the gravity center, friction, and warped surfaces, in particular the hyperbolic paraboloid and its geometric and physical properties. Figure 3 shows the development of one of the exercises by a student. Figure 3 (left) shows the construction at its intermediate point where you can see the importance of the base and how the Pringles chips on the side of the ring are supported thanks to the friction and normal force they exert between them. Figure 3 (right) shows the finished ring.

Figure 3. The Pringles challenge.

The development of each of the 3 challenge sessions is specified below:

- Session 1. A descriptive geometry session is carried out explaining the warped surfaces, including the hyperbolic paraboloid, its mathematical formulation and its properties.
- Session 2. The concept of friction is introduced and exercises are carried out in class with different objects using a ramp and throwing different objects over it to demonstrate that the challenge is met.
- Session 3. The students bring at least 2 Pringles cans and the basic concepts are explained to them to be able to carry them out. These include:
    1. Object stacking: we explain to them what happens when objects are stacked, how much space can protrude from an object to another and how it affects the gravity center of composites.
    2. Friction: The side walls of the ring are supported by friction between the Pringles and the normal force they exert on each other.
    3. Gravity center of composites: One of the keys to overcoming the challenge is to generate a good base of the ring and force the gravity center of composites to remain in that area.

6.1.4. The Letters Challenge

The objective of this challenge is to encourage students to design and manufacture the initial letter of their name with 3.2 × 3.2 cm strips by using joints. The present challenge involves reformulating the concept of joints for rigid body mechanics that is usually explained in traditional physics lessons, see for example [43] (Table 5.1, pp. 210–211), and regarding the specific needs of designers who are using wood joints, see for example [44]. Thus, the classic roller, pin or rocker type joints are replaced by "butt", "miter" joints, and a great diversity of wood joints as we can see in [44]. The challenge consists of designing the letter with certain restrictions, which are:

1. The letter must include at least 3 different joints (from an initial list provided by the teachers), in this challenge the use of one of the joints will be mandatory and assigned through a first draw.
2. The letter must be self-supporting, that is, it must be able to stand upright and must be able to be transported, by grabbing it from a higher point and moving it from one point to another without dismounting.
3. The use of glue or any other element to anchor or fix the joints is not allowed.

- Session 1. A basic design extension session is held in which the concept of structure is explained from the design point of view and real objects and sculptures are analyzed.
- Session 2. A physics session is carried out in which the equations of equilibrium and the calculation of the free-body diagram are introduced. The same real objects and sculptures from session 1 are used for analysis from the physical point of view.
- Session 3. Students are proposed to do research and analysis. They must check the websites of famous designers, select a product and obtain:
    1. The free-body diagram of the selected product.
    2. Dimensions and information from their designer.
    3. Proposal for design improvements, both from a physical and design point of view.
- Session 4. Combined session with the Design and Physics teachers in which an elevator pitch is held: a 2-min short presentation in which each student explains his/her results from session 3 to the rest of their peers. This session is used to share the results but also used as an evaluation for the teacher.
- Session 5. Basic design extension session in which the different types of joints used in design are explained, the challenge is defined and the draw for the mandatory joint to be used is carried out for each student.
- Session 6. Physics session in which the equations of equilibrium are explained and their calculation from the free body diagram.
- Sessions 7 and 8. Work session in the combined classroom with design and physics teachers in which students design their letters, calculate their free body diagram and equilibrium equations to guarantee the viability of the challenge.
- Session 9 and 10. Workshop session. The students go to the workshop and mechanize their letters. Figure 4 shows the students working in the workshop.

Figure 4. The letters challenge. Pictures of the building process.

- Session 11. Combined session with the Design and Physics teachers in which an elevator pitch is held: a 2 min short presentation in which each student presents his/her letter design, calculations, manufacturing problems and they perform the self-support test on-site, moving the letter from a table to the center of the classroom. Figure 5 shows the resulting letters placed in a vertical position on a table. In the elevator pitch session, each student takes his/her letter with one hand from a point chosen by him/her and moves it to the central table where the student will explain his/her design and manufacturing process. Figure 6 shows details of some designs in which we can see how the student, in the case of the letter "P", has used the calculation of the center of gravity to calculate the angle at which the base should be cut in order to keep the P in balance.

Figure 5. The letters challenge. Letters made by students.

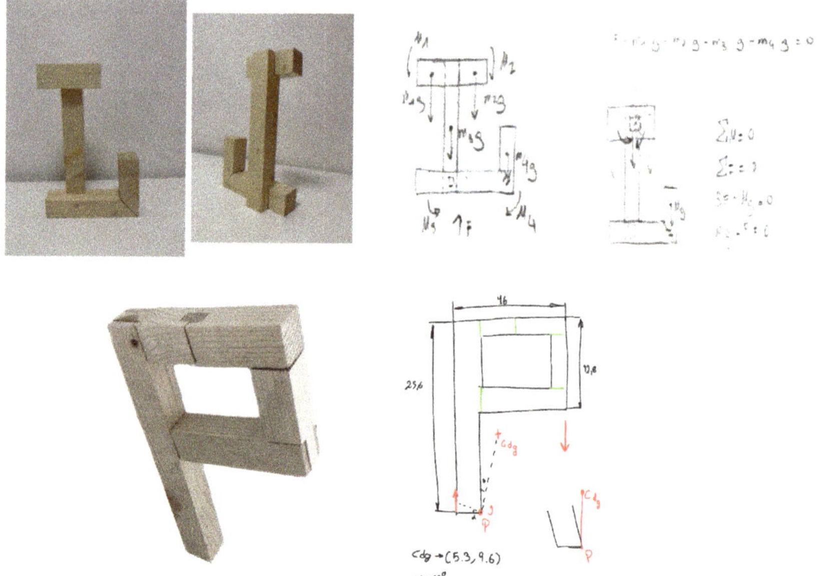

Figure 6. The letters challenge. Examples of results and calculations.

6.1.5. Milestones

At the end of act IV and act V, a Milestone week was carried out. In the Milestone, a topic is proposed on which the students have to provide solutions. In the Milestone of Act IV, the topic Community "Alfara del Patriarca" was proposed, this is the name of the town where the ESET (Technical School of Engineering) of CEU Cardenal Herrera University is located. In this Milestone, the students had to propose design solutions for the inhabitants of the town and the space where they coexist, with the aim of improving common spaces, coexistence, the well-being of the elderly, etc.; and always under the project implementation from the learning and skills obtained in Act IV: Space.

In the Milestone of Act V, Structure, the topic was The Trip, in which the students generated design proposals related to this theme, and in the same way as they did in the previous Milestone, they applied competencies obtained in the previous weeks which were linked, in this case, with the learning outcomes from Act V, Structure. Figure 7 shows one of the resulting works where the students proposed to make a suitcase with drawers. In this case, the students used what they had learned in physics, taking into account the weight limit of the suitcase, the maximum opening of the drawers and the distribution of masses so that in the event that the user had all the drawers open, the suitcase would not overturn, see Figure 7.

MALETEA

Design and prototyping using equilibrium equations:

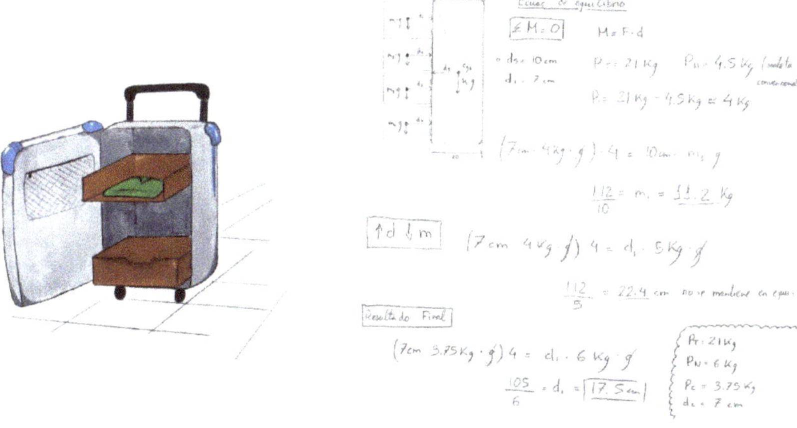

Color selection:

Figure 7. Example of one of the results of Milestone V. The trip.

7. Results

Table 4 shows the answers to the Likert-type questions.

The questionnaire was answered by 95.8% of the enrolled students.

If we focus on the first six questions related to knowledge of physics, in question 1 related to the center of gravity, 100% of the students agree or strongly agree. In question 2, Newton's laws, 91% agree or strongly agree, in question 3, moment of a force, also 91% agree or strongly agree. In question 4, friction, 95% agree or strongly agree. In question 6 related to the free-body diagram, 82.6% agree or strongly agree, and in question 7 related to the equations of equilibrium, 69% agree or strongly agree.

Regarding the usefulness of the challenges for learning physical knowledge, in question 11 related to the challenge of the Pringles ring and in question 12 related to the equilibrium challenge, 82.6% of the students agree or strongly agree, in question 13, understanding how the joints work in the letter, 100% agree or strongly agree and in question 14, the letter exercise and the equilibrium equations, 82.6% agree. Finally, question 18 asks if the exercise of the letter helped them understand the importance of the gravity center, in which 100% of the students agree or strongly agree.

Table 4. Student's questionnaire responses, Likert-type questions.

Question	SD	D	N	A	SA
1	0	0	0	8	15
2	0	0	2	12	9
3	0	0	2	13	8
4	0	0	1	11	11
5	0	0	4	9	10
6	0	0	4	15	4
7	0	2	5	9	7
8	0	0	0	4	19
9	0	0	1	6	16
10	0	0	1	5	16
11	0	1	3	12	7
12	0	0	5	8	10
13	0	0	0	5	18
14	0	0	4	15	4
15	0	0	5	12	5
16	0	0	6	9	8
17	0	0	3	13	7
18	0	0	0	10	13
19	0	0	0	9	14
20	0	0	3	9	11

With regard to learning related to other subjects, in questions 7 and 15, related to the usefulness of integral calculus in product design, 69% and 73%, respectively, agreed or strongly agreed. In question 8, related to knowledge of traditional joints and their physical restrictions, 100% agreed or strongly agreed, and in questions 9 and 20, related to cutting and graphic representation, 95.6% and 86.9%, respectively, agreed or strongly agreed. In questions 10 and 19, related to the aesthetic and added impact of the joints, 95.6% and 100%, respectively, agreed or strongly agreed.

Here are some of the answers given to open question 21:

- "These are exercises that help you understand the syllabus in a very dynamic way. It's a good way to learn theory while putting it into practice".
- "When we carry out this type of projects, we are learning and understanding the lesson explanations".
- "The most interesting exercise is the Pringles ring since with common food you can realize the importance of shape for calculating balance. On the other hand, the balancing exercise was more about calculation of structure. Finally, the letter seems to me a very complete exercise to bring together all the subjects in the same work".
- "When it comes to doing it in a practical way, you learn better".
- "They are examples that perfectly convey the theory taught".
- "They help us to really realize the usefulness of physics in any type of project".
- "I think it was an exercise that allowed us to understand in a practical way how friction and gravity act and how to use these factors to our advantage".
- "These exercises are a very good way to learn, since you understand what you are studying because you can check it at the same time".
- "Very useful exercises, since you can see how to apply physics and mathematics in specific things such as for example the pringles exercise, this helps you see how the content taught in class is reflected in reality and helps you understand the contents much better".

8. Discussion

The EXPLORIA project implemented in the bachelor of Engineering in Industrial Design and Product Development produces a great impact on the learning process. As can be seen in the questionnaire results, the vast majority of students agree or strongly agree that learning all the subjects in an integrated way through challenges and milestones has been very helpful and above all, being participants in the learning process through knowledge discovery.

The feedback given by the students in the questionnaire reinforces the idea that steam-based learning improves the understanding of basic concepts through their implementation in projects and challenges. We can see that a first-year student has not yet received a project subject and therefore he/she will need more knowledge to be able to develop a project, although we think that the students are able to develop their own ideas and put into practice what they have learned so far. So, this is where they are able to apply what they have learned, and their knowledge can be consolidated.

The evaluation of this type of project is an important handicap since, in the end, teachers must assign grade reports for each subject. In this case, in each project or challenge, teachers must independently take the part corresponding to their subject and grade it. The hardest part found by teachers is separating the subject grading from the project applicability or, being able to qualify it as a design project itself. It is important to know how to grade students at the point of learning in which they are. It is not the same grading a first-year student who develops the trip challenge as grading a final-year student. Regarding student evaluation and grades, in our previous work [37], we compared the grades obtained by students in mathematics from the first semester with respect to last year when the improvement was significant, not only in the number of pass grades, but also in the grades obtained. The results of the physics subject are similar compared to those of the previous year: only one of the students failed, and there was an increase in the level of grades obtained. However, this comparison is not included as a result in the paper since the teacher for the year 20/21 was a different one. Understanding the concepts of mathematics and physics applied to projects should solve the problems noted in the research studies by [23,26,27,34,35]. As demonstrated in our previous work [37], the inclusion of this methodology improves students' perception of mathematics by solving the problem indicated in [23,26] and indirectly in the subject of physics, as indicated in [27]. The improvement in physics and mathematics has direct consequences in the rest of the subjects of the same course, and so will have in higher courses, as demonstrated in [34,35].

Regarding the results of the post-test, it can be noted that the proposed challenges help students to understand the concepts of physics, see questions 11–14, 18, and to apply them to design, questions 1–6. As can be seen also in the post-test results, with the questions related to the letter challenge, which involved all the subjects, it has not only served to help in learning the concepts of physics, but also for the rest of the subjects, see questions 15–20. However, although questions 1–20 show how useful the proposed challenges can be, the most important thing is their perception of the activities, see question 21, since, as demonstrated in our previous work [37], the affective domain is key to generate knowledge that is maintained over time and that students do not forget what they have learned. Observing the comments they make, this positive attitude can be denoted with respect to the challenges proposed since they see the physical concepts reflected in the product designs, generating a positive reinforcement that strengthens their learning.

The construction of the learning process could be a problem for the university because a significant effort is required on the part of the teachers since in order to carry out the integration of subjects, a broad knowledge of the rest of the subjects is needed. In this paper, the challenges of the pringles ring exercise and the letter exercise have required the physics teacher to understand ruled surfaces and wood joints. The latter are especially relevant since to achieve an integration of the physical part with the design part, the concept of traditional joints explained in rigid solid mechanics has been adapted to the wood joints used in design.

Another important limitation of the present methodology is that the teachers' schedule is not fixed and is determined by the learning sequence. Additionally, if a student fails a part of the test, it is not easy to pass that part because the whole learning process is intertwined. This could lead to organizational problems for the university.

9. Conclusions and Further Developments

This article shows the design and evaluation of the EXPLORIA project, based on STEAM learning in the bachelor of engineering of product design. The development of an integrated competency map of the learning process, in which the subjects are no longer considered as isolated contents and the elaboration of an integrated learning process in which the competencies and learning outcomes of the subjects are considered as a whole, that is, a global and complete learning, have allowed a comprehensive learning of the basic and specific subjects of the first year, which has offered great advantages. Students can see the application of basic subjects in problems related to their profession which generates positive emotions and reinforces the learning process and the motivation to continue learning.

On the negative side, this integral learning process requires an extra effort from the teaching staff since to be able to achieve perfect integration they need to know the rest of the subjects and we also find problems at organizational level, since the schedules are not fixed but influenced by the learning sequence.

As future works, we intend to implement the EXPLORIA methodology in other bachelors related to STEAM learning, such as Architecture.

Author Contributions: Conceptualization, N.M., P.A., T.F., P.D.R. and S.B.; methodology, N.M., P.A., T.F., P.D.R., S.B. and A.M.C.; validation, N.M., P.A., T.F., P.D.R. and A.M.C.; formal analysis, N.M.; investigation, N.M., P.A., T.F., P.D.R. and S.B.; resources, S.B.; writing—original draft preparation, N.M., P.A. and T.F.; funding acquisition S.B. All authors have read and agreed to the published version of the manuscript.

Funding: This research received no external funding.

Institutional Review Board Statement: Not applicable.

Data Availability Statement: Not applicable.

Acknowledgments: Authors wish to thank the management team of the Fundación San pablo CEU and the Technical School of Engineering for the support given to the teachers in order to promote new methodologies in classrooms that are far away from the classical methodologies.

Conflicts of Interest: The authors declare no conflict of interest.

Abbreviations

The following abbreviations are used in this manuscript:

STEM	Science Technology Engineering Mathsi
STEAM	Science Technology Engineering Art Maths
AGV	Automated Guided Vehicle
AMR	Autonomous Mobile Robot
ESET	Technical School of Engineering

References

1. Irwanto, I.; Saputro, A.D.; Widiyanti; Ramadhan, M.F.; Lukman, I.R. Research Trends in STEM Education from 2011 to 2020: A Systematic Review of Publications in Selected Journals. *Int. J. Interact. Mob. Technol.* **2022**, *16*, 19–32. [CrossRef]
2. Pisa, Programa para la Evaluación Internacional de los Alumnos. Spanish Report. Informe Español. Ministerio de Educacion, Cultura y Deporte. 2018. Available online: https://www.oecd.org/pisa/aboutpisa/ebook%20-%20PISA-D%20Framework_PRELIMINARY%20version_SPANISH.pdf (accessed on 26 May 2021).
3. Sanders, M. STEM, STEM education, STEM mania. In *Technology Teacher*; Association Drive Suite 201; International Technology Education Association (ITEA): Reston, VA, USA, 2009.
4. Ludeña, E.S. La educación STEAM y la cultura maker. Padres y Maestros. *J. Parents Teach.* **2019**, *379*, 45–51.
5. Yakman, G. $ST\Sigma@M$ Education: An overview of creating a model of integrative education. In *PATT-17 and PATT-19 Proceedings*; de Vries, M.J., Ed.; ITEEA: Reston VA, USA, 2008; pp. 335–358.
6. Williams, J. STEM Education: Proceed with caution. *Des. Technol. Educ. Int. J. Spec. Ed.-Stem-Underpinned Res.* **2011**, *16*, 26–35.
7. Wells, J.G. STEM education: The potential of technology education. *Counc. Technol. Eng. Teach. Educ.* **2019**, *16*, 195–229.
8. Pitt, J. Blurring the boundaries—STEM education and education for sustainable development. *Des. Technol. Educ. Int. J.* **2009**, *14*, 37–48.
9. Yakman, G.; Lee, Y. Exploring the exemplary STEAM education in the U.S. as a practical educational framework for Korea. *J. Korea Assoc. Sci. Educ.* **2012**, *32*, 1072–1086. [CrossRef]
10. Kim, E.; Kim, S.; Nam, Lee, D.T. Development of STEAM program Math centered for Middle School Students. *Des. Technol. Educ. Int. J.* **2012**, *1*, 1–5.

11. Parlamento Europeo, Consejo de la Unión Europea. Recomendación 2006/962/CE del Parlamento Europeo y del Consejo, de 18 de Diciembre de 2006, Sobre las Competencias Clave Para el Aprendizaje Permanente. 2006. Available online: https://eur-lex.europa.eu/LexUriServ/LexUriServ.do?uri=OJ:L:2006:394:0010:0018:ES:PDF (accessed on 26 May 2021).

12. Sousa, D.A.; Pilecki, T. *From STEAM to STEAM: Using Brain-Compatible Strategies to Integrate the Arts*; Corwin: Thousand Oaks, CA, USA, 2013.

13. Vergara, D.; Paredes-Velasco, M.; Chivite, C.; Fernandez-Arias, P. The Challenge of Increasing the Effectiveness of Learning by Using Active Methodologies. *Sustainability* **2020**, *12*, 8702. [CrossRef]

14. Ugras, M. The Effects of STEM Activities on STEM Attitudes, Scientific Creativity and Motivation Beliefs of the Students and Their Views on STEM Education. *Int. Online J. Educ. Sci.* **2018**, *10*, 165–182.

15. Konopka, C.L.; Adaime, M.B.; Mosele, P.H. Active Teaching and Learning Methodologies: Some Considerations. *Creat. Educ.* **2015**, *6*, 1536–1545. [CrossRef]

16. Abou-Hayt, I.; Dahl, B.; Rump, C. A Problem-Based Approach to Teaching a Course in Engineering Mechanics. In *Educate for the Future: PBL, Sustainability and Digitalisation*; Guerra, A., Chen, J., Winther, M., Kolmos, A., Eds.; Aalborg Universitetsforlag: Aalborg, Denmark, 2020; pp. 499–509.

17. Santyasa, I.W.; Rapi, N.K.; Sara, I.W.W. Project Based Learning and Academic Procrastination of Students in Learning Physics. *Int. J. Instr.* **2020**, *13*, 489–508. [CrossRef]

18. Zavala, G. Integration of physics, mathematics and computer tools using challenge-based learning. In Proceedings of the 2020 IEEE Global Engineering Education Conference (EDUCON), Porto, Portugal, 27–30 April 2020; pp. 1387–1391.

19. Mills, J.E.; Treagust, D.; Engineering education—Is problem-based or project-based learning the answer. *Australas J. Eng. Educ.* **2003**, *3*, 2–16.

20. Pospiech, G.; Eylon, B.S.; Bagno, E.; Lehavi, Y. Role of Teachers as Facilitators of the Interplay Physics and Mathematics. In *Mathematics in Physics Education*; Pospiech, G., Michelini, M., Eylon, B.S., Eds.; Springer: Cham, Switzerland, 2019; Volume 12, pp. 269–291.

21. UNESCO. *Engineering: Issues Challenges and Opportunities for Development*; UNESCO: Paris, France, 2010.

22. Hidalgo, S.; Maroto, A.; Palacios, A. *¿ Por Qué se Rechazan las Matemáticas? Análisis Evolutivo y Multivariante de Actitudes Relevantes Hacia las Matemáticas*; Revista de Educación: Madrid, Spain, 2004; Volume 334, pp. 75–95.

23. Pedrosa, C. Actitudes Hacia las Matemáticas en Estudiantes Universitarios. Ph.D. Thesis, Universidad de Cordoba, Córdoba, Sapin, 2020.

24. Espinosa, E.O.C.; Mercado, M.T.C.; Mendoza, J.R.R. Actitudes hacia las matemáticas de los estudiantes de posgrado en administración: Un estudio diagnóstico. *REXE Rev. Estud. Exp. Educ.* **2012**, *11*, 81–98.

25. Hill, D.; Bilgin, A.A. Pre-Service Primary Teachers' Attitudes towards Mathematics in an Australian University. *Creat. Educ.* **2018**, *9*, 597. [CrossRef]

26. Mato-Vázquez, D.; Soneira, C.; Muñoz, M. Estudio de las actitudes hacia las Matemáticas en estudiantes universitarios. *Números Rev. didáCtica Mat.* **2018**, *97*, 7–20.

27. Ataide, A.R.P.D.; Greca, I.M. Epistemic views of the relationship between physics and mathematics: Its influence on the approach of undergraduate students to problem solving. *Sci. Educ.* **2013**, *22*, 1405–1421. [CrossRef]

28. Bishof, G.; Zwolfer, A.; Rubesas, D. Correlation between engineering students' performance in mathematics and academic success. In Proceedings of the 122nd ASEE Anual Conference & Exposition Seattle, Seattle, WA, USA, 14–17 June 2015.

29. Adebisi, A.; Olanrewaju, O. Mathematics Skills as Predictors of Physics Students' Performance in Senior Secondary Schools. *Int. J. Sci. Res.* **2013**, *2*, 391–394.

30. Kamal, N.; Rahman, N.N.S.A.; Husain, H.; Nopiah, Z.M. The Correlation Between Electrical Engineering Course Performance and Mathematics and Prerequisite Course Achievement. *Soc. Sci. Humanit.* **2016**, *24*, 97–110.

31. Rossdy, M.; Michael, R.; Janteng, J.; Andrew, S.A. The Role of Physics and Mathematics in Influencing Science Students' Performance. In *Proceedings of the Second International Conference on the Future of ASEAN (ICoFA)*; Mat Noor, A., Mohd Zakuan, Z., Muhamad Noor, S., Eds.; Springer: Cham, Switzerland, 2017; Volume 1, pp. 399–406.

32. Ojonugwa, T.; Umaru, R.; Sujaru, K.O.; Ajah, A.O. Investigation of the Role of Mathematics on Students' Performance in Physics. *J. Res. Educ. Sci. Technol.* **2020**, *5*, 101–108.

33. Pospiech, G. Framework of Mathematization in Physics from a Teaching Perspective. In *Physical Review Special Topics—Physics Education Research*; Springer: Cham, Switzerland, 2019; Volume 1, pp. 1–33.

34. Imran, A.; Nasor, M.; Hayati, F. Relating grades of maths and science courses with students' performance in a multidisciplinary engineering program. A gender inclusive case study. *Procedia Soc. Behav. Sci.* **2012**, *46*, 3989–3992. [CrossRef]

35. Perdigones, A.; Gallego, E.; Garcia, N.; Fernandez, P.; Perez, E.; del Cerro, J. Physics and Mathematics in the Engineering Curriculum: Correlation with Applied Subjects. *Int. J. Eng. Educ.* **2014**, *30*, 1509–1521.

36. Jose, J. An Exploration of the Effective Use of Bloom's Taxonomy in Teaching and Learning. In Proceedings of the International Conference on Business and Information (ICBI), Virtual, 16 November 2021; p. 100.

37. Romero, P.D.; Montés, N.; Barquero, S.; Aloy, P.; Ferrer, T.; Granell, M.; Millán, M. EXPLORIA, a new way to teach maths at university level as part of everything. *Mathematics* **2021**, *9*, 1082. [CrossRef]

38. Makrakis, V.; Kostoulas-Makrakis, N. Bridging the qualitative–quantitative divide: Experiences from conducting a mixed methods evaluation in the RUCAS programme. *Eval. Program Plan.* **2016**, *54*, 144–151. [CrossRef] [PubMed]

39. Jebb, A.T.; Ng, V.; Tay, L. A Review of Key Likert Scale Development Advances: 1995–2019. *Front. Psychol.* **2021**,*12*, 1590. [CrossRef]
40. Hilario, L.; Mora, M.C.; Montés, N.; Pantaleon, D.; Barquero, S. The transport challenge. A new way to learn Mathematics through physics in university studies. *Comput. Educ.* 2022, *under review.*
41. Montes, N.; Herraez, A.; Armesto, L.; Tornero, J. Real-time clothoid approximation by Rational Bezier curves. In Proceedings of the IEEE International Conference on Robotics and Automation, Pasadena, CA, USA, 19–23 May 2008; Volume 1, pp. 2246–2251.
42. Vrontissi, M.; Castellón, J.J.; D'Acunto, P.; Monzó, Ll.E.; Schwartz, J. Constructing equilibrium: A methodological approach to teach structural design in architecture. In Proceedings of the IV International Conference on Structural Engineering Education, Madrid, Spain, 20–22 June 2018; pp. 1–10.
43. Hibbeler, R.C. *Engineering Mechanics*, 14th ed.; Pearson Prentice Hall: Hoboken, NJ, USA, 2016.
44. Sumiyoshi, T.; Matsui, G. *Wood Joints in Classical Japanese Architecture*; Kajima Institute: Tokyo, Japan, 1991.

applied sciences

Article

Educational Case Studies for Pilot Engineer 4.0 Programme: Monitoring and Control of Discrete-Event Systems Using OPC UA and Cloud Applications

Erik Kučera *, Oto Haffner, Peter Drahoš and Ján Cigánek

Faculty of Electrical Engineering and Information Technology, Slovak University of Technology in Bratislava, 812 19 Bratislava, Slovakia
* Correspondence: erik.kucera@stuba.sk

Abstract: The current trend in industry is the digitalisation of production processes using modern information and communication technologies, a trend that falls under the fourth industrial revolution, Industry 4.0. Applications that link the world of information technologies (IT) and operational technologies (OT) are in particular demand. On the basis of information from practice, it can be stated that there is a shortage of specialists in the labour market for the interconnection of PLCs with information and communication technologies (cloud, web, mobile applications, etc.) in Slovakia and neighbouring countries. However, this problem is beginning to affect other countries in Europe as well. The main objective of the work was to prepare case studies suitable for educational purposes, which would address the modelling and control of a virtual discrete-event system using a PLC program and its subsequent interfacing to a cloud application. Within the scope of the work, three case studies were prepared to demonstrate the control of discrete-event system using different programming systems and their communication with the developed cloud applications. These applications are to be used for data monitoring and emergency intervention of the discrete-event system. The characteristics of the prepared case studies, which combine operational and informational technologies, predestines them for use in the sphere of education of engineers for digitalisation of production processes. They can also be helpful in research on the creation of digital twins, which represent a type of symmetry between real and virtual systems.

Keywords: discrete-event system; Industry 4.0; digital factory; system control; cloud computing; engineering education; OPC UA; Node-RED

Citation: Kučera, E.; Haffner, O.; Drahoš, P.; Cigánek, J. Educational Case Studies for Pilot Engineer 4.0 Programme: Monitoring and Control of Discrete-Event Systems Using OPC UA and Cloud Applications. *Appl. Sci.* **2022**, *12*, 8802. https://doi.org/10.3390/app12178802

Academic Editor: Shan Wang, Zehui Zhan, Qintai Hu and Xuan Liu

Received: 30 July 2022
Accepted: 30 August 2022
Published: 1 September 2022

Publisher's Note: MDPI stays neutral with regard to jurisdictional claims in published maps and institutional affiliations.

1. Introduction

The digitalisation of production systems with the help of modern information and communication technologies is one of the trends of our time. These technologies are becoming indispensable tools for the necessary digital transformation of industrial processes. They can strengthen the performance of a company, optimise available capacities and improve the quality of final products. Recently, the merging or convergence of information and operational technologies has begun to be mentioned, a trend linked to the Fourth Industrial Revolution, which brought about the concept of Industry 4.0 [1].

To teach interdisciplinary knowledge in study programmes focused on applied and engineering informatics, it is necessary to update the curricula [2]. Educating engineers for digitised production for a world that requires new, multifunctional professions is not an easy task. It is therefore necessary to prepare new educational materials also in cooperation with practice [3].

PLCs (Programmable Logic Controllers) are an important control element in manufacturing systems and there are many specialists on the market who can interface and program such industrial computers. However, very few of them know how to connect PLCs to the IT world (e.g., cloud, augmented reality, mobile or web applications, etc.). Therefore, in this paper, we will deal with interfacing a PLC with a cloud solution and implementing a service application. This paper and project responds to the demand of practice and consists of three case studies that will be used for the development of educational modules for engineers for Industry 4.0.

2. Background

The presented topic is closely related to the convergence of information and operational technologies.

Information technology is an integral part of our times. It make work easier and save time for us people in today's world in several fields of work such as education, healthcare, transport and others. That is why information technology is constantly advancing. Thus, by information technology (IT), we can imagine that it deals with computer technology including hardware, software, telecommunications and generally anything related to the transmission of information or systems that facilitate communication [4].

Operational technologies, as opposed to information technologies, focus on the control and diagnostics of physical devices. The term operational technology (OT) is hardware and software that detects or causes change through direct monitoring and/or control of industrial equipment, assets, processes and events [5].

Very briefly, it could be said that the basic difference between information technology and operational technology is that information technology deals with information and is often associated with software, and operational technology deals with machines and is often associated with hardware. Until recently, these technologies worked separately, but the Industrial Internet of Things (IIoT) has changed a lot of things and blurred the imaginary line between where IT ends and OT begins. Complex machines have begun to integrate with network elements and are driven by advanced analytics software [6].

The idea of the convergence of IT and OT (Figure 1) is not new, but the concept is only now beginning to be put into practice. With the advent of new technologies, we can generally improve efficiency, reduce errors and costs, but also improve workflows and gain competitive advantages. In short, we would say that we are trying to bring physical technology into the digital realm. We are able to do this thanks to advances in, for example, machine-to-machine communication, as well as the introduction of sophisticated sensors and IoT elements that can be mounted on physical devices. These devices can use wireless communication over standardised network protocols to communicate relevant data from each physical system back to a central server for monitoring and analysis. The results of this analysis can then be relayed back to the physical system to enable more autonomous operation, increase accuracy, improve maintenance and improve uptime [4]. So the question certainly arises as to why convergence between IT and OT is important. There are of course a number of reasons, whether it is the aforementioned blurring of the imaginary line between IT and OT or creating a strategy that will improve operational performance. The convergence of IT and OT is a prerequisite for the development of cyber-physical systems, which are a pillar of the Industry 4.0 concept [6]. One of the important concepts for Industry 4.0 brought about by the convergence of IT and OT is the digital twin. The digital twin represents a kind of symmetry between the real system and its virtual model. New technology areas such as Industrial Internet of Things (IIoT) and the Internet of Services (IoS) are also closely linked to convergence of IT and OT.

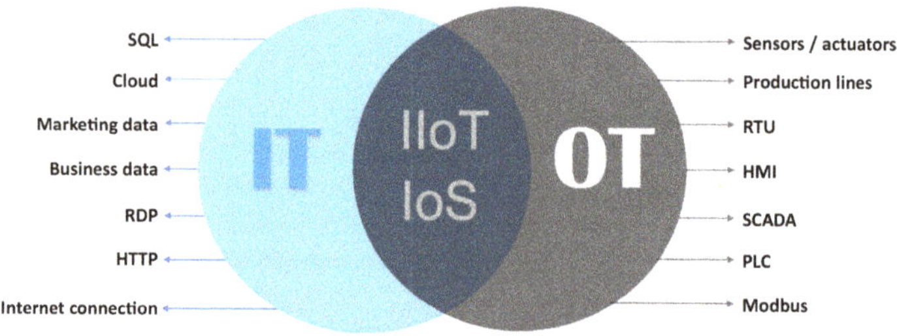

Figure 1. IT and OT convergency.

3. Related Works

In this section, we will address the need to educate the next generation of engineering experts, of which there is still a shortage.

3.1. Urge for Education of Industry 4.0 Technologies

Research and computer laboratories need to be updated in engineering programs to teach interdisciplinary knowledge. The process must include both complex modules and elements from the outside world. This trend in teaching methods is evident in the use of individual HW and SW modules for modelling, testing, and the creation of optimal production lines, cognitive robots, communication systems, and virtual reality models to demonstrate the functionality of individual processes as well as to assess process reconfiguration and the effect of smart features and embedded control systems on the design of production processes [7].

Universities emphasise their role as testbeds of innovation and teachers of the next generation of technologists. The current levels of industrial development and technical advancement are greatly influenced by traditional education. However, it is crucial to take into account how Industry 4.0 will affect higher education institutions if higher education is to impart the necessary skills and knowledge to future generations. As a result, integrating Industry 4.0 concepts into engineering curricula is one of the top concerns for universities and other academic institutions [8].

Given the prevalence of cyber-physical systems, the inclusion of sensors in nearly all industrial components and machinery, and the analysis of all significant data [9], Engineering 4.0 education should concentrate on developing the skills necessary for the digitalisation of the manufacturing sector [10]. Industry 4.0 calls for the development of new, multifaceted professions [11]. These new professionals will need to increase their knowledge of information technology and the procedures that must and should be followed in order to implement it. The report "High-tech skills and leadership for Europe" estimates that there was a shortage of 500,000 IT professionals in Europe by 2020 [12].

The importance of artificial intelligence and computer vision has increased recently due to their contributions to intelligent manufacturing systems [13].

Due to technological advancements and the realisations of Industry 4.0, smart factories require a high degree of precision and accuracy in the measurements and inspection of industrial gears. Machine vision technology enables image-based inspection and analysis for such demanding applications. The level of human expertise in computer vision is rising quickly. Due to the comprehensiveness of image information, visual sensors—unlike other types of sensors—can theoretically span the entire amount of information needed for autonomous driving, the recording of traffic violations, numerous medical tasks, and so forth. The need for experts in computer vision is growing every day. It is crucial to teach experts in various application domains the fundamentals of computer vision at the same time [14].

3.2. Training Programmes for Industry 4.0 and Education 4.0 Concept

In the context of the information presented in the previous subsection, the requirement for specialised training of engineers for the introduction of innovations within the Industry 4.0 concept has arisen in several countries.

The majority of the industry is developing at an incredibly fast rate thanks to Industry 4.0. The education sector has been driven to the Education 4.0 revolution in order to keep up with the industries. The use of automated processes has caused education to shift away from traditional teaching methods and toward a more contemporary model, leading to co-relational research with business that has the potential to alter the entire paradigm of education. It is necessary to evaluate and put into practice the technological developments of this era. The paper compares the revolutions that were led in the fields of industry and education by explaining the ideas of Education 4.0 alongside the study of Industry 4.0 [15,16].

In order to achieve a high degree of production flexibility (individualised mass production), higher productivity rates through real-time monitoring and diagnosis, and a lower rate of material wastage in production, Industry 4.0 was originally a future vision described in the high-tech strategy of the German government. It is conceived upon information and communication technologies such as cyber-physical systems, Internet of Things, Physical Internet, and Internet of Services. The authors of this work present a road map with three pillars that outlines the improvements/changes that will be made to the curriculum development, lab concept, and student club activities. They discuss how they used this road map at the Turkish German University in Istanbul as well [17].

How the business or industry operates and develops has been completely altered by the emergence of Industry 4.0. Its expanding emphasis on automation, decentralisation, system integration, cyber-physical systems, etc., can be attributed to this. The staff is expected to develop computational, cognitive, and adaptive thinking abilities, primarily in the field of information technology, data analytics, etc. The result of this study made in Saudi Arabia is an outcome that the universities that created the foundation for societal trends or future talents must adapt and update their current programs, facilities, and infrastructure. There are opportunities and challenges associated with this transformation of higher education in line with the Industry 4.0 vision. Of course, there are many variables at play, and it takes a fair evaluation to strategically plan this transformation. This research examined the transition from traditional universities to Universities 4.0 and identified the difficulties that lie ahead. There are risks and opportunities associated with the modernisation of higher education in line with the concept of Industry 4.0. SWOT-AHP methodology was subsequently used in this study to analyse this transformation and investigate its likely effects. The importance of practical knowledge and exposure to digital technologies at the university level is highlighted by this study [18].

Due to the requirements for implementing the Industry 4.0 concept, employers will need to hire people with new skills and abilities, particularly in the digital realm. All types of schools, including nontechnical ones, need to change their teaching and educational methods, and rather than narrowly specialising in one area, education should focus on a much broader overview. People also need to be educated in systemic and interdisciplinary thinking at all types of schools. The study's primary objective is to pinpoint areas where educational content should be concentrated in the future in light of Industry 4.0. A questionnaire survey was carried out in Slovak businesses as one of the research methods. According to the research findings, changes in workforce qualification structures related to the implementation of the Industry 4.0 concept should have a positive impact on boosting companies' competitiveness and boosting production effectiveness. Based on the findings, it is suggested that the anticipated positive changes be implemented as structural changes within the Slovak educational system and the enforcement of vocational training in businesses [19].

Ingenjör4.0, a Swedish initiative, is a noteworthy educational project. It is a cutting-edge, web-based upskilling program that 13 Swedish universities collaborated to create. The modules can be mixed and matched by the participant, making it simple to tailor the upskilling to the particular requirements of the company and the individual. For business professionals, Ingenjör4.0 offers a singular, cutting-edge, extensive, and lifelong learning experience. It is targeted at professionals with engineering backgrounds as well as other professionals with an interest in connected and smart production, including operators, technicians, and management [20].

It was the Swedish initiative Ingenjör4.0 that started the discussion at our university on the possibility of implementing a similar programme in Slovakia.

3.3. State-of-the-Art Summary and Task Definition

Today's industry requires professionals in many academic and practical fields. Educational institutions and universities are required by practice to incorporate new Industry 4.0 trends and methods into the current curriculum. This will ensure that future graduates are not caught off guard by changing expectations in the industry. Cyber-physical systems are just one of many major agents of change in the education of engineers and technical professionals.

Education in the field of automation (operational technologies) and in the field of computer science (information technologies) was until recently, and often still is, carried out strictly separately and in different study programmes. However, the convergence of IT and OT described in the previous section of the paper brings about the necessity of new perspectives on the curriculum, whereby knowledge in the fields of ICT, automatic control and mechatronics must be synergistically combined.

Therefore, the presented work deals with the creation of case studies usable for teaching experts for the field of digitalisation of manufacturing processes with emphasis on modern forms of communication in manufacturing systems (from the field level to the cloud) using virtual lines, softPLC and freely available platforms for IIoT (e.g., Node-RED [21]).

4. Basic Theory

This section introduces the basic theoretical aspects and software tools that are used in educational case studies in this article.

Since the studies deal with the control of systems, an environment for running the control system is needed. An industrial computer that has been ruggedised and adapted for the control of manufacturing processes, such as assembly lines, machines, robotic devices, or any activity that requires high reliability, ease of programming, and process fault diagnosis is known as a programmable logic controller (PLC) or programmable controller. Any industry-PC can be turned into an automation system by a PLC runtime system, also known as a runtime system, which allows you to cyclically automate and control technical operations, process sequences, and similar procedures [22].

Node-RED, which is based on flow-based programming, is used in the studies. According to the flow-based programming paradigm, applications are networks of "black box" processes that communicate by passing messages across predefined connections, where the connections are determined externally to the processes. Without requiring internal modification, these "black box" processes can be connected in an infinite number of ways to create various applications. Therefore, flow-based programming is by nature component-oriented. Node-RED is a programming tool for tying new and intriguing connections between hardware components, APIs, and online services. By offering a browser-based flow editor, Node-RED makes it simple to connect flows using the variety of nodes available in the palette. Then, flows can be instantly deployed to the runtime. The event-driven, non-blocking model of Node.js, on which the lightweight runtime is based, is fully utilised. Because of this, it can run both in the cloud and at the network's edge on inexpensive hardware such as the Raspberry Pi [23].

The on-demand availability of computer system resources, particularly data storage (cloud storage) and processing power, without direct active management by the user is known as cloud computing. Functions in large clouds are frequently dispersed among several locations, each of which is a data centre. Cloud computing relies on resource sharing to achieve coherence and typically uses a "pay-as-you-go" model, which can reduce capital expenses but also result in unexpected operating expenses for users who are not aware of them [24].

5. Research Methods

The article deals with the problem of modelling, simulation and control of a discrete-event system. A controlled discrete-event system in real practice consists, for example, of a series of production lines. In a school environment for educational purposes, various physical models (e.g., from Fischertechnik [25] or LEGO [26]) are used. Virtual discrete-event systems are also a suitable alternative, which can be implemented using, for example, Factory I/O [27], and this application is used in the present work.

The use of a virtual/simulated discrete-event system brings several advantages. Since the output of the project is in the form of case studies usable for the field of education of professionals in the field of applied and engineering informatics, it was necessary to ensure that such materials are easily accessible to those interested. The virtual model does not need to be purchased, serviced and does not occupy any physical space, and thus can be offered to a wider group of interested parties.

In order to control the virtual system, we need a control system. In the case of discrete-event systems, which manufacturing systems inherently are, programmable logic controllers (PLCs), which are small computers that have input/output pins, are used. We can plug in appropriate sensors or actuators (e.g., controlled drives). We can also simulate the PLC operation on a computer (softPLC), so it is not necessary to own a real hardware PLC to run the created case studies, which is again very advantageous for educational purposes. PLC runtime is implemented by the PLC runtime engine, which is mostly a standard part of PLC editors, such as Siemens TIA Portal, CODESYS or OpenPLC. We can connect PLC runtime and Factory I/O using a communication protocol.

6. Educational Case Studies

This section contains a description of realised case studies.

6.1. Case Study No. 1: OpenPLC Linked with Node-RED and Microsoft Azure

In the first case study (Figure 2), we would like to demonstrate the use of the freely available open-source application system OpenPLC [28] instead of the traditional paid PLC editors and runtimes. However, OpenPLC does not support the modern communication standard OPC UA [29], but only the conventional Modbus protocol. Although OpenPLC can communicate with Factory I/O using this protocol and implement control processes using it, we also want to send data to the cloud and possibly provide it to other clients via OPC UA. Therefore, we need a Node-RED intermediary (middleware), which is not needed for control, but we also have it there for sharing data to the cloud via MQTT and also via an OPC UA server to which we can connect a wide variety of clients. In our case, we will be testing the OPC UA client UAExpert.

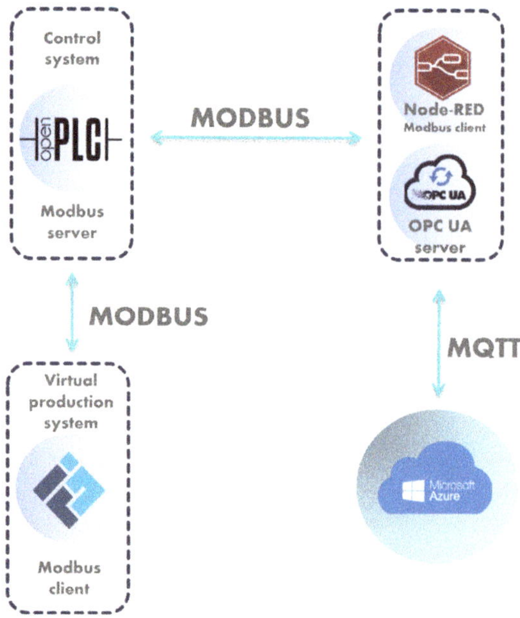

Figure 2. OpenPLC linked with Node-RED and Microsoft Azure

6.1.1. Discrete-Event System Specification and Behaviour

Discrete-event system could generally be defined as a system that can take on multiple states, with transitions between states being event-driven. In this case, we consider **a virtual model of a discrete-event system** in the form of a manufacturing system consisting of a production belt and a machining centre (Figure 3).

The model consists of several parts that should dynamically respond to events that occur in the system. At the **input (emitter), pieces of material (semi-finished products)** are generated at certain time intervals and need to be transported to the **machining centre**. Here, they are transformed by the machining process into **products** ready for dispatch to customers. The material and products are moved by **conveyor belts**. We have six of them (C1–C6). Seven **retro-reflective sensors** (R1–R7) ensure the functionality of the conveyors. Using the emitter, we send the material onto the belt, where it is detected by the sensor R1 and then the belt C1 is started. When the product is detected by sensor R3, the corner belt C2 is triggered and then the belt C3, which takes the material to the machining centre. Here it is processed into the final product, which is passed on to belt C4, which is triggered when sensor R5 detects it. Subsequently, sensor R6 starts belts C5 and C6. The final product is recorded by sensor R7, which should also provide a count of the number of finished products. It is advisable to use an aligner to ensure that the product enters the machining centre correctly. It is also necessary to ensure that there are no collisions between pieces of material in the system and also between finished products.

Figure 4 shows an overall view of the virtual model of the production line.

To create the above production line at Factory I/O, we needed a number of parts. They will be listed in the following paragraphs.

1. **Belt conveyor** (Figure 5)—used for transporting light loads. They are available in lengths of 2, 4 and 6 m and in analogue (we can adjust the speed of the conveyor) and digital versions;
2. **Curved belt conveyor**—used for transporting light loads and available in analogue and digital versions;
3. **Aligners**—metal structures that are attached to the conveyor to prevent the product from falling during transport. There are four types;
4. **Chute conveyor**—mostly used for dispatching items from conveyor belts;
5. **Raw Material**—metal or plastic material for the manufacturing of lids or bases. In our case we understand it as a semi-finished product that needs to be machined into a finished product;
6. **Retroreflective Sensor and Reflector** (Figure 6)—the sensor is used together with the reflector, it detects the presence of an object on the belt.
7. **Emitter** (Figure 7)—it is the entry point of the production line, which ensures the supply of production parts/raw materials to it. Raw materials are automatically generated at time intervals according to the emitter settings.
8. **Remover** (Figure 8)—removes one or more items from the scene.
9. **Machining Center**—a robot used for the production of pedestals.

6.1.2. Control of Discrete-Event System

In order to control the discrete-event system, we use the open-source OpenPLC editor and runtime in this case study. In OpenPLC, we use the Ladder logic. It is necessary to define the variables that are required for the creation of the control program. The variables will also be sent to the cloud later.

We use global **RetroreflectiveSensor** variables from 1 to 7 to sense the product on the belt, based on which we turn the belt on or off, or count the number of semi-products or products. The product values and subsequent calculations are handled through global variables **Result1** to 3. **Result1** counts how many semi-finished products have been dropped on the belt. **Result2** expresses how many finished end products there are and **Result3** expresses how many products are currently in production. For running the production conveyor belts, we use **Conveyor** variables 1 to 6. We also use various timers and mathematical functions. We use timers to ensure that there are no collisions. And with sensors we also count the time of the products on the belt to avoid further collisions again.

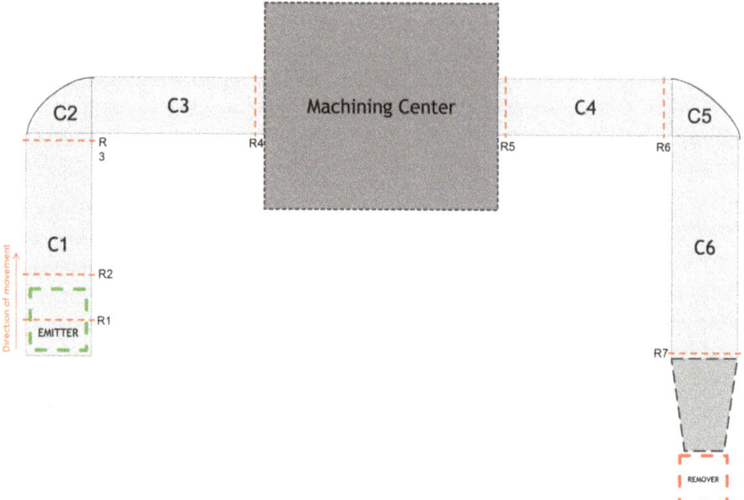

Figure 3. Scheme of production system.

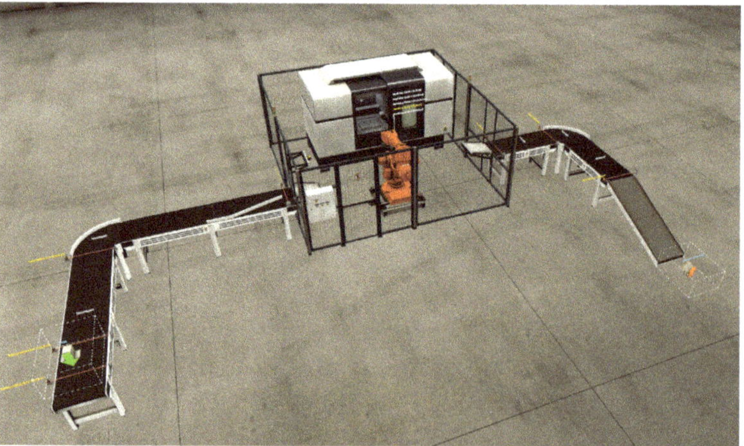

Figure 4. Overall view of the virtual production line.

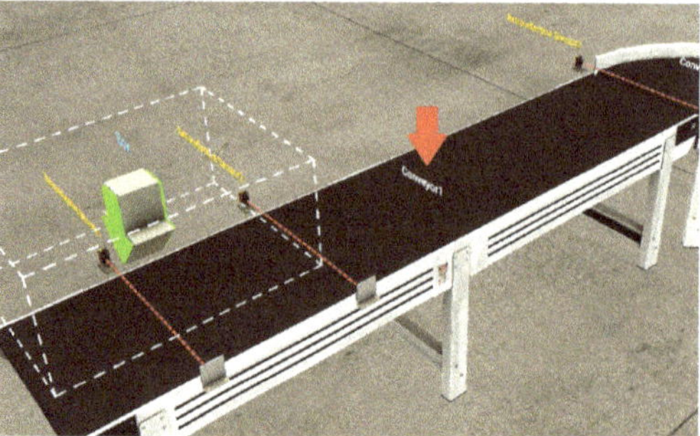

Figure 5. Belt conveyor.

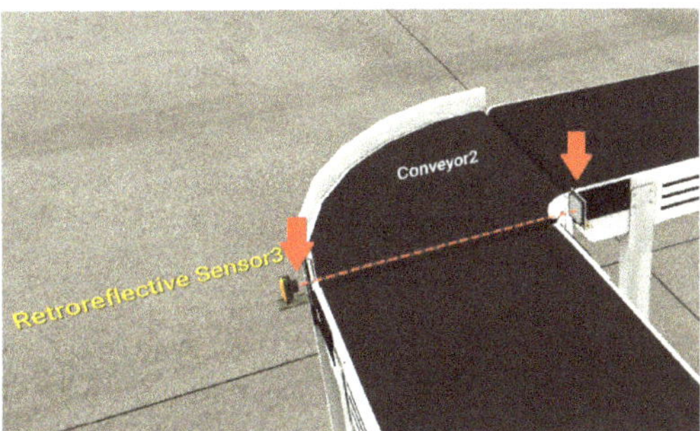

Figure 6. Retroreflective sensor and reflector.

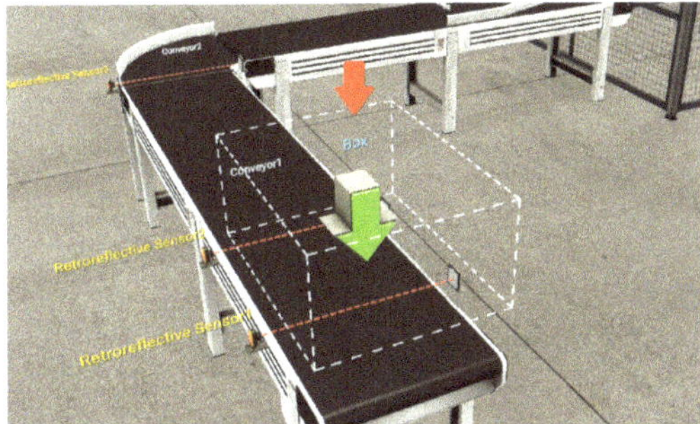

Figure 7. Emitter.

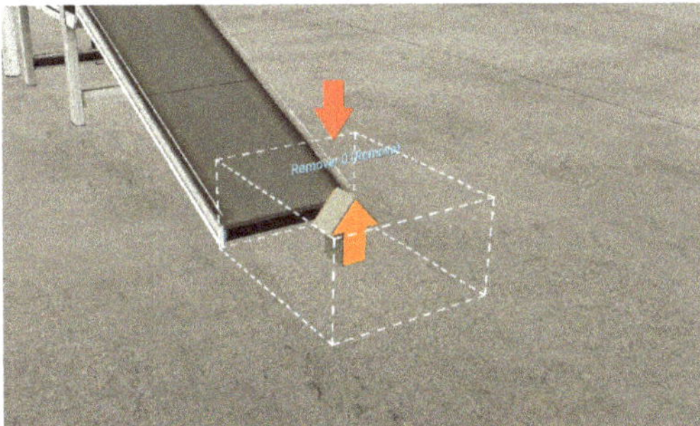

Figure 8. Remover.

6.1.3. Communication between OpenPLC Runtime and Node-RED Middleware

The communication takes place first between OpenPLC and Factory I/O (Figure 9). The communication is provided using the Modbus protocol because OpenPLC does not support more modern protocols (e.g., OPC UA). In this case, OpenPLC behaves as a Modbus server and Factory I/O behaves as a Modbus client. Since we want to send the data to the cloud or provide it to other clients using the OPC UA server, we need an intermediary in the form of Node-RED. Thus, Node-RED is not needed for production control itself, but we use it for the possibility of sharing data for the OPC UA server and to the cloud. We communicate from Node-RED to Microsoft Azure cloud and back using the MQTT communication protocol.

In order to properly connect the OpenPLC runtime with Node-RED, it is necessary to understand how the Modbus protocol works and what is a server and what is a client in our case. OpenPLC can even work as a server and a client at the same time, which we will not actively use. Modbus offers four types of transmitted data.

- **Discrete Input**—A single bit (BOOL) that is used for binary input (e.g., from sensors). In our case, these are addresses of type %IX. It can only be written by the Modbus server;
- **Coil**—A single bit (BOOL), which is mostly used for binary output. In our case it is addresses of type %QX. It can be written not only by the server but also by the client;
- **Input Register**—A 16-bit read-only register. It is kind of like Discrete Input, except it is not BOOL, but it is a 16-bit INT that can be unsigned or signed;
- **Holding Register**—A 16-bit register designed for both read and write. It is kind of like Coil, except it is not a BOOL, but it is a 16-bit INT that can be unsigned or signed.

As mentioned, OpenPLC can act as both a Modbus client and a Modbus server. In these modes it works simultaneously, but a different address is reserved for each. For the OpenPLC server mode we are using, it has the following addresses (see Table 1).

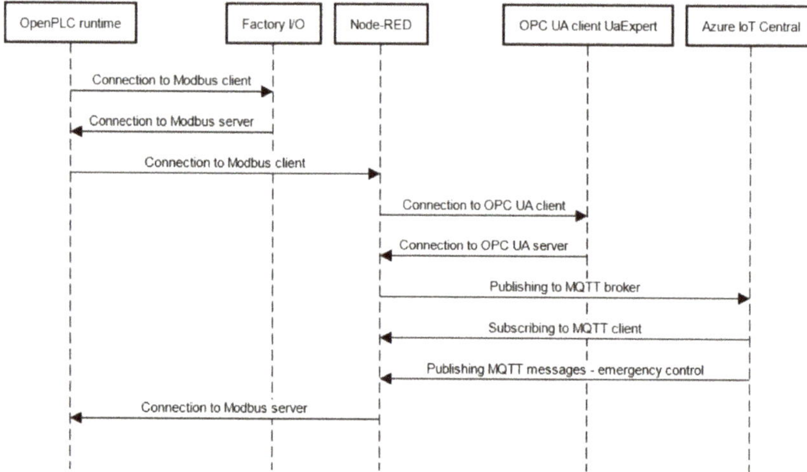

Figure 9. Sequence diagram for the first case study.

Table 1. Modbus server variables in OpenPLC engine [30].

Modbus Table	Usage	PLC Address	Modbus Data Address	Data Size	Range	Access
Discrete Output Coils	Digital Outputs	%QX0.0–%QX99.7	0–799	1 bit	0 or 1	RW
Discrete Input Coils	Digital Inputs	%IX0.0–%IX99.7	0–799	1 bit	0 or 1	R
Analog Input Registers	Analog Input	%IW0–%IW1023	0–1023	16 bits	0–65,535	R
Analog Output Holding registers	Analog Outputs	%QW0–%QW1023	0–1023	16 bits	0–65,535	RW

Next, you need to set the Factory I/O on the Modbus client. We set the localhost where we are running OpenPLC, which is 127.0.0.1. Next, we need to set the digital inputs to be used on Coils. It would be more logical to use Inputs, but since Modbus client can only write to Coils, we have to use Coils for the inputs. And last we need to set up I/O Points and there we set the inputs and outputs according to how much space we need.

In Factory I/O, the input and output variables need to be assigned correctly to the Factory I/O components (Figure 10). The addresses must be identical to those in OpenPLC.

The control program from OpenPLC editor needs to be loaded into OpenPLC runtime and run. Figure 11 shows how values are read from the OpenPLC runtime and how OpenPLC connects to Node-RED using Modbus protocol. The ability to communicate via Modbus can be obtained by installing the *node-red-contrib-modbus 5.14.1* library.

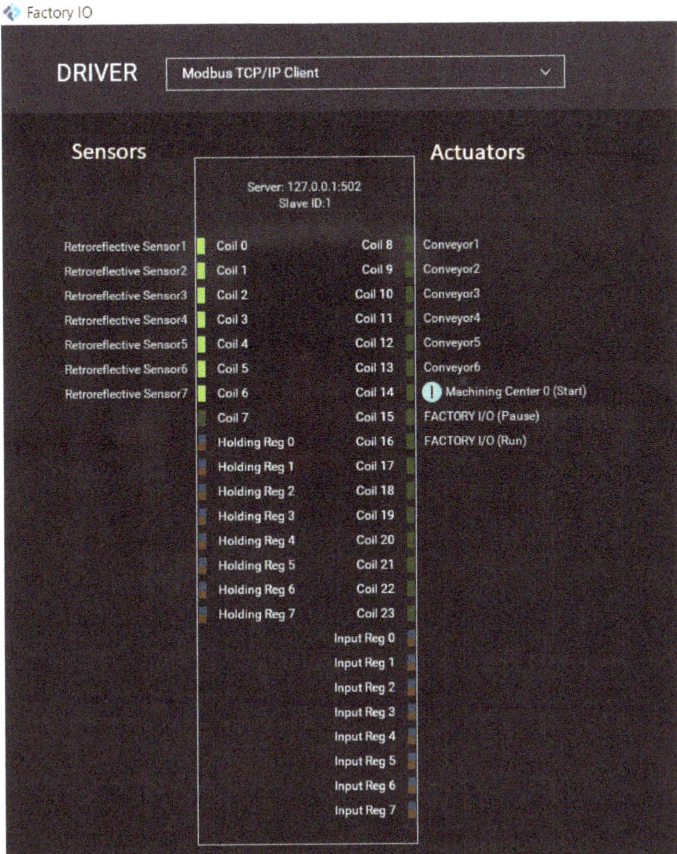

Figure 10. Factory I/O—Modbus client.

We read **inputs (from sensors)** via the node called **Modbus Read-%QX0.0-7** and read **outputs (actuators)** via **Modbus Read- %QX0.7+** node. We start with inputs from address 0, i.e., %QX0.0. The outputs start from address %QX1.0. The Modbus node always reads a whole byte, which is alright in this case since we have exactly eight input variables. The variables are of type BOOL (true/false). Modbus Read works as a client and we need to connect it to the server. We will connect it to a server that we have called OpenPLC local and set the corresponding address 127.0.0.1 and port 502. Next, it is needed to specify that we are going to read coils, which is a standard Modbus protocol command (*FC1: Read Coil Status*). In the case of our input variables, we set the address to 0, since we are reading from %QX0.0 (so in the case of %QX1.0, it would be 8). We set the quantity to 1, since we are reading 1 byte. We set the poll rate to 2 s, which means that the value is read every 2 s.

We used a similar procedure on **Modbus Read- %QX0.7+**, where we read output variables from PLC address %QX1.0 and our Modbus address is 8.

In PLC program there are also 3 values of INT type, which are stored in registers that have different addresses than the coils (these are of BOOL type). These are, for example, the number of finalised products. These values are read using the **Modbus Read Holding** node.

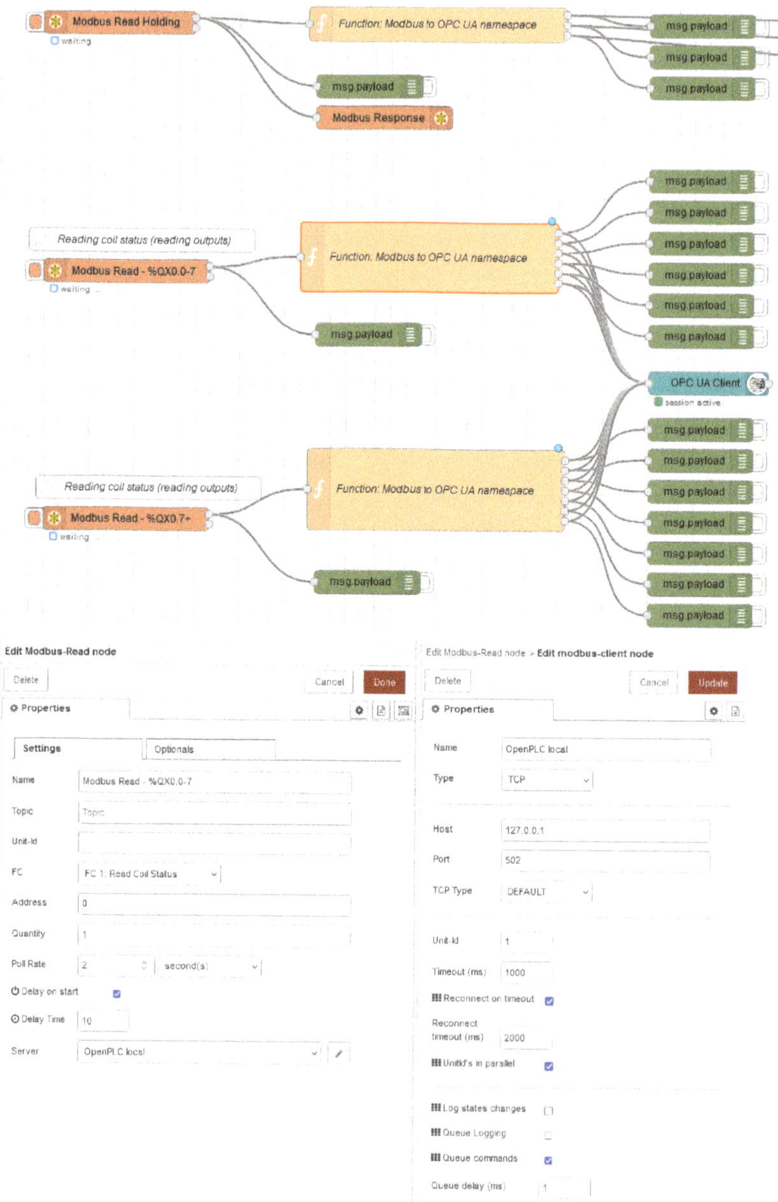

Figure 11. Node-RED as Modbus client.

6.1.4. OPC UA Server and Client

Since we want to offer the values from the production line to other users who may have OPC UA clients, it was necessary to implement an OPC UA server in Node-RED to provide these data.

As already mentioned, OpenPLC runtime cannot function as an OPC UA server, as it is a free open-source tool. This is the domain of more advanced PLC solutions such as

CODESYS or Siemens TIA Portal. Therefore, the OPC UA server will be created using Node-RED, to which the data arrive from OpenPLC runtime via Modbus protocol.

Library *node-red-contrib-opcua 0.2.256* was used to create OPC UA server.

To **create a server**, it is necessary to use the **OPC UA server node**, where the port (in our case 53,880) and optionally its name are set. We use the default name. It is also possible to set authentication options, as OPC UA communication standard supports multiple security profiles. We use anonymous access for clarity (and since we are running on localhost).

The creation of the server is also related to the creation of **address space**, i.e., variables that will initially be empty or have a predefined value, and later we will fill these variables with values that OpenPLC runtime sends using Modbus protocol.

To keep our inputs and outputs clearly separated in the address space, we have created folders *FIOOutputs* and *FIOInputs* in it.

It should be noted that the individual directives that we send to the OPC UA server node are sent using the *Inject* node type. Thus, these directives need to be set to execute automatically when the Node-RED program is started, and it is logical that the timing needs to be set so that the directives that create the folders are executed first, and then the variables are created. This will successfully create an OPC UA server with the desired address space.

We will use the following directives (but there are several supported): **addFolder**, **setFolder** and **addVariable**.

The directives are bound to the *msg.payload* of messages in Node-RED and the content itself to the msg.topic of messages. This can be seen in Figure 12, which shows the creation of the *FIOOutputs* folder. In *msg.payload* there is a command to add the folder and in *msg.topic* there is the defined namespace and folder name.

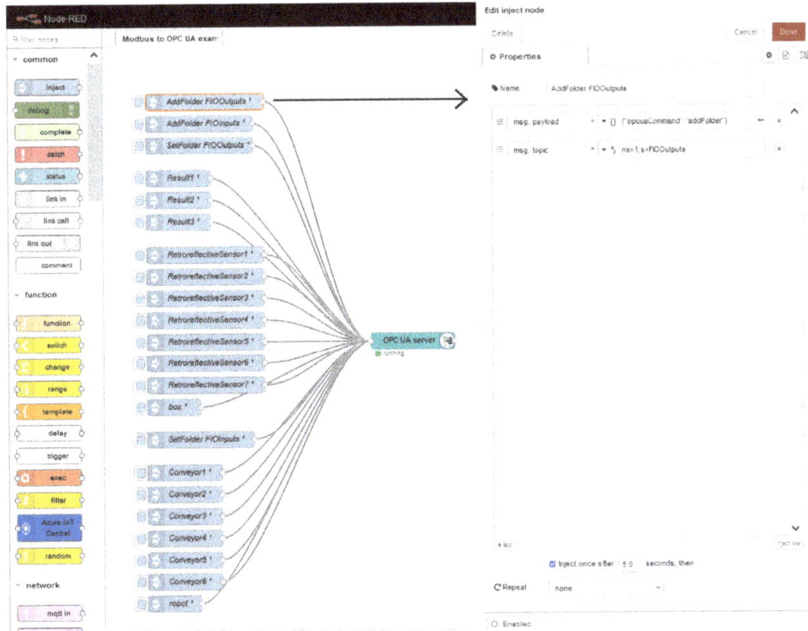

Figure 12. Creating OPC UA server and its address space in Node-RED.

Adding the *RetroreflectiveSensor1* variable would look like this, and it is obvious that the variable's data type is also set:

- `msg.payload:  {"opcuaCommand":"addVariable"};`

- `msg.topic: ns=1;s=RetroreflectiveSensor1;datatype=Boolean`.

In order to be able to populate variables with data, we needed to create a custom function in JavaScript, since in Node-RED it is not necessary to use only the built-in nodes, it is also possible to create your own.

Notice in Figure 11 the node named **Function: Modbus to OPC UA namespace**, i.e., the node representing our own function. Specifically, we will describe the node that is connected to the node *Modbus Read- %QX0.0-7*. The *Modbus Read- %QX0.0-7* node sends an array of 8 values (of type BOOL) to the *Function: Modbus to OPC UA namespace* node. Figure 13 shows the contents of the node. For each value received from Modbus, we have pre-prepared an empty variable (*msg0, msg1, msg2,...*) where we store the elements of the array. In their payload we store the value itself (of type BOOL), but the more interesting part is *msg.topic*. Here, according to the documentation of the OPC UA client installed in Node-RED, we need to define which variable in the OPC UA address space the variable will go to. Thus, we have defined the namespace ID (abbreviated *ns*), the variable name and the data type that matches the input variables obtained from Modbus server.

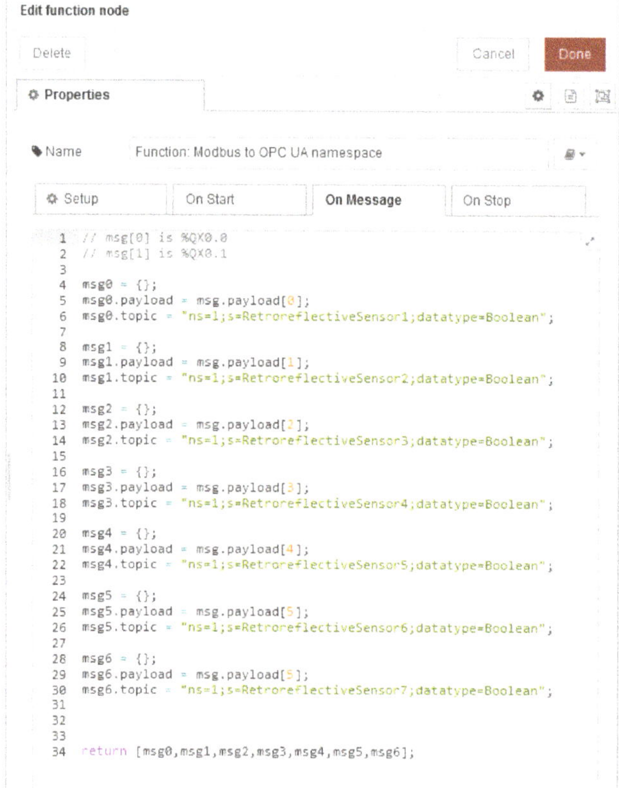

Figure 13. Function for filling OPC UA server with values obtained using Modbus protocol from OpenPLC—*Function: Modbus to OPC UA namespace*.

Then the *Function: Modbus to OPC UA namespace* node is connected to **OPC UA client** node, which stores the values in our OPC UA server.

We can also provide values from our OPC UA server to clients other than the client installed in Node-RED. Theoretically, this could be a client that does not have to run only on a computer, but also on a smartphone, tablet, in the cloud, etc.

6.1.5. Application in Microsoft Azure Cloud

One of the main objectives of the work was the implementation of a cloud application, which could be used to monitor discrete-event production system—to monitor the values of selected variables, to visualize the data appropriately, to process them efficiently and, if necessary, to intervene in the system.

At the beginning, it was necessary to determine what should be in the cloud application. The **functional requirements** were then determined:

- Display of read values and action buttons in a dashboard;
- Communication with Node-RED using the MQTT protocol;
- Display of the number of manufactured (final) products handed over for dispatch;
- Displaying the number of semi-finished products (pieces of raw material) that have entered production;
- Displaying the number of semi-finished products/finished products that are currently on the conveyors (or in the system as such);
- Graphical representation of the current temperature in the production hall;
- Simple processing of the current temperature values in the production hall in order to raise an alarm if the temperature rises above a certain value;
- Possibility of emergency intervention in the system-suspension and start-up of the production line.

The characteristics defined above are intended to describe what the application system (cloud application) being created will be able to accomplish. However, it is necessary to identify a second set of requirements that will not address the functionalities of the application. These will be the so-called qualitative, i.e., **non-functional requirements**:

- The cloud application should have a simple and intuitive user interface;
- Individual displayed variables should be part of a suitable object model, assuming appropriate use of Microsoft Azure cloud components;
- The application should allow for easy extensibility by displaying additional variables, or the possibility of adding additional production lines, each with its own panel;
- In terms of language internationalisation, the application should use English.

After the design, it was necessary to move on to the actual implementation. The most effective way is to use PaaS (Platform as a Service) services, i.e., to use ready developer tools to create your own cloud applications. The original plan was to bundle together multiple PaaS services, such as Azure IoT Hub for monitoring and connecting devices [31], Power BI for data visualisation [32], Azure Stream Analytics for data analytics [33], and Azure Functions for event callbacks [34], for example. During the course of the work, it was determined that the most effective solution would be to use the relatively new comprehensive aPaaS (application Platform as a Service) service **Azure IoT Central** [35], which combines the aforementioned functionalities.

Microsoft Azure IoT Central is used to connect and manage devices on a large scale and provides reliable data for business statistics, so it can be classified as an enterprise resource planning (ERP) system. It incorporates multiple PaaS services to create easily configurable, comprehensive and secure IoT solutions. A web-based user interface allows you to quickly connect devices, monitor device status, create rules, and manage millions of devices and their data throughout their lifecycle [35].

Azure IoT Central works similarly to Azure IoT Hub based on device twins, with each device based on a template. **Device template** is a so-called blueprint that defines the characteristics and behaviour of a device type. We then connect these devices to our application. For example, we can define the telemetry that a device sends so that IoT Central can create visualisations that use the right physical units and data types.

The device template we created is called **template-factoryio**. It contains the device models that we define for integration with our application. Each model has a unique ID and we also implement capabilities or semantic type for each model. The semantic type enables IoT Central, which can make some kind of assumption about how to treat the value.

The capabilities we can assign to our models are:

- *properties*—data fields that represent the state of the device;
- *telemetry*—telemetry (measurements) from sensors;
- *command*—methods that users can execute on the device (e.g., control commands).

The structure of our device named *conveyor-system1* is as follows, where the first entry is the system name and the second entry is the capability type: turnOn (*command*), turnOff (*command*), produced (*telemetry*), entered (*telemetry*), on_line (*telemetry*), temperature (*telemetry*), location (*property*).

The device connection is handled using SAS (Shared access signature) authentication. We connect our device model in the cloud to Node-RED using scope ID, Device ID and Primary key data.

We can then view the data that our cloud application receives. Notice Figure 14, where we can see three values that our cloud is receiving—*Totally produced*, *Entered on line* and *Currently on line*. *Totally produced* means how many products have been produced and submitted for shipment. *Entered on line* is the number of semi-finished products (pieces of raw material) placed on the conveyor belt. *Currently on line* is the number of actual semi-finished and finished products on the belt. There is also a map that can show where a given production is taking place. Next to it is a graph that shows us the air temperature in that factory. Since our production line is not real, we just simulate the temperature by generating random values in Node-RED in the range of 18 to 25. We can also send a signal from Node-RED during the application run that the air temperature is 100 degrees Celsius. In this case, we have implemented an **event (alarm)** on the Azure IoT Central side, which ensures that an email is sent to the authorised personnel. This demonstrates the basic form of **data processing and evaluation** based upon which the event is executed. Events in Azure IoT Central can be selected from predefined types or custom functions can be programmed using Azure Functions (serverless architecture).

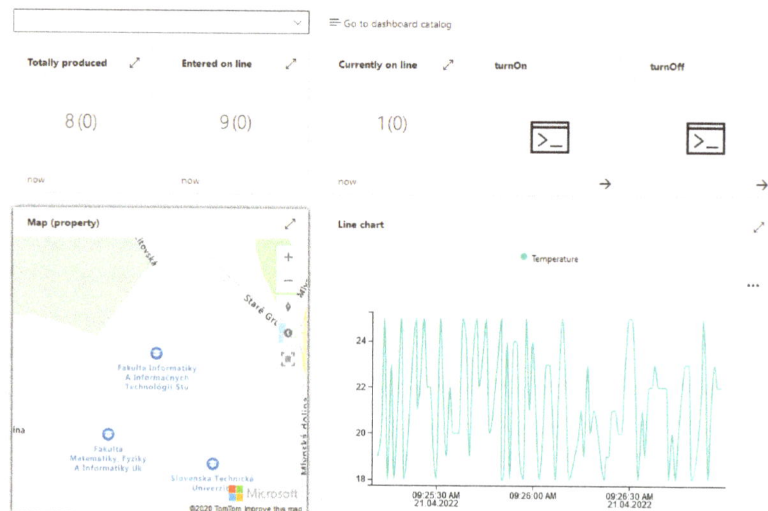

Figure 14. Dashboard in Azure IoT Central.

We will briefly describe how the data are sent in Node-RED, using the example of the generation of temperature values (Figure 15). We start by using the *timestamp* node, which provides us with the cyclic execution of a given program flow. Next, we use the *random* node (named *Temperature* in Figure 15), which generates random integers from 18 to 25. This node goes into its own *temperature* function, where we insert an identifier into the *msg.payload* of our message that is identical to the identifier of the temperature variable in the cloud on the given device model (twin). We also have an *inject 100* node ready, which we can manually push to send a temperature value of 100 degrees Celsius to simulate a fire on the production floor. Finally, the whole branch goes into the *Azure IoT Central* node where the device ID on the cloud and the access key is defined. In addition to this, the communication protocol is also defined here, in our case it is MQTT. One can also communicate over AMQP or HTTPS. We use the *azure-iot-central 1.6.0* library.

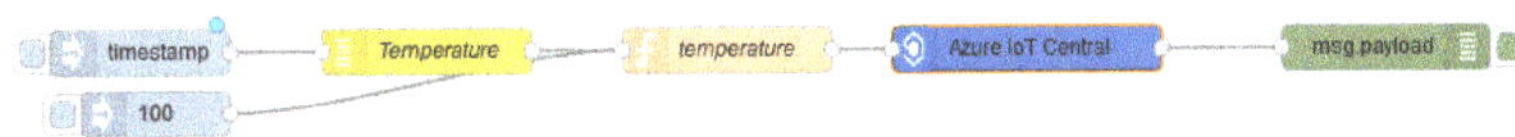

Figure 15. Node-RED data sending flow (air temperature).

We can also **intervene** in our virtual production system using the cloud application. We demonstrated this by implementing the ability to emergency **pause and start the line**. In Figure 14, we can see the buttons in the upper right corner that provide this.

First of all, it is needed to configure in Azure IoT Central node what commands this node listens for. In our case, these will be *turnOff* and *turnOn*. Another important element is the JavaScript functions that will respond to these commands from the cloud application (Figure 16). These functions must be registered in the context of our Node-RED flow (*flow context*). This is carried out (in the case of line suspend) using command `flow.set('turnOff',turnOff);`. The Azure IoT Central node takes care of the rest. After the JavaScript functions, we have other nodes connected to provide us with pause or start links. The command is sent to OpenPLC runtime using Modbus protocol, so we use nodes of type *Modbus Write*.

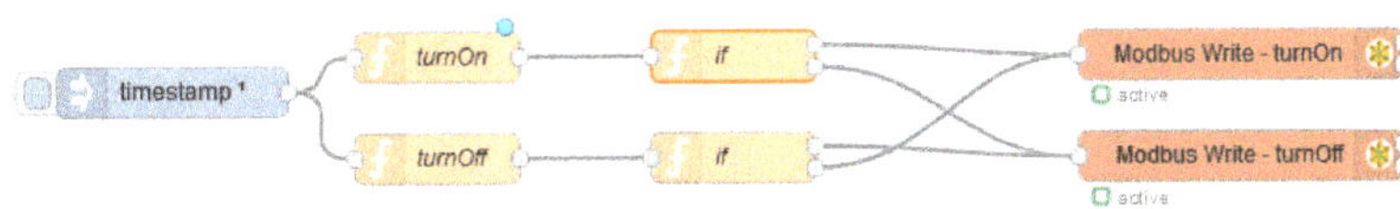

Figure 16. Flow in Node-RED ensuring the pause and start of conveyor belt.

6.2. Case Study No. 2: CODESYS Linked with Node-RED and Microsoft Azure

In the second case study (Figure 17), instead of using a free and open-source PLC editor, we decided to use a comprehensive automation solution called CODESYS, which provides us with a connection to Factory I/O using the modern OPC UA communication standard. OPC UA server will run through CODESYS runtime and Factory I/O will act as OPC UA client. For control purposes, we do not need an intermediary (middleware) in the form of Node-RED. However, we will use Node-RED for capturing data and sending it to the cloud and it will also be used to provide emergency stop and start of production line that can be made by the user through the cloud application. The cloud application will again be implemented using Microsoft Azure and the communication will be secured by MQTT protocol. CODESYS can be used for academic purposes, but the limitation is that the runtime can only run continuously for 1 h in free mode, which would be very limiting in production.

The specification and behaviour of discrete-event system is the same as in the first case study.

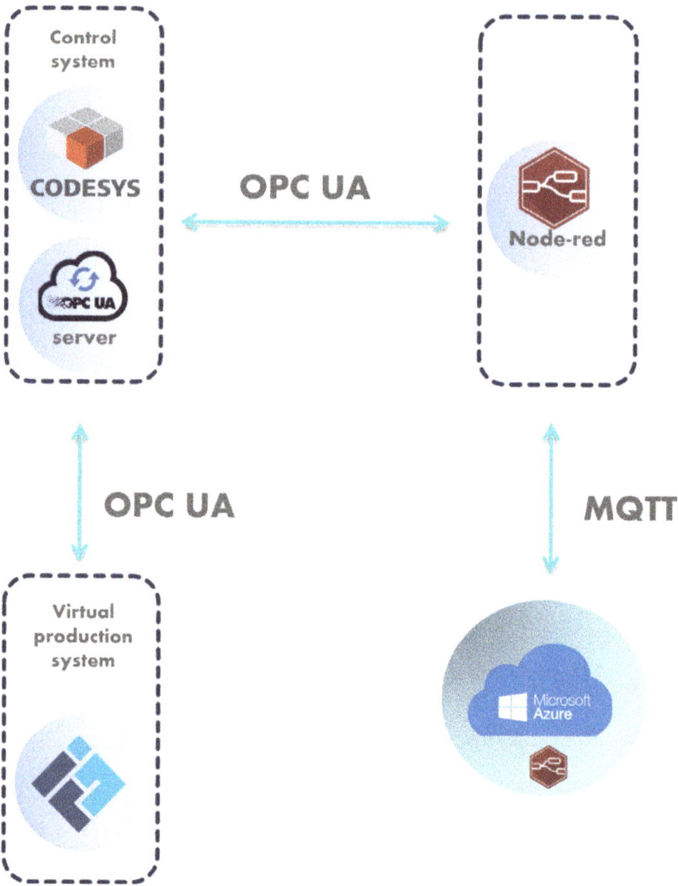

Figure 17. CODESYS linked with Node-RED and Microsoft Azure.

6.2.1. Control of Discrete-Event System

As in the first case study, we need to control discrete-event system. In our first case study, we used open-source and runtime OpenPLC. Now it is replaced by CODESYS. We use the Structured text language to declare variables. The control program is written by Ladder diagram.

The main variables in the control program are the same as in the first case study. However we have also added additional variables that are related to the runtime measurement of the control system. In addition, compared to the first case study, we have the CYCLE_TIME block, which is of type "Function module". We inserted this CYCLE_TIME using the external library OSCAT_BASIC version 3.3.4.0. We use it to measure how long the control system has been running for us.

6.2.2. Communication

The most important aspect of communication (Figure 18) in this case is the communication between CODESYS and Factory I/O, as it provides the control process. The communication is provided by the modern OPC UA protocol. CODESYS will now behave as OPC UA server and Factory I/O will behave as OPC UA client. Thanks to the existence of OPC UA server, we can view the variables in any OPC UA client. We will use Node-RED again in this case to share data to the cloud, but it is not needed for system control. We will communicate with the cloud via MQTT protocol.

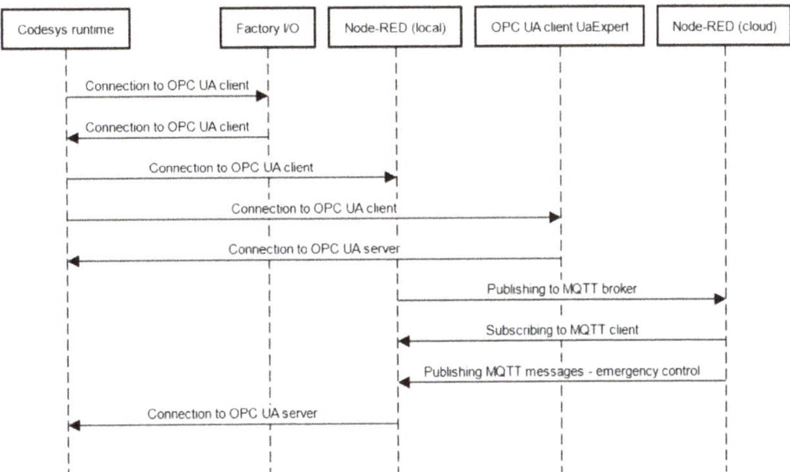

Figure 18. Sequence diagram for second case study.

6.2.3. OPC UA Server in CODESYS

CODESYS is relatively easy to work with. It is important to note that we need to name the variables so that Factory I/O can easily identify them and filter out the ones it needs. The entire OPC UA address space also contains a large number of different configuration and status variables that come with the CODESYS runtime. That is why all our useful variables have *FIO* prefix. We do this in order to be able to recognize our variables and make them easier to read, and most importantly, to be able to retrieve them easily in Factory I/O. In the Symbol configuration of the CODESYS project we select these variables, which actually specifies that they will be offered by the OPC UA server. Then we initialize CODESYS Control Win PLC, which is a softPLC running under Windows. Then, in the CODESYS project, we connect to this runtime and load a program into it. The connection is made by scanning the network and selecting the control unit, which can be the aforementioned softPLC or also a classic hardware PLC unit.

In Symbol configuration, we can click on our entire PLC_PRG program to mark all variables (Figure 19). In this way, all the variables we use are available in the OPC UA server. We then build and run the program.

In Factory I/O we open our scene and set the OPC Client DA/UA as driver in the configuration. We will specify *opc.tcp://localhost:4840* as the server and set "FIO" as the variable filter, this will obtain our variables from the OPC UA address space as we mentioned above. We assign all the variables and the communication between Factory I/O and CODESYS is implemented (Figure 20).

6.2.4. Connection between CODESYS and Local Node-RED

After creating and running OPC UA server using CODESYS, we want to obtain the data using OPC UA to Node-RED, which runs on localhost (like OPC UA server). This will be provided by OPC UA Client node, where we set the same address as in the previous text (opc.tcp://localhost:4840). This way Node-RED acts as the OPC UA client. We accept all the variables we want to work with (Figure 21). On the left in Figure 21, we see nodes of type *Inject* querying each variable every second, defining the variable name in the *msg.topic* of the given message. The variable name is quite complex in the CODESZS OPC UA address space. We can see the names, and also the data, clearly in any OPC UA client, for example. There we can read that the ID of one of the variables is *ns=4;s=|var|CODESYS*

Control Win V3 x64.Application.PLC_PRG.FIO_I_RetroreflectiveSensor1 and based on this ID we can query the variable in Node-RED.

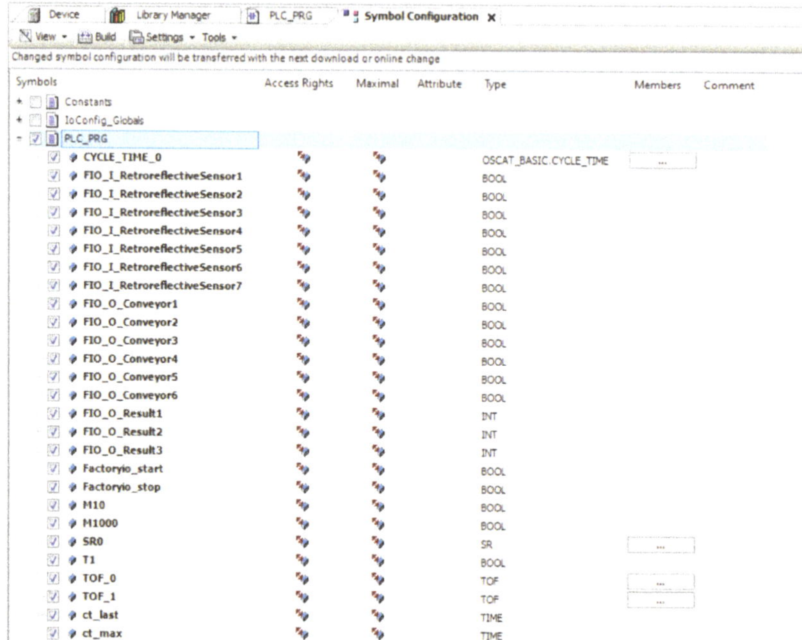

Figure 19. Symbol configuration.

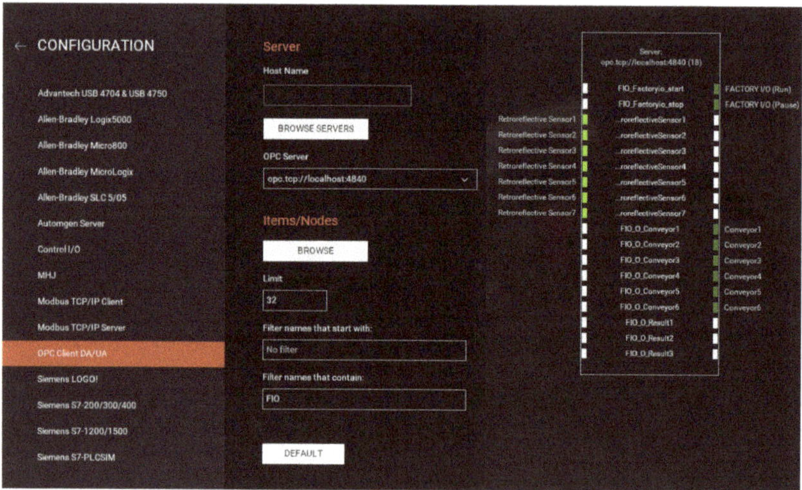

Figure 20. OPC UA Client Factory I/O.

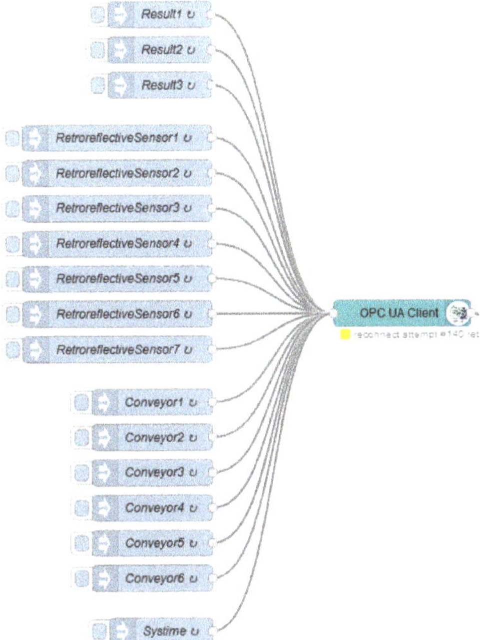

Figure 21. Local Node-RED—OPC UA client.

It should be clarified the reason for running Node-RED on localhost. Similar to the previous case study, we want to use the cloud for discrete-event system monitoring and emergency intervention. Since our OPC UA server is running locally and we do not have a public (static) IP address, we will use localhost Node-RED to send data to the cloud and receive data from the cloud.

We will send our variables to the cloud using MQTT protocol. We will use *MQTT-in* node (acting as a subscriber) and *MQTT-out* node (acting as a publisher). These nodes will connect to the broker (server) that is implemented in the cloud. Specifically, this is the Aedes MQTT broker [36]. We will read our variables in the local Node-RED using OPC UA client, from which we extract individual data (variables) using our own *Filter* functions and send them to the cloud using MQTT. When sending data via MQTT, we need to set the corresponding *MQTT topic* for it. For example, for us it looks like *inputs/FIO_I_RetroreflectiveSensor1*.

6.2.5. Node-RED Dashboard in Microsoft Azure Cloud

In the first case study, we implemented a dashboard for system monitoring and emergency intervention using Azure IoT Central aPaaS service. In the second case study, we opted for a different approach. Node-RED also provides the possibility of implementing a dashboard, so as suitable option turned out to be to deploy Node-RED also in the cloud and create a dashboard using it.

Deploying Node-RED to the cloud is relatively easy, as it is actually an application running on the Node.JS runtime. So we used a virtual machine, specifically the Azure Virtual Machine IaaS service [37]. Azure Virtual Machine is one of several types of scalable on-demand compute resources that Azure offers. Before we create one, we need to think about a few things such as the application name, where the resources are stored, virtual machine size, operating memory, maximum number of virtual machines, operating system, configuration, and related resources. Further, it gives us the flexibility of virtualisation without having to buy and maintain the physical hardware on which it runs. We used a Linux-based operating system, specifically Ubuntu Server 20.04 (Focal).

Thus, we send data from the local Node-RED to the Node-RED in the cloud using the MQTT protocol. Here, we read the data, i.e., receive it using the MQTT protocol, and then display it in the dashboard. We use the *node-reddashboard 3.1.6* library and *node-red-contrib-ui-led 0.4.11* to display the data in the dashboard, since the nodes that allow us to do this are not part of the base installation and need to be installed separately. Table 2 gives us a more detailed description of Figure 22, where we can display data in the dashboard using these nodes.

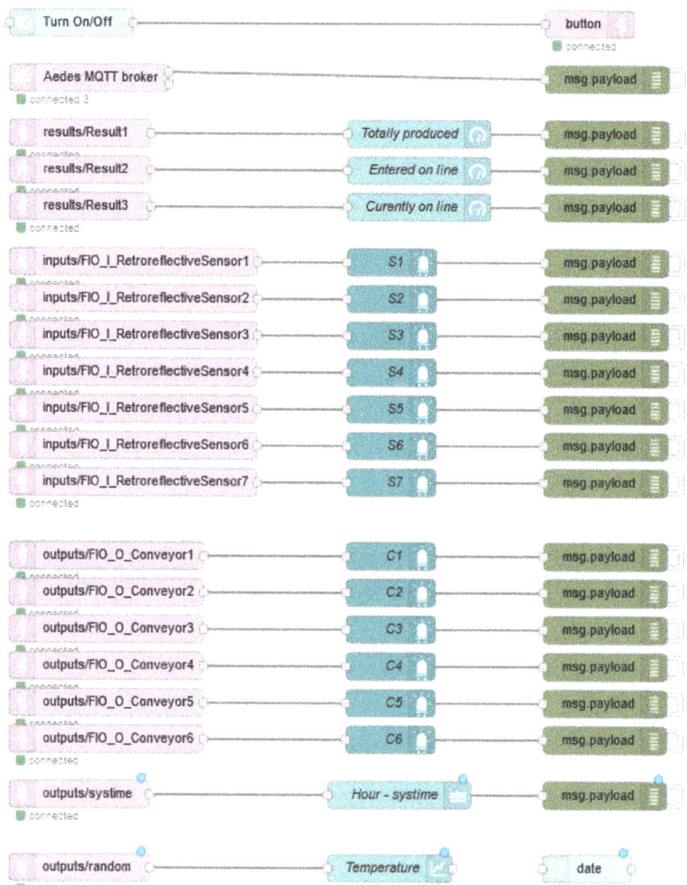

Figure 22. Node-RED in cloud.

Table 2. Variables in local Node-RED.

Node Type	Node Name
led	S1–S7, C1–C6
gauge	Totally produced, Entered on line, Curently on line
text	Systime
chart	Temperature
button	Turn On/Off
date picker	date

Our final dashboard is shown in Figure 23. It is important to note that we can not only monitor the variables, but again we can also intervene in the production process in an emergency.

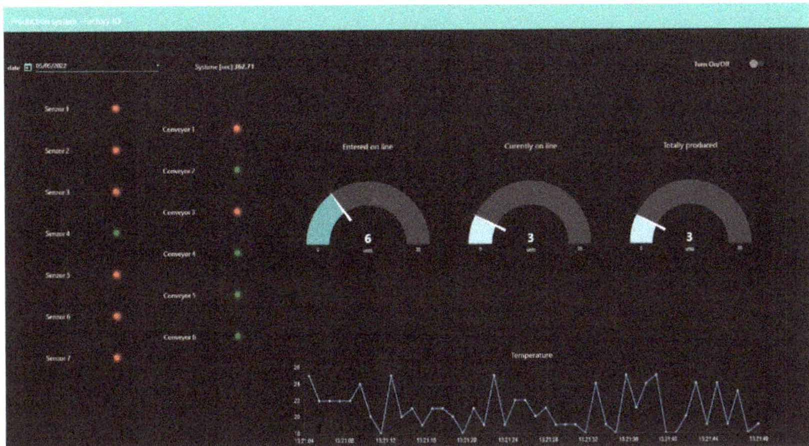

Figure 23. Dashboard in Node-RED (cloud).

6.3. Case Study No. 3: OpenPLC Linked with Node-RED Acting as a Software IIoT Gateway

In the third case study (Figure 24), we will again use the free OpenPLC tool. However, let us imagine a situation where we want to use OpenPLC to control a system that can be remote (accessible via a local network or Internet). OpenPLC only supports the old protocol Modbus, but we want to access this remote system via the modern OPC UA standard for security reasons. Network elements are used for this - for example IIoT gateways, which, in simple terms, provide translation of Modbus messages into the OPC UA address space (and vice versa). In our case study, we will show a software implementation of an IIoT gateway using Node-RED, which can be implemented, for example, using a Raspberry Pi microcomputer. Thus, Node-RED will act as an OPC UA server and at the same time as a Modbus client. Factory I/O will be the OPC UA client and OpenPLC runtime will be the Modbus server.

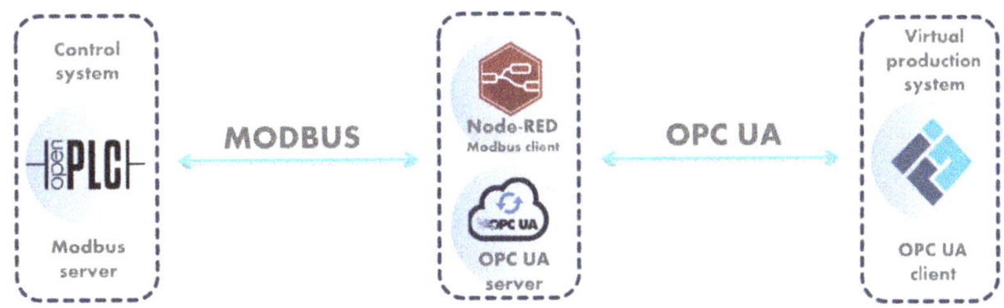

Figure 24. OpenPLC linked with Node-RED acting as a software IIoT gateway.

Of course, such communication cannot be considered as real-time, so such a solution is only suitable for non-time-critical systems (e.g., in a smart home). The control of systems over a network, where substantial delays can occur, is the subject of a special area of control theory focusing on networked control.

For simplicity and greater clarity, we have chosen only a fragment of the production event system for demonstration (Figure 25). It is a simple system that contains parts from both case studies. In the case study, we will use OpenPLC, which will communicate using Modbus. Then the Modbus data will be converted using Node-RED to communicate via OPC UA to Factory I/O.

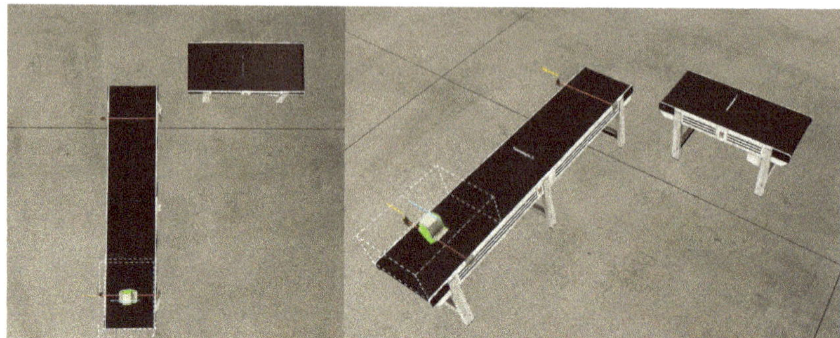

Figure 25. Fragment of discrete-event system.

6.3.1. Control of Discrete-Event System

As in the previous case studies, we will need to control discrete-event system. In this third case study, we use open-source editor and runtime OpenPLC. In OpenPLC, we again use Ladder logic. However, in this case study we use only four global variables named **Sensor_1** and **Sensor_2**, **Conveyor_1** and **Conveyor_2**.

Sensor_1 checks if the product is on the conveyor belt, based on which we switch on the **Conveyor_1** belt. **Sensor_2** stops **Conveyor_1** by sensing the product on the belt. We have the **Conveyor_2** variable for the purpose of checking if we are communicating, so it is set to TRUE by default. So we can think of the whole process as moving the product from point A to point B, where it stops.

6.3.2. Communication

The communication will be between OpenPLC and Factory I/O. However, we want to communicate with Factory I/O using the modern OPC UA protocol. As we already know from previous case studies that OpenPLC cannot communicate with modern protocols, so we will need a so called intermediary (middleware). The intermediary in this case will be Node-RED. OpenPLC will communicate with Node-RED using the Modbus protocol and Node-RED will further communicate with Factory I/O using the OPC UA protocol. Node-RED will provide us with the encapsulation of Modbus data to OPC UA address space as we mentioned above.

6.3.3. Creation of OPC UA Server in Node-RED

In the local Node-RED, we first need to create an OPC UA server using the OPC UA server node (Figure 26) and the procedure already mentioned in the previous section. At the beginning, we create folders in the address space in order to have separate inputs and outputs in the address space. Next, we create 4 variables. These variables are the same as in OpenPLC and we named them **fio_Sensor_1_Q1-0**, **fio_Sensor_2_Q1-1**, **fio_Conveyor_1_Q0-0** and **fio_Conveyor_2_Q0-1**.

6.3.4. Communication from OpenPLC to Factory I/O

First we need to describe how the communication from OpenPLC runtime towards Factory I/O works. From OpenPLC, Factory I/O needs to read the values of the outputs. Therefore, we need to read the outputs using the **Modbus Read- %QX0.0-7** node, since the PLC program in OpenPLC evaluates whether a given Conveyor should be started or not. We connect this *Modbus Read* to our server, which we have called OpenPLC local, where we have again (as in the first case study) set the corresponding address 127.0.0.1 with port 502. Next, we again set the quantity to 1 (because we are reading 1 byte), the address to 0 (the outputs are available from address %QX0.0). This is a reading, so we set the function to *FC: Read Coil Status*.

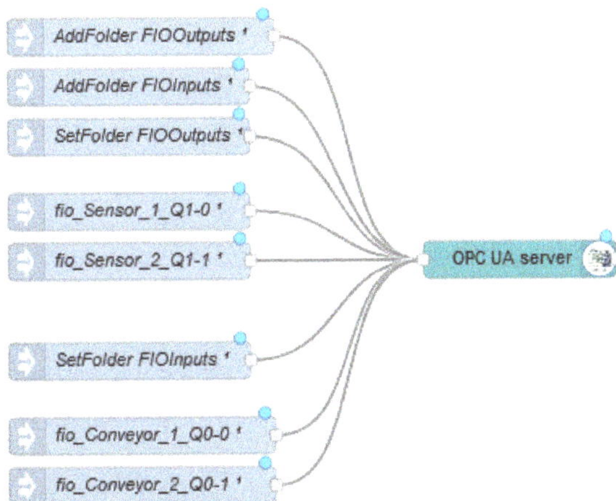

Figure 26. OPC UA server.

Once we have read the values, we need to obtain them to OPC UA server so that Factory I/O can read them from it. Using the **Conveyors: Modbus to OPC UA namespace** custom function (Figure 27), we assign a specific *msg.topic* to the Modbus messages to ensure that the Modbus data are placed in the correct variable in the OPC UA address space. Sending the data to the OPC UA server is handled by the *OPC UA client* node.

The nodes at the bottom of Figure 27 are for program testing purposes only and specifically return the current value of the inputs from the Modbus server (addresses %QX1.0 and %QX1.1).

In Factory I/O, the input and output variables of the address space of the OPC UA server must be correctly assigned to the Factory I/O components (Figure 28).

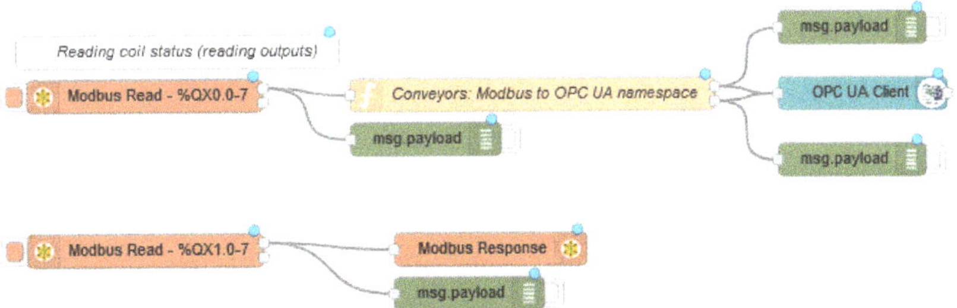

Figure 27. Communication from OpenPLC to Factory I/O.

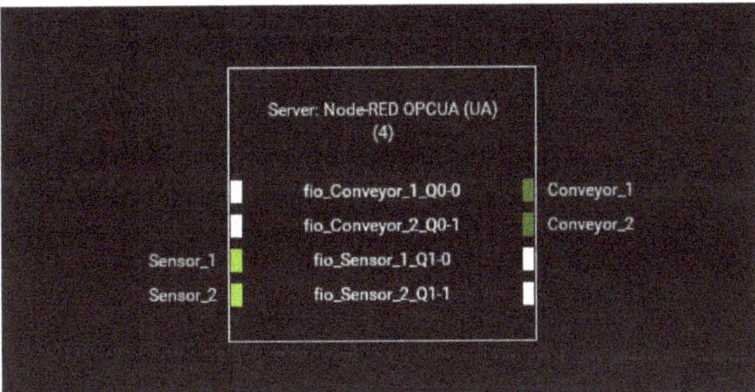

Figure 28. Factory I/O—connection.

6.3.5. Communication from Factory I/O to OpenPLC

In this subsection, we will show how the communication from Factory I/O towards the OpenPLC runtime takes place. Factory I/O needs to send values to the OpenPLC runtime from inputs, i.e., from sensors that detect the presence of a product. For this we need the variables **fio_Sensor_1_Q1-0** and **fio_Sensor_2_Q1-1**, which we will send to OpenPLC. In Figure 29 we see on the left two nodes of type *Inject* named after specific variables. These nodes are queried every 500 ms for the values of these variables by the *OPC UA client* node. This node then returns the value of the first or second sensor, and this is detected using the custom functions that follow in the flow (functions starting with the word *if*). Depending on the type of value, the value is written using *Modbus Write* to address %QX1.0 (address 8 in Modbus) or %QX1.1 (address 9 in Modbus) in OpenPLC runtime.

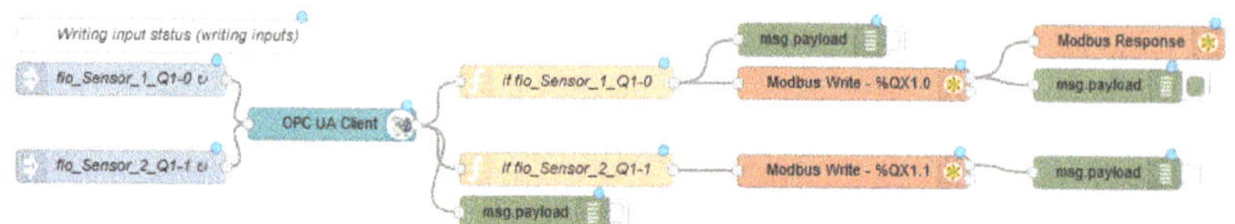

Figure 29. Communication from Factory I/O to OpenPLC.

In a similar way, a program in Node-RED could be extended to control an entire production line.

7. Discussion of Results

The presented work dealt with the current issue of convergence of information and operational technologies, which is related to the digitalisation of production processes. For the development of complex information systems in digital enterprises, it is necessary to use modern development tools (e.g., Node-RED), communication standards (OPC Unified Architecture) and also cloud technologies. Consequently, they need to be effectively interfaced with discrete-event systems. The case studies presented in this paper dealt with the modelling and control of virtual discrete-event systems implemented in Factory I/O and their communication with implemented cloud applications. The addressed issues are thus fully in line with the requirements of the industry regarding the creation of new educational materials for Industry 4.0 engineers. An advanced application for Industry 4.0 (that is also related to the convergence of IT and OT) is a digital twin. Digital twin creates a type of symmetry between real and virtual system.

It is essential for students to acquire knowledge and especially practical skills and competences in the field of technologies related to the digitalisation of products, production and related processes, which they can then use highly effectively as experts in the digitalisation of the enterprise in modern manufacturing companies of a wide range of focus. The evaluation of the module will be made possible by cooperation with partners from industrial practice, who will define the content needs of the implemented modules, which they will also use for the training of their experts. Such an approach will provide highly valuable feedback that will be used to modify both the formal and the content of the modules. This will ensure that the University's students have up-to-date and required knowledge for industrial practice.

The courses will be designed for two groups of people, based on the nature of the convergence of IT and OT:

- People focused on IT—computer engineers, software engineers, programmers, etc.
- People focused on OT—automation engineers, mechatronics engineers, PLC programmers, etc.

The two groups will attend the courses together. The groups will therefore be mixed. This will ensure that they learn to work together and also acquire the necessary soft skills. Everyone will improve in the area (IT or OT) that is their weaker side.

The course can be divided into four parts and consists of a motivational introduction, basic IT concepts, basic OT concepts and finally the core of the course—the convergence of IT and OT.

1. Motivational introduction
   - Importance of IT and OT convergence, key aspects.
2. Basic IT concepts
   - Review of basic programming paradigms, functions, objects;
   - Object-oriented programming;
   - Cloud computing—basic aspects.
3. Basic OT concepts
   - Systems theory, overview of different types of cyber-physical systems, discrete-event systems;
   - Programming languages for PLC;
   - PLC programming.
4. Core of the course—convergence of IT and OT
   - Simulation tool Factory I/O;
   - Industry 4.0 concept, Reference Architectural Model Industrie 4.0 (RAMI 4.0);
   - OPC Unified Architecture;
   - Node-RED: basic programming;
   - Node-RED: communication with cloud and using OPC UA and MQTT;
   - Cloud computing—Microsoft Azure.

8. Conclusions

In conclusion, it can be stated that three educational studies have been created, which will be the basis for the creation of pilot courses for Industry 4.0 education. Their nature is to explain to the participants the convergence of information and operational technologies using practical examples. The case studies focus on the monitoring and control of virtual manufacturing systems, using two types of PLC runtimes, Node-RED flow-oriented programming tool, and cloud technologies.

The application benefits of the article can be summarised in the following points:

- **Modelling and control of virtual discrete-event system**
  A virtual model of a production discrete-event system was created within the work, which was inspired by a real model available at the Institute of Automotive Mechatronics FEI STU in Bratislava, Slovakia. The control programs were created using ladder diagrams and were implemented using two development environments—OpenPLC and CODESYS.
- **Design and implementation of cloud applications**
  Two cloud-based applications were designed and implemented for monitoring and emergency intervention of the production discrete-event system. The cloud applications were implemented in different ways. In the first case, the aPaaS service Azure IoT Central was used and in the second case, the IaaS service Azure Virtual Machine was used. This is an Ubuntu Linux based virtual machine where a dashboard (graphical user interface) created using Node-RED was deployed.
- **Implementation of a software IIoT gateway applicable for edge computing**
  In the last case study, it was necessary to encapsulate data arriving via the Modbus protocol in the address space of OPC UA. This is commonly handled using hardware IIoT gateways or similar network elements. However, our solution is to implement a software IIoT gateway using the flow-oriented programming tool Node-RED, and this type of application can also be implemented in affordable microcomputers (e.g., Raspberry Pi). We can then place such a microcomputer at the edge of the network as an edge device.

- **Creation of the basis for pilot projects for the creation of educational materials for the training of engineers for Industry 4.0—pilot Engineer 4.0 programmes**
  The nature of the prepared case studies, which combine operational and information technologies, predestines them to be used in the sphere of the education of engineers for the digitalisation of production processes.

In conclusion, the results are beneficial and can be further modified for the development of educational materials and technical practice in Industry 4.0.

Author Contributions: E.K. proposed the idea in this paper and prepared the software application; O.H., P.D. and J.C. designed the experiments, E.K. and P.D. performed the experiments; O.H., P.D. and E.K. analysed the data; E.K. wrote the paper; O.H., P.D. and J.C. edited and reviewed the paper; All authors have read and agreed to the published version of the manuscript.

Funding: This research was funded by the Slovak research and Development Agency under the contract no. APVV-21-0125, by the Cultural and Educational Grant Agency of the Ministry of Education, Science, Research and Sport of the Slovak Republic KEGA 016STU-4/2020 and 039STU-4/2021, by the Scientific Grant Agency of the Ministry of Education, Research and Sport of the Slovak Republic No. 1/0107/22.

Institutional Review Board Statement: Not applicable.

Informed Consent Statement: Not applicable.

Data Availability Statement: Not applicable.

Acknowledgments: We would like to thank to Simona Lopatniková for helping with programming the implementation.

Conflicts of Interest: The authors declare no conflict of interest.

References

1. Kamal, S.; Al Mubarak, S.; Scodova, B.; Naik, P.; Flichy, P.; Coffin, G. IT and OT convergence-Opportunities and challenges. In Proceedings of the SPE Intelligent Energy International Conference and Exhibition, Aberdeen, Scotland, UK, 6–8 September 2016.
2. Shahroom, A.; Hussin, N. Industrial Revolution 4.0 and Education. *Int. J. Acad. Res. Bus. Soc. Sci.* **2018**, *8*, 314–319. [CrossRef]
3. González-Pérez, L.I.; Ramírez-Montoya, M.S. Components of Education 4.0 in 21st Century Skills Frameworks: Systematic Review. *Sustainability* **2022**, *14*, 1493. [CrossRef]
4. Information Technology (in German). Available online: https://theastrologypage.com/information-technology (accessed on 26 April 2022).
5. Operational Technology. Available online: https://www.gartner.com/en/information-technology/glossary/operational-technology-ot (accessed on 26 April 2022).
6. What's the Difference between IT and OT? Available online: https://www.3pillarglobal.com/insights/how-does-iiot-bring-it-and-ot-together/ (accessed on 26 April 2022).
7. Huba, M.; Kozák, Š. From E-learning to Industry 4.0. In Proceedings of the 2016 International Conference on Emerging e-Learning Technologies and Applications (ICETA), Stary Smokovec, Slovakia, 24–25 November 2016; pp. 103–108.
8. Leiden, A.; Posselt, G.; Bhakar, V.; Singh, R.; Sangwan, K.; Herrmann, C. Transferring experience labs for production engineering students to universities in newly industrialized countries. In Proceedings of the IOP Conference Series: Materials Science and Engineering (TSME-ICoME 2017), Bangkok, Thailand, 12–15 December 2017; Volume 297, p. 012053.
9. Souza, R.G.d.; Quelhas, O.L.G. Model proposal for diagnosis and integration of industry 4.0 concepts in production engineering courses. *Sustainability* **2020**, *12*, 3471. [CrossRef]
10. Assante, D.; Caforio, A.; Flamini, M.; Romano, E. Smart Education in the context of Industry 4.0. In Proceedings of the 2019 IEEE Global Engineering Education Conference (EDUCON), Dubai, United Arab Emirates, 8–11 April 2019, pp. 1140–1145.
11. Sackey, S.M.; Bester, A. Industrial engineering curriculum in Industry 4.0 in a South African context. *S. Afr. J. Ind. Eng.* **2016**, *27*, 101–114. [CrossRef]
12. Ciolacu, M.; Svasta, P.M.; Berg, W.; Popp, H. Education 4.0 for tall thin engineer in a data driven society. In Proceedings of the 2017 IEEE 23rd International Symposium for Design and Technology in Electronic Packaging (SIITME), Constanta, Romania, 26–29 October 2017; pp. 432–437.
13. Pierleoni, P.; Belli, A.; Palma, L.; Sabbatini, L. A versatile machine vision algorithm for real-time counting manually assembled pieces. *J. Imaging* **2020**, *6*, 48. [CrossRef] [PubMed]
14. Merkulova, I.Y.; Shavetov, S.V.; Borisov, O.I.; Gromov, V.S. Object detection and tracking basics: Student education. *IFAC-PapersOnLine* **2019**, *52*, 79–84.

15. Dr, P.; Kumar, P.; Johri, P.; Srivastava, S.; Suhag, S. A Comparative Study of Industry 4.0 with Education 4.0. In Proceedings of the 4th International Conference: Innovative Advancement in Engineering Technology (IAET), Jaipur, India, 21–22 February 2020. [CrossRef]
16. Hussin, A.A. Education 4.0 Made Simple: Ideas For Teaching. *Int. J. Educ. Lit. Stud.* **2018**, *6*, 92–98. [CrossRef]
17. Coşkun, S.; Kayıkcı, Y.; Gençay, E. Adapting Engineering Education to Industry 4.0 Vision. *Technologies* **2019**, *7*, 10. [CrossRef]
18. Mian, S.H.; Salah, B.; Ameen, W.; Moiduddin, K.; Alkhalefah, H. Adapting Universities for Sustainability Education in Industry 4.0: Channel of Challenges and Opportunities. *Sustainability* **2020**, *12*, 6100. [CrossRef]
19. Grenčiková, A.; Kordoš, M.; Navickas, V. The impact of Industry 4.0 on education contents. *Business Theory Pract.* **2021**, *22*, 29–38. [CrossRef]
20. Produktion2030 Ingenjör4.0. Available online: https://produktion2030.se/en/ingenjor-4-0/ (accessed on 27 December 2021).
21. Chanthakit, S.; Rattanapoka, C. Mqtt based air quality monitoring system using node MCU and node-red. In Proceedings of the 2018 Seventh ICT International Student Project Conference (ICT-ISPC), Nakhonpathom, Thailand, 11–13 July 2018; pp. 1–5.
22. Langmann, R.; Stiller, M. The PLC as a Smart Service in Industry 4.0 Production Systems. *Appl. Sci.* **2019**, *9*, 3815. [CrossRef]
23. Ferrari, P.; Flammini, A.; Rinaldi, S.; Sisinni, E.; Maffei, D.; Malara, M. Impact of Quality of Service on Cloud Based Industrial IoT Applications with OPC UA. *Electronics* **2018**, *7*, 109. [CrossRef]
24. Lakhan, A.; Mohammed, M.A.; Abdulkareem, K.H.; Jaber, M.M.; Nedoma, J.; Martinek, R.; Zmij, P. Delay Optimal Schemes for Internet of Things Applications in Heterogeneous Edge Cloud Computing Networks. *Sensors* **2022**, *22*, 5937. [CrossRef] [PubMed]
25. Indexed Line with Two Machining Stations 24V—Simulation. Available online: https://www.fischertechnik.de/en/products/simulating/training-models/96790-sim-indexed-line-with-two-machining-stations-24v-simulation (accessed on 27 April 2022).
26. Pajpach, M.; Haffner, O.; Kučera, E.; Drahoš, P. Low-Cost Education Kit for Teaching Basic Skills for Industry 4.0 Using Deep-Learning in Quality Control Tasks. *Electronics* **2022**, *11*, 230. [CrossRef]
27. Factory IO. Available online: https://docs.factoryio.com/ (accessed on 27 April 2022).
28. OpenPLC Editor. Available online: https://openplcproject.com/docs/3-1-openplc-editor-overview/ (accessed on 26 April 2022).
29. Philip Samuel, A.K.; Shyamkumar, A.; Ramesh, H. Industry 4.0-Connected Drives Using OPC UA. In *Industry 4.0 and Advanced Manufacturing*; Chakrabarti, A., Arora, M., Eds.; Springer: Singapore, 2021; pp. 3–12.
30. Modbus Address Mapping. Available online: https://openplcproject.com/docs/2-5-modbus-addressing/ (accessed on 26 April 2022).
31. Azure IoT Hub. Available online: https://azure.microsoft.com/en-us/services/iot-hub/ (accessed on 27 July 2022).
32. Power BI. Available online: https://powerbi.microsoft.com/en-us/ (accessed on 27 July 2022).
33. Azure Stream Analytics. Available online: https://azure.microsoft.com/en-us/services/stream-analytics/ (accessed on 27 July 2022).
34. Azure Functions. Available online: https://docs.microsoft.com/en-us/azure/azure-functions/ (accessed on 27 July 2022).
35. Azure IoT Central. Available online: https://azure.microsoft.com/en-us/services/iot-central/ (accessed on 27 July 2022).
36. Aedes MQTT Broker. Available online: https://github.com/moscajs/aedes (accessed on 27 July 2022).
37. Azure Virtual Machines. Available online: https://azure.microsoft.com/en-us/services/virtual-machines/ (accessed on 27 July 2022).

Review

A Systematic Review on Design Thinking Integrated Learning in K-12 Education

Tingting Li and Zehui Zhan *

School of Information Technology in Education, South China Normal University, Guangzhou 510631, China
* Correspondence: zhanzehui@m.scnu.edu.cn

Abstract: Design thinking is regarded as an essential way to cultivate 21st century competency and there has been a concomitant rise of needs and interest in introducing K-12 students to design thinking. This study aimed to review high-qualified empirical studies on design thinking integrated learning (DTIL) in K-12 education and explore its future research perspectives. After a systematic search in online database via a keyword search and snowballing approach, 43 SSCI journal papers with 44 studies were included in this review. The results indicate that: (1) There has been a growing popularity of integrating design thinking into K-12 education over the past decade, and most empirical studies target middle school students with small group size and a short period; (2) Studies tend to pay more attention to STEM related curriculum domains by incorporating non-unified design thinking models or processes, and the core concepts of design thinking in K-12 education have been frequently valued and pursued including prototype, ideate, define, test, explore, empathize, evaluate, and optimize; (3) The mostly evaluated learning performances are design thinking, followed by emotional/social aspect, subject learning performance and skill. For evaluation, qualitative assessments are used more frequently with instruments like survey/questionnaire, portfolio, interview, observation, protocol analysis, etc. (4) interventions with non-experimental study, formal classroom setting, collaborative learning, and traditional tools or materials have been mainly applied to the open-ended and challenging activities in real situated DTIL. Overall, the 43 papers suggest that design thinking shows great educational potential in K-12 education, however, the empirical evidence that supports the effectiveness of DTIL is still rather limited. Research gaps and future directions derived from reviewed papers are also discussed.

Keywords: design thinking; systematic review; K-12 education; educational method; transdisciplinary issue

Citation: Li, T.; Zhan, Z. A Systematic Review on Design Thinking Integrated Learning in K-12 Education. *Appl. Sci.* **2022**, *12*, 8077. https://doi.org/10.3390/app12168077

Academic Editor: Mirco Peron

Received: 12 July 2022
Accepted: 8 August 2022
Published: 12 August 2022

Publisher's Note: MDPI stays neutral with regard to jurisdictional claims in published maps and institutional affiliations.

1. Introduction

Originating from design field, design thinking has attracted considerable interest from practitioners and academics alike, as it offers a novel approach to innovation and problem-solving [1,2]. In the artificial world, design plays an indispensable role in the progress of human society by realizing "the transformation of existing conditions into preferred ones" [3]. In the field of design, design thinking is associated with the understanding of expertise in design, such as what constitutes expertise in design, and how to assist novice students to gain that expertise so that they can become expert and outstanding designers [4]. According to Jonassen's typology [5], design problems are usually among the most complex and ill-structured kinds of problems that are encountered in practice. Therefore, with their creative ways of solving problems, expert designers are seen as one group of innovative problem solvers, then they and their thinking have something important to offer for wider areas [1].

As an innovative problem-solving method, the concept of design thinking has gradually expanded from a professional concept to a more general one. Nowadays, "Design Thinking" is identified as an exciting new paradigm for dealing with problems in many sectors such as IT, Business, Education and Medicine [6]. Practitioners and researchers have put forward their views on design thinking. For example, in the fast-moving world of business, design thinking is defined as "a discipline that uses the designer's sensibility and methods to match people's needs with what is technologically feasible and what a viable business strategy can convert into customer value and market opportunity" [7]. A similar definition was proposed by the Kelley brothers, in which design thinking is seen as "a way of finding human needs and creating new solutions using the tools and mindsets of design practitioners" [8]. Researchers also explore design thinking's role and teaching in school. According to [9], design thinking is rooted in abductive reasoning and is the next competitive advantage. Some researchers in this cluster advocate that design thinking could (and should) be learned and adopted by non-designers [2]. For this reason, there is a growing interest in teaching design for business or management settings [1,10].

In addition to being valued in the adult world and higher education, design thinking has also been introduced into K-12 education as an innovative teaching approach [11]. K-12 is an American expression for kindergarten to 12th grade school students. For younger students, design thinking is usually introduced by integrating with other subjects, such as science, engineering, technology, STEM, etc. [11–14]. Consequently, rooted in design thinking, design thinking integrated learning (DTIL) refers to a new paradigm in non-professional design fields that aims to develop students' innovative problem-solving ability through design practice [15,16]. Through DTIL, students are expected to develop practical and thinking abilities, as well as domain knowledge. Moreover, design thinking is applied to prepare students for their future life and career by fostering competences such as creativity, collaboration, communication, and critical thinking, or the so called 21st century skills [16]. To help students better engage in the design thinking process and understand the core tenets of design thinking, more profound, different design thinking models are employed. Typical models, like the Stanford d.school's five iterative stages (Empathize, Define, Ideate, Prototype, Test) [17], the IDEO process model (Discovery, Interpretation, Ideation, Experimentation, and Evolution) [18], the four-step Double Diamond model (Discover, Define, Develop and Deliver) [19], etc., have been adopted in education.

Overall, as an innovative methodology, design thinking has been employed within an extensive field as high-leverage practice and K-12 education is no exception. However, there are also some persistent questions, such as the "design-science gap", whereby projects focus more on building successful design products rather than on the learners and relevant scientific principles [20–22], and the question of whether design thinking training can really boost creativity, or just generate unfounded confidence not accompanied by real gains in creativity [23]. More evidence using empirical research is required to show whether 'design thinking' is an effective approach for K-12 education and how it is applied in the teaching context. Therefore, a systematic literature review is needed to present the current research status of design thinking in K-12 education.

Some researchers have synthesized the work related to design thinking. Razzouk and Shute [16] identified the features and characteristics of design thinking and discussed its importance in promoting students' problem-solving skills in the 21st century. By referring to the processes and methods that designers use to approach problems, a design thinking competency model was constructed, but the search in their review was not limited to experimental studies. Micheli et al. [2] concentrated on design thinking in management discourse, by reviewing the knowledge and conceptualizations of design thinking, they identified 10 principal attributes, including abductive reasoning, ability to visualize, blending analysis and intuition, creativity and innovation, gestalt view, interdisciplinary collaboration, iteration and experimentation, problem solving, tolerance for ambiguity and failure, user-centeredness and involvement. Besides, eight tools and methods of design thinking were summarized. Zhang, Markopoulos and Bekker [12] presented a review of literature that referred to K-12 children's emotion while involving in design related learning activity. Rusmann and Ejsing-Duun [11] summarized the design competence framework in the K-12 school context: reasoning, problem setting, empathy, ideation, modelling, and process management, among which, the skills of reasoning and process management were stated to be needed at every step of the design process. But they did not assess the quality of the reviewed papers and included both review articles and primary theoretical or empirical studies.

Considering these limitations, our study aims to systematically review DTIL research in more detail than previous reviews. In particular, our review focuses specifically on the empirical studies that apply DTIL for the educational levels from kindergarten to secondary education (K-12). The purposes of this study are: (1) to systematically review high-quality empirical studies on DTIL in K-12 education, and (2) to explore future research perspectives of design thinking based on the reviewed papers. The following research questions (RQ) formed the basis of this review:

RQ1: What is the current research status of DTIL in K-12 education?
RQ2: What kind of curriculum domains are taught in DTIL?
RQ3: How to evaluate students' learning in DTIL?
RQ4: What intervention approaches are employed in DTIL?

2. Method

In order to conduct a reliable systematic review on this topic, we followed the three-stage guideline provided by Tranfield et al. [24]: stage I—planning the review, stage II—conducting a review, stage III—reporting and dissemination. Additionally, in stage II, we mainly used the snowball method [25] to select relevant studies.

Under stage I, as discussed in the previous section, we identified the need and goal of this review based on the scoping study surrounding the field.

Under stage II, to select useful and high-quality papers, we performed a keyword search in an online bibliographic database ISI Web of Science. The main reasons for selecting this database were: (1) to retrieve authoritative research articles that were of sufficiently high quality for analysis, (2) to draw more representative and reliable conclusions of this systematic literature review.

After the initial search, we utilized a snowball method [25] using the references in the selected articles before 14 March 2022 to search more literature. A snowballing approach refers to using the reference list of a paper or the citations to the paper to identify additional papers for a literature review [25]. The following inclusion criteria were used to select papers:

(a) Full peer-reviewed English paper published in SSCI journals.
(b) Empirical study conducted for K-12 students, the term "empirical study", which means either quantitative or qualitative, and not a literature review, framework, or proposal.
(c) Papers involved in DTIL (using "design thinking" in any part of the paper, such as title, abstract, keywords, or main text).
(d) The DTIL featured in the study targets students rather than pre-service/in-service teachers.

2.1. Keyword Search

Considering searching for relevant and mainstream papers that lie within the scope of this study, we used representative search terms to capture articles in the field. The keywords consisted of two clusters: design thinking and K-12. We adopted the keywords such as K-12, middle school, high school, and children, etc., to ensure that the target age levels of this study could be obtained. Initially, we used the search string "design thinking" AND (K-12 OR K12 OR middle school OR high school OR secondary school OR primary school OR elementary school OR kindergarten OR preschool OR children OR child) to identify SSCI journal papers in the ISI Web of Science. The results returned 99 papers. We quickly analysed the titles, abstracts, and methods sections of these papers with reference to the above four inclusion criteria. As a result, 27 papers were left.

2.2. Snowball Approach

After finishing the keyword search, we employed the first-round snowballing approach with the 27 papers as seeds to search papers in the ISI Web of Science. A total of 1485 references from backward snowballing and 236 citations from forward snowballing (a total of 1721 papers) were found. Using the above four inclusion criteria, 12 new papers were selected.

In the second round, 12 papers were identified as new seeds for the snowballing approach from the results of the aforementioned searches. A total of 703 references from backward snowballing and 123 citations from forward snowballing (a total of 826 papers) were examined. As a result, three new papers were singled out.

We then launched the third-round snowballing approach. This time three papers were used as new seeds and we retrieved 186 papers (173 references from backward snowballing and 13 citations from forward snowballing). Only one new paper was selected for review.

At the end, the fourth-round of snowballing was conducted, and no more papers were found, thus the iterative process of snowballing approach could be ended.

After four rounds of the snowballing approach, as depicted in Table 1, 43 papers were selected as samples for the subsequent literature review.

Table 1. Results of paper search.

Selection Strategy	Papers Resulting from the Search	Selected
Keyword search	99	27
First-round snowballing approach	1721	12
Second-round snowballing approach	826	3
Third-round snowballing approach	186	1
Fourth-round snowballing approach	37	0
Total	2869	43

Under stage III, to compare the features of the sample papers, a coding form was created (see Tables A1 and A2 in Appendix A). To address the above four research questions, information on the author, publishing year, educational level, sample size, duration, subject, and the implemented design thinking model is presented in Table A1. Table A2 summarizes each study's research design, course type, group size, design task/challenge, learning tools/materials, dependent variable, and evaluation instrument. The coding of the papers was initially done by one author, another author was responsible for checking, and the two authors studied together in ambiguous cases. Finally, the coding was analysed to draw conclusions.

3. Results

3.1. What Is the Current Status of DTIL in K-12 Education?

3.1.1. Distribution of Articles on Design Thinking over Time

The yearly publication distribution of the 43 articles on empirical studies on design thinking illustrated in Figure 1 reveals an upward trend from 2010 till now, consistent with the growing popularity and importance of design thinking [2]. Although we did not limit the publication period of papers in the literature retrieval process, in the K-12 education context, design thinking has attracted increasing attention since about ten years ago. Specifically, only 2 (4.7%) of the 43 papers were published before 2015, and related research has increased year by year. Especially, after 2017, it shows a steady upward trend, and subsequent years' publications peaked in 2021 (12 or 27.9%).

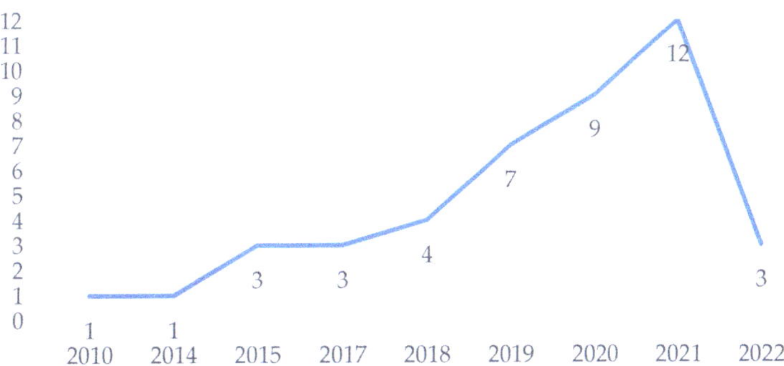

Figure 1. The publication year of design thinking papers.

3.1.2. Sample Group Level, Size, and Duration

Considering one paper included two experiments, we identified a total of 44 studies from the 43 papers. Figure 2 shows the results for the different categories of educational levels. Researchers conducted studies across various educational levels. In total, design thinking in the middle school setting accounted for the largest proportion (23 or 52.3%), or more than half of the reviewed studies. There were 10 (22.7%) studies for elementary school students, seven (15.9%) studies for high school students, two (4.5%) for kindergarten children, and two (4.5%) studies for mixed level students; one is mixed with kindergarten and primary school students, another is middle and high school across categories.

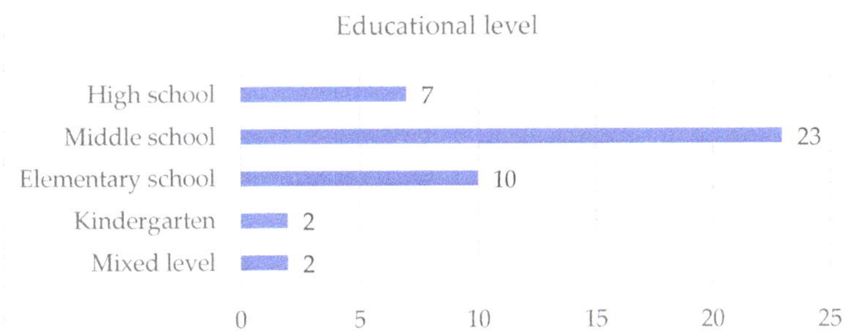

Figure 2. The educational level of design thinking papers.

Among the 44 empirical studies, 43 mentioned the number of participants. Nineteen (44.2%) of the papers were for studies that recruited less than 40 participants (see Figure 3). Only 11 (25.6%) of the papers involved more than 100 participants. This indicates that the sample sizes were not large in the K-12 educational research of design thinking. Promisingly, three papers distanced themselves from a small sample size: [26] surveyed 613 high school students who experienced a 16-week engineering design training course; [27] collected data from 576 children that participated in STEM Learning via 3D technology-enhanced makerspaces; [28] analysed data from 350 eighth and ninth graders.

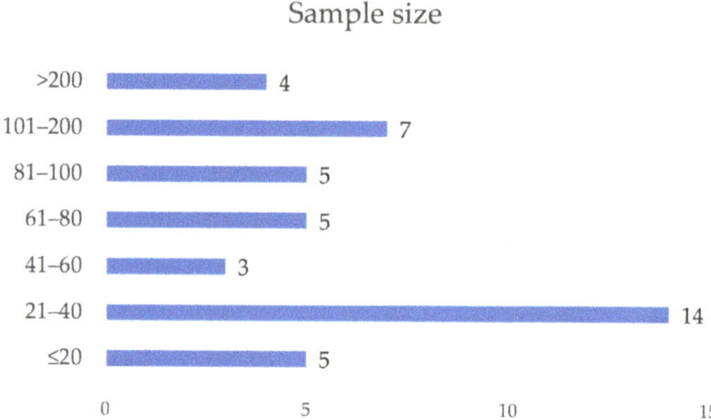

Figure 3. Sample size in the reviewed papers.

Among the 44 studies, 38 reported duration of the research on design thinking. As can be seen in Figure 4, the largest proportion of studies (9 or 23.7%) were conducted over a duration less than one month. While the second largest proportion of studies (8 or 21.1%) lasted for less than 2 months. Besides, five (13.2%) of the studies were one-off tasks, which lasted for less than 1 day. Only four studies were conducted for more than 6 months. Overall, the duration of most studies was comparatively short (i.e., typically for one semester).

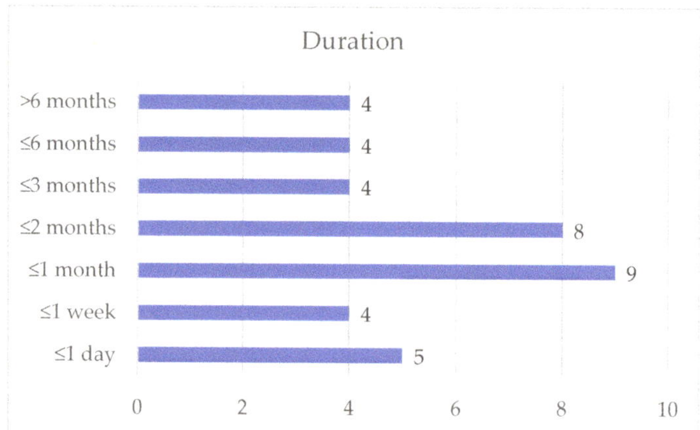

Figure 4. Duration of the reviewed papers.

3.2. What Kind of Curriculum Domains Are Taught in DTIL?

3.2.1. Distribution of Studies on Curriculums

We examined the curriculum domains of the reviewed studies. Overall, there were four main categories of curriculum domains involved in the research, including STEM-related, design-related, non-STEM multidisciplinary and other curriculums. As shown in Figure 5, among the 44 studies, 41 reported the subjects or curriculums, and science (11 or 26.8%) was the subject/curriculum matter most often researched, followed by STEM/SETAM in eight (19.5%) studies, and engineering in seven (17.1%) studies. Five (12.2%) studies covered multiple disciplines (non-STEM), three (7.3%) studies mentioned design related subjects/curriculums, such as design, design and research, design, and technology. The technology curriculum also had three (7.3%) studies. And the number for the remaining four curriculums was one (2.4%), including science and engineering, technology and engineering, robotics, and geography. Overall, subjects or curriculums related to STEM, such as Technology and engineering (1), technology (3), STEM/STEAM (8), science and engineering (1), science (11), robotics (1), engineering (7), accounted for the largest proportion (32 or 78.0%). In other words, in the K-12 educational context, design thinking was mostly integrated with STEM/SETAM subjects or curriculums. To compare the application of subjects by different educational levels, the number of empirical papers is shown in Figure 6. This shows that engineering was most popular in the high school level, while science was most popular in the middle school and elementary school levels. Engineering and STEM were applied in kindergarten. STEM/STEAM was applied in the two mixed level papers.

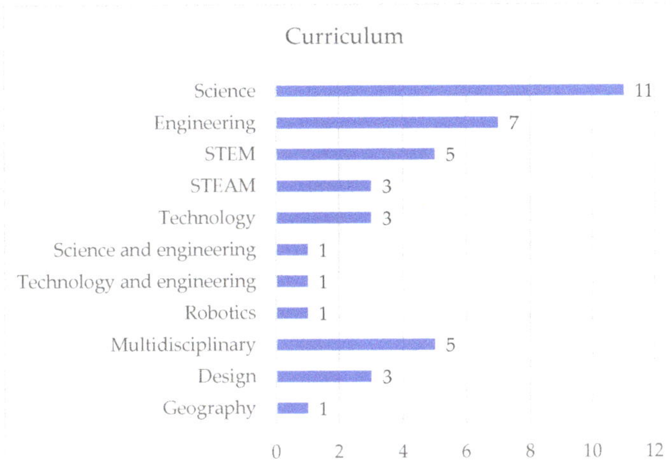

Figure 5. Curriculum of the reviewed papers.

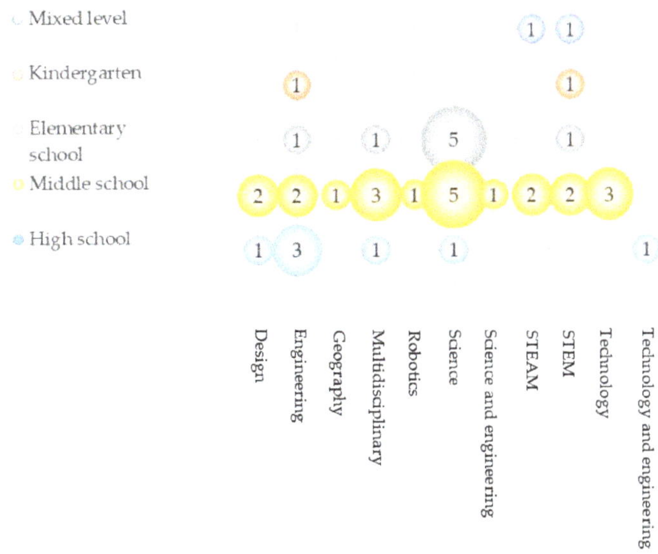

Figure 6. The number of empirical papers by educational level and curriculum.

3.2.2. Design Thinking Model Implemented in the Empirical Studies

Among the 44 studies, 37 mentioned the design or design thinking phases or elements. The authors described design thinking in different ways and associated a variety of attributes with the concept, and different terms and sequences of action were also employed [2]. Thus, models varied in the sample of empirical studies. For example, the typical five-stage model proposed by Stanford was adapted to integrate within the K-12 educational context (7 or 18.9%), see for example [29–33]. Some studies drew on other design thinking models, such as the model proposed by IDEO, see for example [27].

There also existed varied vocabulary in expression of the models [11]. For example, prototype was expressed in words as "modelling", "making", "create", "producing", "building". Ideate was also described by "brainstorming", "develop possible solutions", "generate design ideas", "product ideation", etc. Define also meant "clarify problems and constraints", "identify problem", "understanding and defining the problem", "Understand" etc. It was not easy to identify all the models that applied. Despite this, some elements among these were employed more regularly, suggesting a level of concurrence [2]. Therefore, we summarized the mostly mentioned design thinking competences in the selected studies, including prototype, ideate, define, test, explore, empathize, evaluate, and optimize. Explanation and codes are also listed (see Table 2).

Table 2. The frequently mentioned design thinking competences among the empirical studies.

Design Thinking Competence	Frequency in the Data Set	Codes	Explanation
Prototype	32	Prototype Modelling Build Create Make Fabrication	Creating the original or early solution model, it can be a sketch, or other physical or virtual structure that designed.
Ideate	31	Ideate Design Brainstorming develop possible solutions	Generating alternative ideas that may lead to solutions.
Define	18	Define Understand Problem definition Identify problem Clarify problems and constraints	Actionable problem statement based on insights into the problem situation.
Test	15	Test	Experimenting and gathering feedbacks.
Explore	14	Explore Collect information Data collection Discovery Field Studies Observe	Questioning and collecting information to gain deep understanding of the problem.
Empathize	13	Empathize Human-centeredness Needs-finding Sensitizing Feel	Carrying out design around the needs of users, highlighting human-centred design.
Evaluate	9	Evaluating Appraising	Checking if the design meets the user's needs.
Optimize	8	Optimization Improve Evolution Iteration Redesign	Refining solution based on user's feedbacks.

3.3. How to Evaluate Students' Learning in DTIL?

3.3.1. Dependent Variables

All the dependent variables assessed across all the studies are included in Table 3. The measured competences mainly covered four aspects of students' learning in the DTIL context: subject learning performance, design thinking, emotional/social aspect and skill. (1) Subject concepts like scientific knowledge (e.g., [34–37]) were discussed. (2) The design thinking concept, such as the understanding of design thinking (e.g., [35,38–40]), and design thinking practice, such as the design thinking process (e.g., [21,26,31,41–46]) or the design thinking work (e.g., [31,43,45,47–49]), were assessed. (3) Emotional/social aspects, like attitude, desire, engagement, or collaboration, etc, were tested in some reviewed studies (e.g., [27,49–52]). (4) Other skills, such as creativity, productive thinking, problem solving, and critical thinking, were also evaluated (e.g., [23,26,29,34,53,54]).

Among these dependent variables assessed (see Figure 7), 31 (or 70.5%) studies tested the design thinking, 17 (or 38.6%) discussed the emotional/social aspect, nine evaluated the subject learning performance, and eight tested skills (or 18.2%).

Table 3. Dependent variables evaluated in the empirical studies.

Dependent Variables	Construct
Subject learning performance	Subject concept
	Subject skill
Design thinking	Design thinking concept
	Design thinking process
	Design thinking work
Emotional/social aspect	Attitude
	Interest
	Satisfaction
	Desire
	Acceptance
	collaboration
Skill	Creativity
	Critical thinking
	Problem solving
	Productive thinking

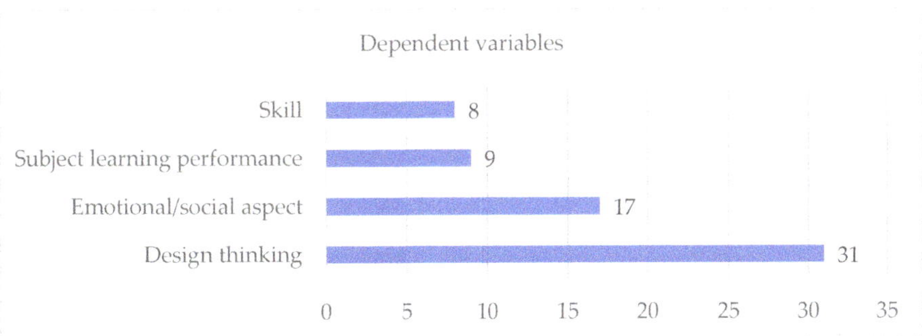

Figure 7. Dependent variables of the reviewed empirical studies.

3.3.2. Evaluation Instruments

Despite the variability in students' learning performance assessed for DTIL, multiple measurement instruments emerged from the literature, including the qualitative and quantitative assessment types. The results of each instrument were revealed in Figure 8. Among the 44 studies, 11 (25.0%) just adapted one kind of measurement instrument. Some employed more than one assessment instrument to collect multidimensional evidence of students' learning performances. Thirteen (29.5%) used two measurement instruments, and twenty (45.5%) combined three or more evaluation methods.

For the measurement instruments, 23 (52.3%) of the studies adopted a survey/questionnaire for measuring the students' learning performances in the design thinking integrated context, which is the one with the largest proportion (see Figure 8). The second-largest method adopted was portfolios (14, or 31.8%), followed by interview, observation, protocol analysis, and test categories, which all accounted for a frequency of 11 (25.0%). Next was design work evaluation and video recording, and the frequency was 10 (22.7%) for both. Audio recording (6 or 13.6%) and journal (5 or 11.4%) were also included in the measurement instruments. Explanation and examples are also presented in Table 4. Because some studies used a combination of evaluation methods, the sum of the numbers was greater than 44 (or 100%). Overall, qualitative measurement instruments were used more than quantitative ones.

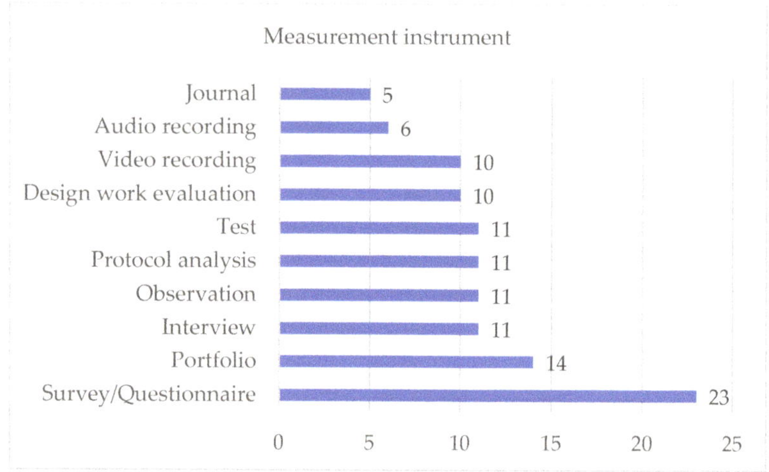

Figure 8. Measurement instruments adopted in the empirical studies.

Table 4. Measurement instruments adopted in the empirical studies.

Measurement Instrument	Explanation and Examples
Survey/Questionnaire	Surveys or questionnaires are often used for investigating skills or emotional/social dispositions towards DTIL (e.g., [27,32,35,55,56]).
Portfolio	Collecting and evaluating students' design products purposefully and systematically (e.g., [26,33,52]).
Interview	Researcher adopted interview to probe participants' understanding of DTIL (e.g., [30,41,57]).
Observation	Observation is usually employed to explore participants' procedural performance in greater detail (e.g., [33,41,58]).
Protocol analysis	Protocol analysis is often adopted to understand the thought process of individual or groups in a natural way, the object of its analysis includes the coded verbal communication, or the thought process being asked to speak out (e.g., [41,46,59]).
Test	To estimate students' mastery of relevant knowledge, test or examination is usually adopted (e.g., [23,35,43,60]).
Design work evaluation	Design work is seen as a direct way to reflect students' learning outcome, and it is widely implemented in the evaluation of DTIL (e.g., [31,35,48]).
Video recording/Audio recording	In DTIL, researcher recorded the design activities by video or audio so that to understand the participants' learning process more fully (e.g., [38,44,61,62]).
Journal	Journals or diaries from participants is analysed to help understand the process by which learning occurs (e.g., [27,29,38]).

3.4. What Intervention Approaches Were Employed in DTIL?

3.4.1. Study Design

According to the existing taxonomy framework [63], we divided the research designs of these empirical studies into three types: experimental (using random assignment), quasi-experimental (no random assignment but with a control group or multiple measures), and non-experimental (no random assignment, no control group or multiple measures). Among the 44 studies, most used a non-experimental design (31 or 70.5%) (see Figure 9). These studies usually presented one or more cases where an intervention was applied and measured student performance through various instruments. Seven (or 15.9%) studies used quasi-experimental designs, and six (or 13.6%) used experimental designs.

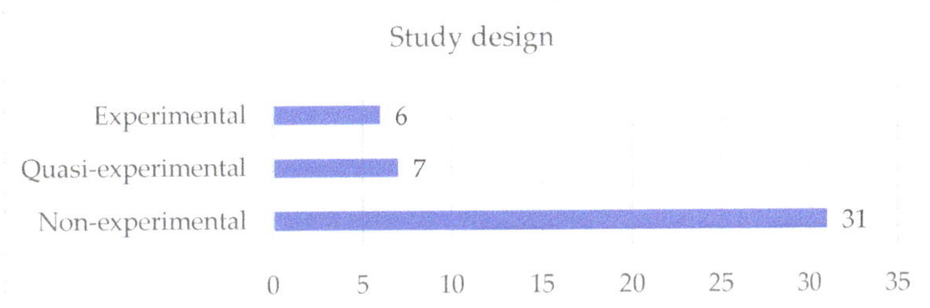

Figure 9. Study design of the reviewed empirical studies.

3.4.2. Course Type

Apart from some unclear research contexts, we summarize the course types from 40 studies. Results (see Figure 10) show that 34 (or 85.0%) studies conducted empirical research in the formal K-12 school courses, like the regular or elective courses offered during the conventional teaching schedule. Six (or 15.0%) were conducted in the informal course environment, for example, conducting the experiment during a period of school holidays [23], or the study was conducted as an afterschool program which was not applicable within the conventional school hours [37,53].

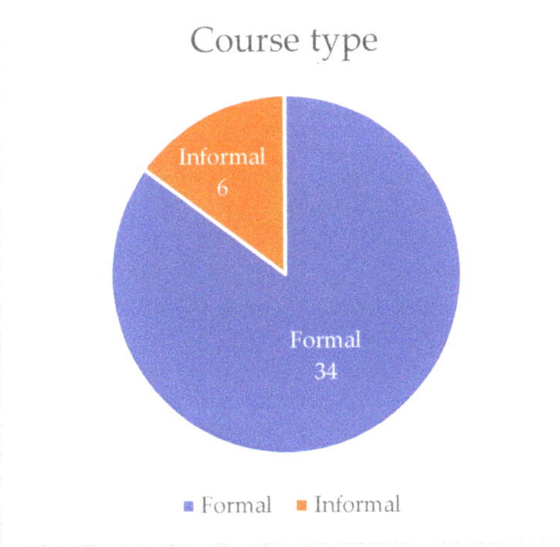

Figure 10. The frequency of studies by course type.

3.4.3. Grouping

Twenty-nine of the forty-four studies mentioned grouping in the empirical research. Four studies mentioned that students used group learning but did not specify the size:

one study used both individual and collective design activities [64], and students in another three studies were organized in groups [48,55,57]. In the remaining 25 studies (see Figure 11), the group size varied in the research. Six studies employed a group size of 3–4 students per group [30,33,34,39,45,52], five employed a group size of three students per group [21,44,46,51,59], followed by four students per group (4) [29,50,54,61] and individual group (4) (e.g., [41,65]). Next was the size of 2–4 (2) [36,47]. There were also four studies with group sizes of 2–3 [62], 4–5 [38], 4–6 [58], and 5–6 [53], respectively.

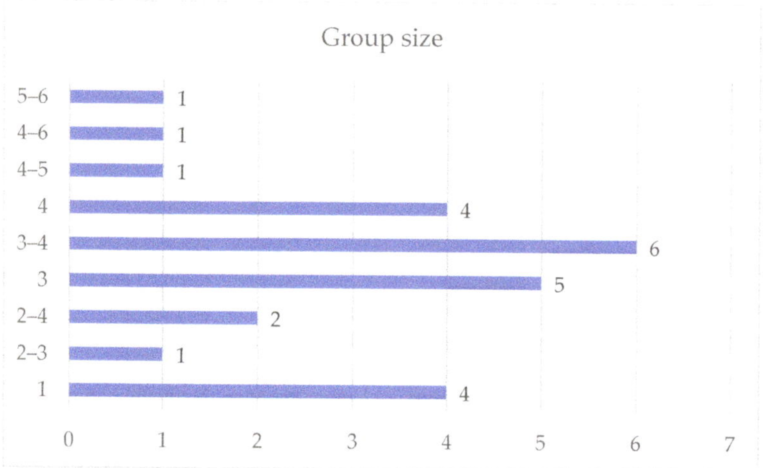

Figure 11. Group size of the empirical studies.

3.4.4. Design Thinking Task/Challenge

The design tasks or challenges used in these empirical studies are summarized in Appendix A. It can be found that these tasks or challenges are usually real situated, open-ended, challenging, and cooperative.

On one hand, most of these design activities were based on authentic situations, which are close to the real-life experience of students. Challenges were carried out around topics related to social life, such as to identify and redesign systems that existed at school [38]; the design of an escape room for the local fire department to allow participants to playfully and interactively improve awareness of fire safety in and around the house [30]; working revolved around interviewing senior citizens and creating prototypes that met their needs on the background of the aging society [32]; to create a secure environment for the elderly without taking away their freedom [40]; or designing a heat retaining food container for street food vendors at a taxi depot [46]; designing a water filter system for the city's wastewater management plant to help prevent the pollution of a local river [34]; to develop sustainable food products for peer group [55]; designing and constructing shoes [21]; the design theme on preventing bullying in the social context of the class [58]; or targeting on more local or social problems that could be solved with new products, services, or other solutions [33,50,66], etc.

On the other hand, for all design activities, there were no unified answers. Students were encouraged to design as many creative solutions as possible and choose the relatively optimal solution to solve the problem. Many tasks were engineering, such as the marshmallow tower activity and the trebuchet design activity [45], the design of toys [39,43], the engineering design activities of musical instrument, simple machines, and bio-inspired flower [47], or the design of water filters, bridges, circuits in maglev vehicles and windmills, and pollinators and knee braces [62], etc. In another study, students were asked to design a 3D model and print a 3D artefact [57].

In addition, the design activities were usually challenging and cooperative. The problems addressed by designers are wicked problems, such as the design of complex systems or environments for living, working, playing, and learning, and the wicked problems theory of design was explained in [67]. In school education, an important goal is to develop cooperative citizens, which helps to equip students with the ability to collaboratively solve wicked problems in social life. Therefore, to help students solve these complex and wicked problems, some activities were game-based and aimed to create interesting design situations for learners. Especially for the younger children, the fun of activity design is especially important. Thus, some activities were role-playing, such as the design of castle according to an imaginary engineering situation [64]; or storytelling, such as reading the narrative content of the books and solving the engineering problems presented in the books, and finally rewriting the story [52]. Others also employed computer games to study students' learning based on the design process [60,68].

3.4.5. Design Tools and Materials

Among the 44 studies, 20 mentioned the design tools or materials used. Six provided students with conventional design tools or materials, such as paper, pencil, glue, tapes, scissors, etc., [36,42,44,46,50,51]. Specially, one of these studies employed a reverse engineering teaching context, in which students learned from already designed products rather than started from scratch [51]. By dissecting the product's components and structures, it helps students understanding how those components function and work together, and then students can redesign or build their own product to properly solving a design challenge with micro-innovation or applying the scientific concept they had learned during the learning process.

Fourteen studies employed digital platforms as teaching aids of designing and making in the class. For example, some studies organized design activities with open-source hardware and software. Kim, Seo, and Kim [52] provided students with Arduino Leonardo-based device to solve the engineering problems, by which artifacts can be made through assembly of various blocks with sensors and actuators, and also be driven by the programming language based on Scratch and app inventor. In [49], Raspberry Pi with programmable microelectronics, adapted versions of SNaP and Tiles toolkits, and Google's Design Sprint Kit techniques were combined to help IoT design across generations. Another platform, such as the mobile telepresence robot called KT, which is controlled over the internet through a web interface, was used to design human-centred robots that served a need in the local environment and allowed remote peers to explore the local spaces [61]. Christensen et al. [40] also explored middle school students' design literacy who had experience with digital technology in maker settings (e.g., FabLabs).

Lin, Chang, and Li [31] studied virtual reality (VR) teaching application on engineering design creativity. In their study, the junior high school students experienced design teaching with VR devices used as teaching tools. The VR box (for a 360° view), zSpace built-in software Franklin's Lab (for detailing motor operation), Cyber Sciences-3D (for explaining circuits), and Leopoly-3D (for modelling simulation) (created by zSpace, Inc., San Jose, CA, USA) were used to design an electric model vehicle capable of automatically avoiding obstacles.

There were other tools, meant for prototyping or concept understanding, for example, the 3D design and printing tools, that were employed in some reviewed papers. In [26], the design stage used computer-aided design to create a three-dimensional model of the toy. Forbes et al. [27] studied children' STEM learning performance in 3D technology-enhanced makerspaces by using iPad and 3D design software and print device (Makers Empire, created by Makers Empire Pty Ltd., Adelaide, Australia). Leinonen et al. [57] discussed the digital fabrication in elementary school education. The 3D model design by Tinkercad software (created by Autodesk, Inc., San Francisco, CA, USA) and 3D artefact printing by Ultimaker printer were adopted to aid the design activities, such as designing and printing a name tag, a floor plan, and a game piece. In [43], the teacher explained the mechanical functions and STEM knowledge via physical and 3D virtual simulation models, and then in the inquiry experiment stage, students were asked to assemble the LEGO to design a movable toy with various mechanical structure types.

Besides 3D modelling tools, other design platforms were also employed. Four studies involved students in the poster design activities, and three of them used poster design games on computer [60,65,68], while another one used WPS Writer® to complete the new year poster design [48].

A tool dedicated to assisting design communication was also mentioned. Won et al. [56] reported on the ways that middle school students appropriated a social networking forum (Edmodo) as a part of the iterative design process in an informal learning environment. In their study, the forum was utilized for the purpose of collaborative design by interaction in progression of the design process.

4. Discussion

4.1. The Current Status of DTIL

4.1.1. Research Trend

Among the selected 43 empirical papers, most were published after 2017 and most (12) in 2021 (see Figure 1). It was concluded that the growing trend of high-quality empirical studies on DTIL in K-12 education was significant in the past five years. "The proper study of mankind is the science of design, not only as the professional component of a technical education but as a core discipline for every liberally educated man" [3]. According to the rapid development of empirical research, we can predict that more evidence-based research will focus on this topic in the future, and the potential value of design thinking in K-12 education will be further explored.

4.1.2. Educational Level

Considering the uneven distribution of the age groups in K-12 education, more research is needed to engage the younger (like kindergarten and primary school students) and older learners (high school students) into DTIL. As shown in Section 3, the research covered students of all levels in K-12 education, aged between 3 and 18 years. Students from kindergarten to high school were all involved, and among them, the papers studied middle school students' learning accounted for the largest proportion (23 or 52.3%). It shows that the distribution of existing studies was uneven across educational levels. As a methodology for innovative problem solving, the design thinking intelligence is not necessarily something that comes naturally, it needs to be nurtured and developed in pedagogical approach. According to [69], design must be nurtured from early beginnings, as building design capital is a visionary and long-term job. Wells [70] also pointed out that, similar to language learning, design thinking is something that should be carefully nurtured from an early age and be included in all areas of education. However, in comparison with the studies conducted in university or other adult learning scenarios, there is still a lot of room for research in K-12 education.

4.1.3. Samples and Duration

Among the 44 studies extracted from the 43 papers, many selected small samples or short periods to carry out the empirical research. The reason may be that the activities integrating design thinking had higher requirements for teachers and students. First, teachers need to spend more time preparing various sources for class teaching, such as the design tools and materials. Second, during the lesson, support and guidance are required for student activities. Especially in the hands-on activities, teachers are also concerned about the safety of their students. Third, the diversification of evaluation, such as the evaluation of the design process and design works, also requires teachers to spend more time on observation, interviews, surveys, etc., to obtain and analyse multi-dimensional learning data. The limited energy of teachers makes it difficult to carry out large-scale class teaching for a long time. Fourth, for students, the design thinking tasks are usually more challenging, which require a longer cycle to constantly iterate and optimize.

However, for DTIL in K-12 education, empirical studies with longer periods and larger samples are still expected to provide more sufficient and convincing evidence to demonstrate the application effect. Despite the difficulties, researchers and teachers could try to solve these problems with relevant technologies, thereby expanding the sample size and extending the experimental period. For example, teachers could try to use intelligent teaching assistants to assist in monitoring students' learning, real-time feedback and forewarning of learning, etc., so that they can better communicate with students, guide them according to students' needs, and realize personalized teaching. Students could use intelligent technologies such as intelligent learning partners to carry out learning, reduce their learning burden and increase their interest, so that they can actively participate in longer-term learning activities. In addition, multiple subjects are encouraged to participate in teaching and research related to DTIL in K-12 education. In addition to research institutions (e.g., university), K-12 schools, and government departments, more social forces are also expected to join, such as communities, science and technology museums, enterprises, etc., to jointly provide support for DTIL, thus ensuring larger scale and longer cycles teaching practice.

4.2. Curriculum Domain in DTIL

4.2.1. Curriculum

More research is needed to explore the integration of design thinking with single or multiple disciplines from a wider subject range. As the social movement of design thinking has called for the involvement of design thinking into K-12 classrooms, we put together the disciplines mentioned in the reviewed empirical studies. The results revealed that the implementation of design thinking in non-design classrooms is increasingly concerned. Although design thinking by its nature originated from design field in an effort to encourage people to think and practice like an expert and prepare future designers [70], it is now become a simplified version of "designerly thinking" or a way of describing a designer's methods that is integrated beyond the design context [7,9,15]. In addition to design-related courses, most researchers employed the design thinking integrated teaching and learning in the STEM related subjects, such as science, STEM/STEAM, engineering, technology, etc., which is consistent with previous findings [11]. While other researchers have extended design thinking to non-STEM subjects, such as geography [38]. This has a lot to do with the rise of STEM education in recent years. There is societal recognition of the role played by STEM education in preparing students for college and career readiness, and design challenge exactly provides a pathway for the STEM disciplines to work together [14]. Additionally, it has adopted the stance that design and technology should be embedded in various school subjects [71].

How is design thinking integrated into these curriculums? Despite the differences in the curriculums themselves, there are some common threads of DTIL. First, for the purpose of DTIL, it mainly responds to the need for interdisciplinary and innovative talent cultivation in the age of intelligence. Therefore, design thinking is integrated by enabling students to learn and connect multidisciplinary knowledge and skills via involving the design process/skills. Second, for the learning content, it is usually embedded inside the open-ended and authentic problem context, in which students' learning content stems from real life, not those hypothetical questions. That is, the integration of design thinking provides ideas for the reframing of teaching content or topics. Third, for the organization of learning, the curriculums adapted DTIL are characterized by student-centred and co-creative. Teachers act as coach to enable active learning, while students can communicate and collaborate with peers and even with stakeholders so that be able to solve problems creatively.

4.2.2. Design Thinking Model

The design thinking models should be aligned with the complex teaching situations. It was found that the core concepts of design thinking are valued and pursued in the field of K-12 education. However, in terms of the design thinking model, there is no panacea for all situations, that is, there is no unified framework or process in academic. For example, when talking about the design process and characteristics of a certain design activity (e.g., the technology and engineering design), it does not mean that there is just one process [14]. Researchers have also pointed out that if teachers rely too much on the confined procedural, pre-determined process structure, it may strangle the qualitative forms of intelligence inherent in design thinking [70]. In fact, in the human-made world, different fields or disciplines have their applicable design processes, the process of design thinking is surely not deterministic [44]. Therefore, the key ideas of the existing studies for design thinking provide a foundation for understanding the capabilities related to design. On this basis, teachers can adjust the design thinking model reasonably according to the actual situation.

In this study, based on the evidence of the reviewed empirical papers, it reveals that the highly concerned aspects of design thinking are the following: prototype, ideate, define, test, explore, empathize, evaluate, and optimize. These key elements provide a holistic outline for students to solve problems with design thinking. Overall, the results of this study are consistent with previous studies. For example, ITEEA [14] elaborated eight key ideas of design in the PreK-12 technology and engineering education, including a fundamental human activity, open-ended and can always be improved and refined, iterative, a range of skills, universal principles and elements, making, optimized by criteria and constraint, and diversity of approaches. In [11], design thinking in the K-12 educational context was noticed, and they organized the multiple codes of design thinking competences into six areas, including reasoning, problem setting, empathy, ideation, modelling, and process management.

4.3. Learning Evaluation in DTIL

4.3.1. Dependent Variables

Knowledge, skills, and dispositions are all covered by the dependent variables reported from the reviewed papers. For example, the existing empirical studies measured students' knowledge of relevant subject concepts, e.g., physics concepts of acceleration, Newton laws of motion, velocity, motion, energy, etc., [37], or the understanding of the mechanical concepts [43]. Some studies also checked students' understanding of design thinking (e.g., [39]). Various skills, like creativity, productive thinking, problem solving, and critical thinking, are evaluated as crucial learning effects. Except for the concepts and skills, students' performance on design thinking practice (including the design process and design work) has been paid close attention, and it is the most frequently evaluated one. This exactly confirms the practical orientation of design, or the non-verbal competence [15] that is required. Besides, many studies surveyed students' dispositions, as there exists a bi-directional relationship between emotion and design-based learning [12]. To sum up, in the DTIL context, design thinking is considered in many studies as both a dependent variable and partly as an independent one. Results of the learning effects indicate optimistic learning outcomes of students; additionally, the competences cultivated by the design thinking activities were often associated with the 21st century skills, like communication, collaboration, creativity, and critical thinking [11,27,29,40].

4.3.2. Evaluation Instruments

Considering the benefits of applying design thinking into other subjects, more objective and apt evaluation instruments need to be developed to emphasize the alignment between design thinking competences and domain knowledge. The evaluation instruments used in the existing empirical studies include survey/questionnaire, portfolio, interview, observation, protocol analysis, design work evaluation, test, video recording, audio recording, and journal. Qualitative and quantitative assessments are both included, and the qualitative assessments are used more frequently. However, it was found that most of them were self-designed measures, such as portfolio, survey, interview, observation, design work evaluation. These assessments were highly subjective. To better serve the trending integration of design thinking into STEM and non-STEM subjects in K-12 education, it calls for more objective and apt design thinking-embedded evaluation instruments that evaluate comprehensive performance for researchers and educators to adapt in their interventions or classrooms. Researchers have pointed out the shortcomings of current evaluation methods, for example, Blom and Bogaers [46] indicated that current verbal protocol analysis methods and theoretical frameworks did not explain how internal and external information sources contribute to novice designers' moment-to-moment thought processes, and they employed a protocol analysis method—Linkography—to investigate the nature of novice designers' thought processes. Furthermore, given that DTIL is usually suggested to be involved in the complex problem-solving situations, more timely evaluation and feedback are needed to help students gain a better learning experience. In the era of intelligence, there is also an urgent need to explore assessment supported by various intelligent technologies.

4.4. Interventions in DTIL

4.4.1. Study Design

Most of the studies tended to report the DTIL via non-experimental case study in K-12 context [16], the number of which far surpassed that of the quasi-experimental and experimental studies. In the educational context, samples were usually restricted by the original class settings, therefore, it was relatively more difficult to carry out rigorous experimental research that emphasizes the randomization of samples. In this instance, the application of case studies does have value and facilitates the introduction of concrete practical experiences. However, the problem that ensues is the difficulty in identifying the certain competences can be developed by participating in design thinking integrated activities [11], so that a compromise approach is to conduct quasi-experimental study design. According to [63], a study with no random assignment but a control group, multiple measures can be classified into the type of quasi-experimental design to examine whether such a design thinking integrated intervention, as the independent variable, influences students' learning, as the dependent variable. In the future, more rigorous research with quasi-experimental or experimental study design is needed to strengthen the learning effect verification.

4.4.2. Course Type

Most studies were conducted in a formal educational context (i.e., classrooms). The high proportion of studies conducted in the formal educational setting appeared to align with the move of involving design thinking into educational context. Besides, it is also consistent with the finding of large number of research conducted in standard courses, such as science, engineering, technology, etc., (see Section 3). In the future, more studies should be conducted for design thinking integrated interventions or activities applied in informal educational scenarios, such as the design workshops or makerspaces, so that it can be a complementary way to highlight the educational value of design thinking.

4.4.3. Grouping

Most research defaults to a collaborative learning approach to design thinking embedded activities. Collaborative learning has the potential to foster students' understanding of various subject domains, help complex problem solving, and master social skills [58,72]. The group sizes typically employed were between two and four people and were usually randomly grouped or arbitrarily appointed by the teachers. It lacked detailed research to explain the specific reasons or rules. A few studies have looked at the impact of gender differences on design thinking embedded learning [31,34,37]. More in-depth studies are needed to explore applicative collaborative strategies, such as group size, grouping methods like role assignments, dynamic grouping or fixed grouping, gender combination, etc.

4.4.4. Design Thinking Task/Challenge

The design thinking embedded activities were usually comprehensive, transdisciplinary, and integrate a variety of learning methods, such as project-based learning [55], problem-based learning [21], inquiry-based learning [36], game-based learning [60], storytelling [64], etc. It was in line with the characteristics of transdisciplinary learning such as STEM. Additionally, process-oriented design (design-no-make) and product-oriented design (design-and-make) were both adopted to produce various creative solutions, object models, smart products, engineering devices, etc. For example, while involved in the activities of poster design [48,65], playground design [41,42], or the design of smart-things ideas for an outdoor park environment [49], students were encouraged to propose creative ideas to meet certain needs. When participated in the product-oriented design tasks, students were also required to construct artifacts, such as designing and constructing a device that was engineered to provide electricity to a third world community scenario [36]. Results of this study should inspire researchers and teachers to develop design thinking embedded activities that are reality situated, open-ended, challenging, and cooperative, to engage younger students to better develop their domain knowledge and thinking skills, activities that are fun and hand-and-brain on will be in greater need. In future, the integration of design thinking in project-based learning and STEM/STEAM education can be more diverse, and it would be more expected if it can be combined with local cultural characteristics. For example, in one study [73], the "Chinese Wooden Arch Bridges' Intelligent Monitoring system " project was developed to explore the integration of cultural education, scientific inquiry and design thinking. In this STEAM project, the integration of design thinking is believed to not only promote transdisciplinary learning, but also help cultural understanding and inheritance.

4.4.5. Learning Tools and Materials

The tools and materials employed are expected to help students formulate and express creative design ideas. For example, the design cards are convenient for students to sort out the design process and carry out reasoning [49]. The 3D design and printing platforms can help students design and visualize prototypes [27,57]. Conventional design and construct tools, as well as the digital fabrication kits, provide students with hands-on opportunities to materialize ideas. According to the reviewed articles, there were more applications of conventional tools and materials than intelligent ones. Researcher argued that the functions of design are making better what is inadequate and delivering a value that has not yet been delivered, thus technology is a key driver of the project as a whole and will play an important role in that [74–76], likewise, design can promote technological literacy. Future study could be conducted to further explore the technology-enabled design thinking activities in K-12 context.

Especially, in the emerging digital lean manufacturing paradigm [77], the advent of Industry 4.0 (I4.0) technologies such as Artificial Intelligence (AI), Internet of Things (IoT), Virtual Reality (VR), and Mixed Reality (MR), etc., have provided significant enlightenment to education. On one hand, new technologies can be used to enable better performance. For example, during the COVID-19 pandemic, technologies make such large-scale online teaching activities possible. In particular, for experimental or hands-on courses, the technology application like the online simulation platform (i.e., IRobotQ3D) [78], the integration of VR and motion capture system [79], etc., can solve the problems of interactivity and presence to a certain extent. On the other hand, the essence of I4.0 is also to better meet the needs of users, which coincides with the human-centred concept emphasized by design thinking. As we enter the era of intelligence, customers require new and customised products more and more often, and companies have hence to be able to easily follow these requirements in order to survive in today's competitive market [79]. The implication for education is that to integrate technology into DTIL activities may help students collect and analyse data more scientifically, open up broader design space, and enrich the ways and forms of design work presenting, so as to promote the cultivation of innovative intelligent manufacturing talents that meet the needs of the era.

In conclusion, the implications and suggestions for DTIL in K-12 context are: (a) The potential value of design thinking in K-12 education needs to be further explored with larger samples and longer durations, in particular, more attention needs to be paid to students at the preK-elementary and high school levels. (b) Design thinking models be aligned with the domain learning situations are needed to expand the application of DTIL in a wider range of disciplines, especially non-STEM disciplines. (c) DTIL is expected to facilitate the acquisition of higher-order learning achievements, correspondingly, more objective and apt evaluation instruments are needed. (d) The educational value of design thinking does not arise spontaneously, more targeted interventions are expected to reach its full potential, such as rigorous study design in the pre-post mode or with control groups, more focus on learning in informal settings, intensive grouping strategies that consider gender, group size, role, etc., pertinent design thinking tasks/challenges what are both fun and challenging, and finally, support design activities with appropriate technical tools or materials.

5. Conclusions and Implication

In the wake of the popularity in design thinking, this literature review analyses 43 high-quality empirical papers on DTIL in K-12 education. The current status and future directions are identified to narrow the gaps. The major findings are summarized as follow.

First, as deduced from the publication trend (see Figure 1), a positive picture of design thinking emerges. However, the empirical evidence to support the effectiveness of DTIL is still rather limited. The results indicate that although samples from preschool to high school are all involved, it shows an uneven distribution of the age groups in K-12 education, which calls for more research to engage the younger learners (like kindergarten and primary school students) into DTIL. If possible, studies with longer periods and larger samples are expected.

Second, DTIL has the potential to deepen students' learning with diversified models in various subject domains, including that of STEM and non-STEM, while the most mentioned applications are the fields of STEM, which is consistent with the result of previous study [12]. In addition, the design thinking models applied varied among studies, which suggests inconsistency in current perceptions of design thinking. On one hand, a prevailing view was to apply design thinking from a thinking model of designer or designer's mindsets (e.g., [40,49–52]), in other words, think of it as a unique problem-solving methodology (e.g., [33,35,37,42,44,48,53]). Our study also supports this view. The theoretical basis of design thinking has been propositioned to be interconnected to Dewey's notion of pragmatist inquiry and aesthetic experience [23,36]. The corresponding training goals were the abilities required to solve the complex and open-ended problems, such as communication skills, cooperation skills, creativity, and critical thinking, etc. On the other hand, however, there were differences in the focus of studies on the application of design thinking, leading to the emergence of different models or competence frameworks. For example, some studies focused on inquiry and experimentation in the early stages of problem solving (such as discover, define, empathy), others focused on modelling or prototyping during the problem-solving process, and still others discussed the impact of emotions, etc.

How do teachers choose or construct a design thinking model or framework that meets the needs of the teaching situation? In fact, it is difficult to construct a universal design thinking model to deal with a variety of complex educational scenarios. Therefore, to meet the teaching needs of more subject domains, future study needs more exploration of the design thinking models that could be compatible with teaching practice in K-12 education. Referring to the existing empirical research, it may be necessary to comprehensively consider related factors such as students, teachers, and the school environments. In addition, although existing research claims that design thinking has great educational value, it is difficult to cover all aspects at once. Therefore, it is a feasible way to develop various competencies by stages or disciplines. Future study could bring design thinking into both STEM and non-STEM curriculums to further verify how design thinking will help students build knowledge and skills in a wider range of domains. Furthermore, teachers can make targeted combinations or adjustments of elements in a design thinking model or framework to better meet teaching needs.

Third, in response to the many existing design thinking models and subject domains, various dependent variables were measured by different instruments. Mostly, the empirical studies evaluated students' performance on design thinking, such as the design processes and design works. Another two types of learning performances that were paid close attention were emotional/social aspects and subject concepts. Besides, the design thinking concept and other skills like creativity, problem solving, critical thinking, etc., were also involved in the measured variables. For assessment, qualitative and quantitative assessments were both included, and the qualitative assessments were used more frequently, which highlights the practice-orientation and process-orientation of design thinking.

Implications for instruction include: (1) for the dependent variables, both process performance and learning outcomes should be emphasized. In particular, it is helpful to explore the development process of students' thinking by analysing the process of students' design activities. (2) Besides, the culture competency also needs to be included in the assessment. Design was seen as an important force driving the development of the artificial world, so one of the core characteristics of design thinking is that it is human-centred [1,7,32,80,81]. Every citizen should have a certain quality of cultural understanding and inheritance, which emphasizes the practical implementation of these values contained in the excellent culture at the behavioural level [82]. On this basis, we can better solve the problems faced by human beings through design thinking activities, to promote the benign development of human society. (3) To verify the design thinking embedded learning more scientifically and efficiently, more objective and apt evaluation instruments are needed. In addition to conventional subject knowledge and skills, for the evaluation of design thinking, existing research has provided some experience for reference. For the understanding of design thinking concepts, it can be assessed by test [35], survey [39], questionnaire [40], or other qualitative method, such as interview, video/audio recording, etc. [38]. For design thinking, evaluation instruments that can be used are portfolio (e.g., [21,33,42,55,61]), observation (e.g., [45,64]), survey (e.g., [31,33,55,56]), protocol analysis (e.g., [21,42,44,46,59,62]), design work evaluation (e.g., [31,35,45,47,58,60]), interview (e.g., [55]), etc. (4) Furthermore, there is more room for development of relevant evaluation methods supported by technology. In the existing studies, evaluation instruments supported by technology are rarely used. In the future, the application of artificial intelligence technology, multimodal data collection and analysis, real-time evaluation and feedback technology in the classroom still needs to be explored to promote the improvement of teaching quality.

Finally, interventions including study design, course type, grouping, design task/challenge and teaching tools and materials are summarized from the empirical studies. Results show that most research reported DTIL by conducting non-experimental studies in formal classrooms. In the reviewed studies, collaborative learning was frequently used, through which students were usually arranged in groups of 2–4 people. Except for cooperative, the design tasks or challenges employed in the empirical studies were also characterized as real situated, open-ended, and challenging. Moreover, conventional tools or materials were used more often, technical tools and materials were relatively lacking. Overall, at present, the development of design thinking intervention in K-12 education is not sufficient. We are inspired to pay more attention to the application of rigorous study design, balance between formal and informal classrooms, intensive grouping strategies, applicative tasks or challenges, and technical materials in the future. For instruction, because the situations vary among schools, the DTIL intervention should vary accordingly. Moreover, in the selection of design materials or tools, it is also necessary to pay attention to the application of technical materials or tools, to help students understand the technological world in which they live and learn to use technology to innovate and shape the world.

6. Limitations

As this study mainly focuses on students' learning in a design thinking integrated context of K-12 education, there are some limitations. First, we selected the reviewed papers only from SSCI journals, which ensured the representativeness and quality of sample but lacked comprehensiveness. Second, the school levels were limited to preschool to high school, while colleges and other adult learners were not included, so that the conclusions are only applicable to K-12 education. Third, teachers were considered crucial for learning; separate study is needed to amply elaborate the role of teachers in DTIL. In future, it is expected that more research on teaching, learning and evaluation of DTIL will be carried.

Author Contributions: Conceptualization, T.L. and Z.Z.; methodology, T.L. and Z.Z.; writing—original draft preparation, T.L.; writing—review and editing, T.L. and Z.Z. All authors have read and agreed to the published version of the manuscript.

Funding: This research was financially supported by Ministry of Education in China Project of Humanities and Social Sciences "The Collaborative Innovation Mechanism for Guangdong-Hong Kong-Macao Greater Bay Area Education based on C-STEAM" (22YJC881901), the Major Project of Social Science in South China Normal University (ZDPY2208), the Major basic research and applied research projects of Guangdong Education Department (#2017WZDXM004). The author gratefully acknowledges the support of "2022 Guangdong-Hong Kong-Macao Greater Bay Area Exchange Programs of SCNU".

Institutional Review Board Statement: Not applicable.

Informed Consent Statement: Not applicable.

Data Availability Statement: Not applicable.

Conflicts of Interest: The authors declare no conflict of interest.

Appendix A

Table A1. General information of the papers.

Paper	Sample Size	Educational Level	Duration	Curriculum	Design Thinking Model (Competence)
Aflatoony, Wakkary, Neustaedter (2018) [33]	39	high school	9 weeks	design and technology	A five-step design process including empathise, define, ideate, prototype and test
Aranda, Lie, Guzey (2020) [54]	26	middle school	6 days	science	Stage 1: plan a design: cognitive memory, divergent thinking, evaluative thinking. Stage 2: redesign: evaluative thinking. Stage 3: communicate to the client: cognitive memory, divergent thinking
Blom and Bogaers (2020) [46]	18	middle school	2 h	technology	N/A
Carroll et al. (2010) [38]	24	middle school	3 weeks	geography	The components of the design thinking process include the following: understand, observe, point of view, ideate, prototype, test
Chin et al. (2019) [68]	197	middle school	5 weeks	multidisciplinary	N/A
Christensen et al. (2019) [40]	246	middle school	2 years	design	Six phases: (1) design problem; (2) field studies; (3) ideation; (4) fabrication; (5) argumentation; (6) reflection
Cutumisu, Chin, Schwartz (2019) [60]	97	middle school	N/A	multidisciplinary	N/A

Table A1. *Cont.*

Paper	Sample Size	Educational Level	Duration	Curriculum	Design Thinking Model (Competence)
Cutumisu, Schwartz, Lou (2020) [65]	80	middle school	5 weeks	multidisciplinary	N/A
Derler et al. (2020) [55]	117	high school	2 years	multidisciplinary	Three phases: (1) exploration, (2) product ideation, and (3) product prototyping and optimisation
English (2019) [21]	34	elementary school	N/A	STEM	The inclusion of design processes involved students in learning through design involving planning, sketching, and testing
Fan, Yu, Lou (2018) [43]	103	high school	15 weeks	technology and engineering	Clarify problems and constraints, collect information, develop possible solutions, predictive analysis, selection solutions, modelling, and testing, evaluating and revising, optimization
Fleer (2021) [64]	13	kindergarten	7 weeks	engineering	Designing, making, appraising
Forbes et al. (2021) [27]	576	Kindergarten and Primary School	N/A	STEM	Five phases of discovery, interpretation, ideation, experimentation, and evolution to scaffold the 'design process' (The IDEO model)
Gennari, Melonio, Rizvi (2021) [49]	8	middle school	N/A	technology	(1) exploration and familiarisation, (2) ideation and conceptualisation, (3) programming and prototyping
Gomoll et al. (2018) [61]	16	middle school	5 weeks	robotics	Ask questions (define problem), imagine (brainstorm ideas), collect information,develop, and test solutions, improve (how did this work? How can we make it better?)
Guzey and Jung (2021) [34]	27	middle school	15 days	science	Cognitive memory, convergent thinking, divergent thinking, evaluative thinking
Kelley and Sung (2017) [47]	91	elementary school	1 year	science	N/A
Kelley, Capobianco, Kaluf (2015) [59]	21	elementary school	1–2 weeks	science	Cognitive processes identified by Halfin's (1973) study of high-level designers: Analysing, computing, defining problem(s), designing, interpreting data, modelling, predicting, questions/hypotheses, testing.
Kijima and Sun (2021) [32]	26	middle school	3 days	STEAM	Five iterative stages (Stanford d.school), including: empathise, define (capture needs), ideate (brainstorm solutions), prototype and test (seek feedback)
Kijima, Yang-Yoshihara, Maekawa (2021) [50]	97	middle and high school	3 days	STEAM	Five stages of—from empathy-building to needs—finding, brainstorming, prototyping, and testing
Kim, Seo, Kim, (2022) [52]	28	elementary school	3 weeks	engineering	Explore, empathize, ideate, create, test

Table A1. *Cont.*

Paper	Sample Size	Educational Level	Duration	Curriculum	Design Thinking Model (Competence)
Ladachart et al. (2021) [51]	38	middle school	4 weeks	STEM	Six aspects of design thinking, namely (1) collaboratively working with diversity, (2) being confident and optimistic to use creativity, (3) orientation to learning by making and testing, (4) mindfulness to process and impacts on others, (5) being comfortable with uncertainty and risks, and (6) human-centeredness
Leinonen et al. (2020) [57]	64	elementary school	2 months	multidisciplinary	N/A
Lin et al. (2020) [48]	62	middle school	7 weeks	technology	Three phases: Inspiration: real-world problem, ideation: design scheme, implementation: digital work
Lin, Chang, Li (2020) [31]	169	middle school	8 weeks	engineering	Five stages of engineering design thinking (empathize, define, ideate, prototype, and test)
Marks and Chase (2019) [35]	78	middle school	4 weeks	science	Iterative make-test-think process
Marks and Chase (2019) [35]	89	middle school	3 weeks	science	Iterative make-test-think process
Mentzer, Becker, Sutton (2015) [42]	59	high school	3 h	engineering	8 stages: problem definition, information gathering, idea generation, modelling, feasibility, evaluation, decision making, communication
Mentzer, Huffman, Thayer (2014) [41]	20	high school	3 h	engineering	Problem definition, gathering information, generating ideas, modelling, feasibility, evaluation, decision, communication
Nichols et al. (2021) [36]	159	elementary school	12 weeks	science	Defining, designing, producing, evaluating
Parikh, Maddulety, Meadows (2020) [53]	70	middle school	5 months	N/A	N/A
Rao, Puranam, Singh (2022) [23]	195	middle school	4 days	science	Four stages called 'feel', 'imagine', 'do' and 'share'
Simeon, Samsudin, Yakob (2022) [37]	89	high school	3 months	science	The five-stage model proposed by the Hasso Plattner Institute of Design at Stanford: empathy, define (the problem), ideate, prototype, and test
Sung and Kelley (2019) [44]	27	elementary school	1 year	science	Identify problem, share and develop a plan, create, and test, communicate results and gather feedback, improve and retest
Tsai and Wang (2021) [28]	350	middle school	N/A	STEAM	Four phases: empathize, define, ideate and prototype
Van Mechelen et al. (2019) [58]	49	elementary school	1 day	N/A	The Collaborative Design Thinking (CoDeT): introduction, sensitizing, scaffolding collaboration, defining a design goal, reflection on collaboration, ideation, grouping and selection, elaboration through making, presentation and peer jury, iteration or wrap up

Table A1. *Cont.*

Paper	Sample Size	Educational Level	Duration	Curriculum	Design Thinking Model (Competence)
Wendell, Wright, Paugh (2017) [62]	N/A	elementary school	1 day	science	Reflective decision making: articulate multiple solutions, evaluate pros and cons, intentionally select solution, retell performance of solution, analyse solution according to specific evidence, purposefully choose improvements
Won et al. (2015) [56]	44	middle school	N/A	STEM	Articulation of the learning phenomenon, design, data collection, actual construction, redesign
Yalcin and Erden (2021) [29]	39	kindergarten	8 weeks	STEM	Five stages used by the Hasso Platter Institute at Stanford: Empathize, Define, Ideate, Prototype, Test
Yu, Wu, Fan (2020) [26]	613	high school	16 weeks	engineering	Observing, predicting, creating, analysing, and evaluating
Zhang et al. (2022) [30]	30	middle school	3 months	Design & Research	Empathize Design User (EDU), Define Design Problem (DDP), Ideate Design Solution (IDS), Make Prototype (MP), Test Prototype (TP)
Zhou et al. (2017) [39]	24	middle school	2 weeks	science and engineering	Nine coding categories: sketching, prototyping, design goals, inference/predictions about design, generate design ideas, design of structure, design of system/process, materials, and collaboration
Zhou et al. (2021) [45]	27	middle school	2 weeks	engineering	Design cycle of planning, building, and testing
Zupan, Cankar, Cankar (2018) [66]	146	elementary school	17.5 weeks	N/A	The process was divided into five interrelated phases: understanding and defining the problem, observation, ideation, prototyping and testing, implementation

Note: N/A—Not available.

Table A2. Study design and intervention.

Paper	Course Type	Task/Challenge	Tools and Materials	Grouping	Study Design	Study Type	Dependent Variable	Measurement Instrument
Aflatoony, Wakkary, Neustaedter (2018) [33]	formal	To use design to change their communities	N/A	3–4	Non-experimental: design thinking based pedagogy in the context of interaction design	O X O	Design thinking	Observation, survey, portfolio
Aranda, Lie, Guzey (2020) [54]	formal	Design a process to both prevent and test for cross-pollination of non-GMO fields from GMO fields. (Genetically Modified Organisms, GMOs)	N/A	4	Non-experimental: engineering design as a tool to improve student science learning	X O	Skill	Protocol analysis
Blom and Bogaers (2020) [46]	N/A	Design a heat retaining food container for street food vendors at a taxi depot	Basic stationary items, including pens, pencils, safety rulers, post-it notes, coloured pencils, paper, paper clips, felt-tip pens and highlighters.	3	Non-experimental: STEM design task	X O	Design thinking	Protocol analysis, video recording
Carroll et al. (2010) [38]	formal	To identify and redesign systems that existed at the school	N/A	4–5	Non-experimental: introducing students both to the design process and to systems in geography	X O	Design thinking, emotional/ social aspect	Journal, audio recording, video recording, portfolio, interview
Chin et al. (2019) [68]	formal	Design digital posters	Computer design games	1	Experimental: EG1: Feedback design-thinking strategies treatment; EG2: Explore design-thinking strategies treatment	R O X1 O R O X2 O	Design thinking	Test
Christensen et al. (2019) [40]	formal	To create a secure environment for the elderly without taking away their freedom	Digital technology	N/A	Quasi-experimental: EG: students who had already received design education in their school (FabLab group); CG: without intervention	X1 O X2 O	Design thinking, emotional/ social aspect	Questionnaire
Cutumisu, Chin, Schwartz (2019) [60]	formal	Design digital posters	Computer design games	1	Non-experimental: digital poster design game (Posterlet)	X O	Design thinking, subject learning performance	Test, design work evaluation
Cutumisu, Schwartz, Lou (2020) [65]	formal	Design digital posters	Computer design games	1	Non-experimental: digital poster design game (Posterlet)	O X O	Design thinking	Test

Table A2. *Cont.*

Paper	Course Type	Task/Challenge	Tools and Materials	Grouping	Study Design	Study Type	Dependent Variable	Measurement Instrument
Derler et al. (2020) [55]	formal	To develop sustainable food products for peer group	N/A	Group	Non-experimental: Project-Based Learning focused on the development of sustainable food products	X O	Design thinking	Portfolio, journal, survey, interview
English (2019) [21]	formal	Design and construct shoes	N/A	3	Non-experimental: problem solving activities (shoes design)	X O	Design thinking	Portfolio, protocol analysis
Fan, Yu, Lou (2018) [43]	formal	To design a movable toy with various mechanical structure types	Physical and 3D virtual simulation models, LEGO	N/A	Non-experimental: project-based engineering design program	O X O	Subject learning performance, design thinking, emotional/ social aspect	Test, questionnaire, design work evaluation, survey, observation
Fleer (2021) [64]	formal	Design castle according to an imaginary engineering situation of Sherwood forest	N/A	Individual and collective	Non-experimental: designerly play	X O	Design thinking	Observation
Forbes et al. (2021) [27]	formal	3D design and printing: Floatable boats, shadow puppets, Headphone cable holders, Spinning tops, Playground sculptures, Habitat for hermit crabs, Herb markers, Designing keyrings, Bag tags	3D design and printing technologies (Ipad and 3D design software and print device)	N/A	Non-experimental: STEM-focussed curricula in 3D technology based makerspace	X O	Subject learning performance, design thinking, skill, emotional/social aspect	Interview, survey, journal, observation
Gennari, Melonio, Rizvi (2021) [49]	N/A	Generate smart-things ideas for an outdoor park environment	Card-based toolkits, microelectronics components: Raspberry Pi, Google's Design Sprint Kit	N/A	Non-experimental: IoT design workshop	X O	Design thinking, emotional/ social aspect	Portfolio, questionnaire, observation, interview
Gomoll et al. (2018) [61]	formal	Design a robot that served a need in their local environment and allowed remote peers to explore their local spaces	A mobile telepresence robot that we called KT, controlled over the Internet through a web interface	4	Non-experimental: human-centred robotics curriculum	X O	Design thinking	Portfolio, audio recording, video recording

Table A2. *Cont.*

Paper	Course Type	Task/Challenge	Tools and Materials	Grouping	Study Design	Study Type	Dependent Variable	Measurement Instrument
Guzey and Jung (2021) [34]	formal	Design a water filter system for the city's wastewater management plant to help prevent the pollution of a local river	N/A	3–4	Non-experimental: engineering design task in teams	O X O	Skill, subject learning performance	Audio recording, protocol analysis, test
Kelley and Sung (2017) [47]	formal	3 engineering design activities: Musical Instrument, Simple Machines, and bio-inspired flower	N/A	2–4	Quasi-experimental: EG1: pretreatment on basic engineering design sketching strategies before the three design activities; EG2: delayed treatment before the third design activity	X1 O X2 O	Design thinking	Design work evaluation
Kelley, Capobianco, Kaluf (2015) [59]	formal	To work in teams to build a prototype for a prosthetic leg to function like a human leg joint and strike the ball; paper football kicker	N/A	3	Non-experimental: engineering design activity	X O	Design thinking	Protocol analysis, video recording
Kijima and Sun (2021) [32]	N/A	Work revolved around interviewing senior citizens and creating prototypes that met their needs on the background of Japan's aging society	N/A	N/A	Non-experimental: design thinking workshop	O X O	Emotional/ social aspect	Survey
Kijima, Yang-Yoshihara, Maekawa (2021) [50]	N/A	Design local solutions addressing global issues	Using basic prototyping materials such as recycled plastic bottles and cardboards, glue, tapes, scissors,	4	Non-experimental: design thinking and STEAM workshop	O X O	Emotional/ social aspect	Questionnaire, interview
Kim, Seo, Kim, (2022) [52]	formal	Reading the narrative content of the books and solving the engineering problems presented in the books, and finally rewrite the story	COBL-S (Arduino Leonardo-based device, supported programming language was developed based on Scratch and app inventor)	3–4	Quasi-experimental: EG: Class activities according to the NE-Maker instructional model; CG: Normal software education class according to the textbook.	O X1 O O X2 O	Emotional/ social aspect	Questionnaire, journal, interview

Table A2. *Cont.*

Paper	Course Type	Task/Challenge	Tools and Materials	Grouping	Study Design	Study Type	Dependent Variable	Measurement Instrument
Ladachart et al. (2021) [51]	formal	Reverse engineering project: to design a bimetal thermostat	A dissected bimetal thermostat, metal, tape, and scissors	3	Non-experimental: design-based reverse engineering	O X O	Emotional/ social aspect	Questionnaire, video recording, protocol analysis
Leinonen et al. (2020) [57]	formal	3D model design and 3D artefact printing	3D model design by Tinkercad software and 3D artefact printing by Ultimaker printer	group	Non-experimental: 3D design and printing activities	X O	Subject learning performance, skill, design thinking	Observation, interview, questionnaire, portfolio
Lin et al. (2020) [48]	formal	Design digital documents for new year party (e.g., to make posters for party promotion)	WPS Writer®	group	Quasi-experiment: EG: using the design thinking approach (class a: project); CG: using traditional teaching methods (class b, according to the textbook)	O X1 O O X2 O	Subject learning performance	Design work evaluation
Lin, Chang, Li (2020) [31]	formal	Design an electric model vehicle capable of automatically avoiding obstacles was developed	The experimental group experienced design teaching with VR devices used as teaching tools	N/A	Experimental: EG: engineering design teaching with VR CG: conventional engineering design teaching	R O X1 O R O X2 O	Design thinking	Survey, design work evaluation
Marks and Chase (2019) [35]	formal	Drop challenge, playground challenge, and a post-design challenge (the boat challenge)	N/A	N/A	Experimental: EG: iterative prototyping (Prototype); CG: content-focused design (Content)	R O X1 O R O X2 O	Design thinking, emotional/ social aspect	Test, survey, design work evaluation
Marks and Chase (2019) [35]	formal	Base-line tower design task,drop challenge, playground challenge, and a post-design challenge (the boat challenge)	N/A	N/A	Experimental: EG: design thinking intervention focused on effective iterative prototyping (Prototype); CG: content-focused intervention (Content)	R O X1 O R O X2 O	Design thinking, emotional/ social aspect	Test, survey, design work evaluation

Table A2. *Cont.*

Paper	Course Type	Task/Challenge	Tools and Materials	Grouping	Study Design	Study Type	Dependent Variable	Measurement Instrument
Mentzer, Becker, Sutton (2015) [42]	formal	To design a playground	A calculator, ruler, a small note pad, graph paper, white paper, pencil, highlighter, sticky notes, and a piece of paper identifying the design task were placed on the table before the student entered the room	N/A	Quasi-experimental: EG1: high school freshmen starting the sequence of engineering courses; EG2: high school seniors who had taken multiple engineering courses; CG: engineering experts	X1 O X2 O X3 O	Design thinking	Audio recording, video recording, protocol analysis, portfolio
Mentzer, Huffman, Thayer (2014) [41]	formal	Playground design	N/A	1	Non-experimental: Engineering design challenge	X O	Design thinking, Emotional/social aspect	Observation, protocol Analysis, video recording, audio recording, portfolio, interview, survey
Nichols et al. (2021) [36]	formal	To design and construct a device that is engineered to provide electricity to a third world community scenario	Materials like LED, water	2–4	Quasi-experimental: EG: design task embedded in an inquiry science unit and a community of inquiry (CoI); CG: design task embedded in an inquiry science unit (Non-CoI)	O X1 O O X2 O	Design thinking, subject learning performance	Protocol analysis, video recording, test, interview
Parikh, Maddulety, Meadows (2020) [53]	informal	Design prototypes for solving a Design Thinking challenge	N/A	5–6	Quasi-experiment: EG: Design Thinking training spread over two action research cycles; CG: received no intervention	O X1 O X2 O O X3 O X4 O	Skill	Portfolio, test
Rao, Furanam, Singh (2022) [23]	informal	Three key design thinking exercises: 'Bag Exercise', 'Cartographer', 'Be a Detective'	N/A	N/A	Experimental: EG: design thinking training programme; CG: usual hands-on science education curriculum	R X1 O R X2 O	Skill	Test

Table A2. *Cont.*

Paper	Course Type	Task/Challenge	Tools and Materials	Grouping	Study Design	Study Type	Dependent Variable	Measurement Instrument
Simeon, Samsudin, Yakob (2022) [37]	informal	Zip line delivery challenge, truss bridge challenge	N/A	N/A	Non-experimental: STEM-Design thinking modules	O X O	Subject learning performance	Test
Sung and Kelley (2019) [44]	formal	Design a Doggie Door Alarm	Normal design tools, such as paper, pencil	3	Non-experimental: engineering design activity for science learning	X O	Design thinking	Audio recording, video recording, protocol analysis
Tsai and Wang (2021) [28]	formal	Design a robot for solving some problems related to natural science or ecological environmental issues	N/A	N/A	Non-experimental: project-based STEAM curriculum	X O	Emotional/social aspect	Questionnaire
Van Mechelen et al. (2019) [58]	formal	The design theme on preventing bullying in the social context of the class	N/A	4–6	Non-experimental: design activities on the theme of preventing bullying in the social context of the class	X O	Emotional/social aspect, design thinking	Observation, portfolio, design work evaluation
Wendell, Wright, Paugh (2017) [62]	formal	Design water filters, bridges, circuits in, maglev vehicles and windmills, and pollinators and knee braces	N/A	2–3	Non-experimental: engineering design tasks	X O	Design thinking	Video recording, portfolio, protocol analysis
Won et al. (2015) [56]	informal	Design of lights powered through motion	Social media technologies	N/A	Non-experimental: integrating learning technologies such as social networking forum (SNF) into design-based learning activities	X O	Design thinking	Survey, portfolio
Yalcin and Erden (2021) [29]	formal	Design thinking STEM activities	N/A	4	Experimental: EG: design thinking STEM activities CG: non-STEM activities	R O X1 O R O X2 O	Skill	Survey, journal
Yu, Wu, Fan (2020) [26]	formal	Design mechanical toy	The design stage used computer-aided design to create a three-dimensional model of the toy	N/A	Non-experimental: engineering project	X O	Subject learning performance, design thinking, skill	Questionnaire, design work evaluation, portfolio

Table A2. *Cont.*

Paper	Course Type	Task/Challenge	Tools and Materials	Grouping	Study Design	Study Type	Dependent Variable	Measurement Instrument
Zhang et al. (2022) [30]	formal	Design an escape room for the local fire department to allow participants to playfully and interactively improve awareness of fire safety in and around the house	N/A	3–4	Non-experimental: DBL (design-based learning) activities	X O	Design thinking, emotional/social aspect	Questionnaire, observation, interview
Zhou et al. (2017) [39]	informal	A total of five toy design activities	N/A	3–4	Non-experimental: toy design workshop	O X O	Emotional/social aspect, design thinking	Questionnaire, survey
Zhou et al. (2021) [45]	informal	Marshmallow tower activity and the trebuchet design activity	N/A	3–4	Non-experimental: design workshop	X O	Design thinking	Observation, design work evaluation
Zupan, Cankar, Cankar (2018) [66]	formal	Identify and define a local or social problem that could be solved with a new product, service, or other solution	N/A	N/A	Non-experimental: use the design thinking method to develop the entrepreneurial mindset	X O	Emotional/social aspect	Interview, observation

Note: N/A—Not available.

References

1. Kimbell, L. Rethinking Design Thinking: Part I. *Des. Cult.* **2011**, *3*, 285–306. [CrossRef]
2. Micheli, P.; Wilner, S.J.; Bhatti, S.H.; Mura, M.; Beverland, M.B. Doing design thinking: Conceptual review, synthesis, and research agenda. *J. Prod. Innov. Manag.* **2019**, *36*, 124–148. [CrossRef]
3. Simon, H.A. *The Sciences of the Artificial*; MIT Press: Cambridge, MA, USA, 1969.
4. Cross, N. Expertise in design: An overview. *Des. Stud.* **2004**, *25*, 427–441. [CrossRef]
5. Jonassen, D.H. Toward a design theory of problem solving. *Educ. Technol. Res. Dev.* **2000**, *48*, 63–85. [CrossRef]
6. Dorst, K. The core of 'design thinking'and its application. *Des. Stud.* **2011**, *32*, 521–532. [CrossRef]
7. Brown, T. Design thinking. *Harv. Bus. Rev.* **2008**, *86*, 84–92. Available online: https://pubmed.ncbi.nlm.nih.gov/18605031/ (accessed on 9 June 2022).
8. Kelley, T.; Kelley, D. *Creative Confidence: Unleashing the Creative Potential Within Us All*; Crown Business: New York, NY, USA, 2013.
9. Martin, R.; Martin, R.L. *The Design of Business: Why Design Thinking is the Next Competitive Advantage*; Harvard Business Press: Boston, MA, USA, 2009.
10. Brenner, W.; Uebernickel, F.; Abrell, T. Design thinking as mindset, process, and toolbox. In *Design Thinking for Innovation*; Springer: Berlin, Germany, 2016; pp. 3–21. [CrossRef]
11. Rusmann, A.; Ejsing-Duun, S. When design thinking goes to school: A literature review of design competences for the K-12 level. *Int. J. Technol. Des. Educ.* **2021**. [CrossRef]
12. Zhang, F.; Markopoulos, P.; Bekker, T. Children's Emotions in Design-Based Learning: A Systematic Review. *J. Sci. Educ. Technol.* **2020**, *29*, 459–481. [CrossRef]
13. Zhou, D.; Gomez, R.; Wright, N.; Rittenbruch, M.; Davis, J. A design-led conceptual framework for developing school integrated STEM programs: The Australian context. *Int. J. Technol. Des. Educ.* **2020**, *32*, 383–411. [CrossRef]
14. International Technology and Engineering Educators Association. Standards for Technological and Engineering Literacy: The Role of Technology and Engineering in STEM Education. Available online: https://www.iteea.org/STEL.aspx (accessed on 9 June 2022).
15. Johansson-Sköldberg, U.; Woodilla, J.; Çetinkaya, M. Design thinking: Past, present and possible futures. *Creat. Innov. Manag.* **2013**, *22*, 121–146. [CrossRef]
16. Razzouk, R.; Shute, V. What is design thinking and why is it important? *Rev. Educ. Res.* **2012**, *82*, 330–348. [CrossRef]
17. Hasso Plattner Institute of Design at Stanford University. Design Thinking Bootleg. Available online: https://dschool.stanford.edu/resources/design-thinking-bootleg (accessed on 9 June 2022).
18. IDEO. Design Thinking for Educators Toolkit (2nd ed.). Available online: https://designthinkingforeducators.com/toolkit/ (accessed on 9 June 2022).
19. Design Council. The Double Diamond: A Universally Accepted Depiction of the Design Process. Available online: https://www.designcouncil.org.uk/our-work/news-opinion/double-diamond-universally-accepted-depiction-design-process/ (accessed on 9 June 2022).
20. Burghardt, M.D.; Hacker, M. Informed design: A contemporary approach to design pedagogy as the core process in technology. *Technol. Teach.* **2004**, *64*, 6–8.
21. English, L.D. Learning while designing in a fourth-grade integrated STEM problem. *Int. J. Technol. Des. Educ.* **2018**, *29*, 1011–1032. [CrossRef]
22. Chase, C.C.; Malkiewich, L.; Kumar, A.S. Learning to notice science concepts in engineering activities and transfer situations. *Sci. Educ.* **2018**, *103*, 440–471. [CrossRef]
23. Rao, H.; Puranam, P.; Singh, J. Does design thinking training increase creativity? Results from a field experiment with middle-school students. *Innovation* **2021**, *24*, 315–332. [CrossRef]
24. Tranfield, D.; Denyer, D.; Smart, P. Towards a Methodology for Developing Evidence-Informed Management Knowledge by Means of Systematic Review. *Br. J. Manag.* **2003**, *14*, 207–222. [CrossRef]
25. Wohlin, C. Guidelines for snowballing in systematic literature studies and a replication in software engineering. In Proceedings of the 18th international conference on evaluation and assessment in software engineering, London, UK, 13 May 2014; pp. 1–10. [CrossRef]
26. Yu, K.C.; Wu, P.H.; Fan, S.C. Structural Relationships among High School Students' Scientific Knowledge, Critical Thinking, Engineering Design Process, and Design Product. *Int. J. Sci. Math. Educ.* **2020**, *18*, 1001–1022. [CrossRef]
27. Forbes, A.; Falloon, G.; Stevenson, M.; Hatzigianni, M.; Bower, M. An Analysis of the Nature of Young Students' STEM Learning in 3D Technology-Enhanced Makerspaces. *Early Educ. Dev.* **2020**, *32*, 172–187. [CrossRef]
28. Tsai, M.-J.; Wang, C.-Y. Assessing Young Students' Design Thinking Disposition and Its Relationship With Computer Programming Self-Efficacy. *J. Educ. Comput. Res.* **2020**, *59*, 410–428. [CrossRef]
29. Yalcin, V.; Erden, S. The Effect of STEM Activities Prepared According to the Design Thinking Model on Preschool Children's Creativity and Problem-Solving Skills. *Think. Ski. Creat.* **2021**, *41*, 100864. [CrossRef]
30. Zhang, F.; Markopoulos, P.; Bekker, T.; Paule-Ruíz, M.; Schüll, M. Understanding design-based learning context and the associated emotional experience. *Int. J. Technol. Des. Educ.* **2020**, *32*, 845–882. [CrossRef]
31. Lin, H.-C.; Chang, Y.-S.; Li, W.-H. Effects of a virtual reality teaching application on engineering design creativity of boys and girls. *Think. Ski. Creat.* **2020**, *37*, 100705. [CrossRef]

32. Kijima, R.; Sun, K.L. 'Females Don't Need to be Reluctant': Employing Design Thinking to Harness Creative Confidence and Interest in STEAM. *Int. J. Art Des. Educ.* **2020**, *40*, 66–81. [CrossRef]

33. Aflatoony, L.; Wakkary, R.; Neustaedter, C. Becoming a Design Thinker: Assessing the Learning Process of Students in a Secondary Level Design Thinking Course. *Int. J. Art Des. Educ.* **2017**, *37*, 438–453. [CrossRef]

34. Guzey, S.S.; Jung, J.Y. Productive Thinking and Science Learning in Design Teams. *Int. J. Sci. Math. Educ.* **2021**, *19*, 215–232. [CrossRef]

35. Marks, J.; Chase, C.C. Impact of a prototyping intervention on middle school students' iterative practices and reactions to failure. *J. Eng. Educ.* **2019**, *108*, 547–573. [CrossRef]

36. Nichols, K.; Musofer, R.; Fynes-Clinton, L.; Blundell, R. Design thinking and inquiry behaviours are co-constituted in a community of inquiry middle years' science classroom context: Empirical evidence for design thinking and pragmatist inquiry interconnections. *Int. J. Technol. Des. Educ.* **2021**. [CrossRef]

37. Simeon, M.I.; Samsudin, M.A.; Yakob, N. Effect of design thinking approach on students' achievement in some selected physics concepts in the context of STEM learning. *Int. J. Technol. Des. Educ.* **2020**, *32*, 185–212. [CrossRef]

38. Carroll, M.; Goldman, S.; Britos, L.; Koh, J.; Royalty, A.; Hornstein, M. Destination, Imagination and the Fires Within: Design Thinking in a Middle School Classroom. *Int. J. Art Des. Educ.* **2010**, *29*, 37–53. [CrossRef]

39. Zhou, N.; Pereira, N.; George, T.T.; Alperovich, J.; Booth, J.; Chandrasegaran, S.; Tew, J.D.; Kulkarni, D.M.; Ramani, K. The Influence of Toy Design Activities on Middle School Students' Understanding of the Engineering Design Processes. *J. Sci. Educ. Technol.* **2017**, *26*, 481–493. [CrossRef]

40. Christensen, K.S.; Hjorth, M.; Iversen, O.S.; Smith, R.C. Understanding design literacy in middle-school education: Assessing students' stances towards inquiry. *Int. J. Technol. Des. Educ.* **2018**, *29*, 633–654. [CrossRef]

41. Mentzer, N.; Huffman, T.; Thayer, H. High school student modeling in the engineering design process. *Int. J. Technol. Des. Educ.* **2014**, *24*, 293–316. [CrossRef]

42. Mentzer, N.; Becker, K.; Sutton, M. Engineering Design Thinking: High School Students' Performance and Knowledge. *J. Eng. Educ.* **2015**, *104*, 417–432. [CrossRef]

43. Fan, S.C.; Yu, K.C.; Lou, S.J. Why do students present different design objectives in engineering design projects? *Int. J. Technol. Des. Educ.* **2018**, *28*, 1039–1060. [CrossRef]

44. Sung, E.; Kelley, T.R. Identifying design process patterns: A sequential analysis study of design thinking. *Int. J. Technol. Des. Educ.* **2018**, *29*, 283–302. [CrossRef]

45. Zhou, N.; Pereira, N.; Chandrasegaran, S.; George, T.T.; Booth, J.; Ramani, K. Examining Middle School Students' Engineering Design Processes in a Design Workshop. *Res. Sci. Educ.* **2019**, *51* (Suppl. 2), 617–646. [CrossRef]

46. Blom, N.; Bogaers, A. Using Linkography to investigate students' thinking and information use during a STEM task. *Int. J. Technol. Des. Educ.* **2018**, *30*, 1–20. [CrossRef]

47. Kelley, T.R.; Sung, E. Sketching by design: Teaching sketching to young learners. *Int. J. Technol. Des. Educ.* **2017**, *27*, 363–386. [CrossRef]

48. Lin, L.; Shadiev, R.; Hwang, W.-Y.; Shen, S. From knowledge and skills to digital works: An application of design thinking in the information technology course. *Think. Ski. Creat.* **2020**, *36*, 100646. [CrossRef]

49. Gennari, R.; Melonio, A.; Rizvi, M. From children's ideas to prototypes for the internet of things: A case study of cross-generational end-user design. *Behav. Inf. Technol.* **2021**. [CrossRef]

50. Kijima, R.; Yang-Yoshihara, M.; Maekawa, M.S. Using design thinking to cultivate the next generation of female STEAM thinkers. *Int. J. STEM Educ.* **2021**, *8*, 1–15. [CrossRef]

51. Ladachart, L.; Cholsin, J.; Kwanpet, S.; Teerapanpong, R.; Dessi, A.; Phuangsuwan, L.; Phothong, W. Ninth-grade students' perceptions on the design-thinking mindset in the context of reverse engineering. *Int. J. Technol. Des. Educ.* 2021. [CrossRef]

52. Kim, J.-Y.; Seo, J.S.; Kim, K. Development of novel-engineering-based maker education instructional model. *Educ. Inf. Technol.* **2022**, *27*, 7327–7371. [CrossRef]

53. Parikh, C.; Maddulety, K.; Meadows, C. Improving creative ability of base of pyramid (BOP) students in India. *Think. Ski. Creativity* **2020**, *36*, 100652. [CrossRef]

54. Aranda, M.L.; Lie, R.; Guzey, S.S. Productive thinking in middle school science students' design conversations in a design-based engineering challenge. *Int. J. Technol. Des. Educ.* **2019**, *30*, 67–81. [CrossRef]

55. Derler, H.; Berner, S.; Grach, D.; Posch, A.; Seebacher, U. Project-Based Learning in a Transinstitutional Research Setting: Case Study on the Development of Sustainable Food Products. *Sustainability* **2019**, *12*, 233. [CrossRef]

56. Won, S.G.; Evans, M.A.; Carey, C.; Schnittka, C.G. Youth appropriation of social media for collaborative and facilitated design-based learning. *Comput. Hum. Behav.* **2015**, *50*, 385–391. [CrossRef]

57. Leinonen, T.; Virnes, M.; Hietala, I.; Brinck, J. 3D Printing in the Wild: Adopting Digital Fabrication in Elementary School Education. *Int. J. Art Des. Educ.* **2020**, *39*, 600–615. [CrossRef]

58. Van Mechelen, M.; Laenen, A.; Zaman, B.; Willems, B.; Abeele, V.V. Collaborative Design Thinking (CoDeT): A co-design approach for high child-to-adult ratios. *Int. J. Human-Computer Stud.* **2019**, *130*, 179–195. [CrossRef]

59. Kelley, T.R.; Capobianco, B.M.; Kaluf, K.J. Concurrent think-aloud protocols to assess elementary design students. *Int. J. Technol. Des. Educ.* **2014**, *25*, 521–540. [CrossRef]

60. Cutumisu, M.; Chin, D.B.; Schwartz, D.L. A digital game-based assessment of middle-school and college students' choices to seek critical feedback and to revise. *Br. J. Educ. Technol.* **2019**, *50*, 2977–3003. [CrossRef]
61. Gomoll, A.; Tolar, E.; Hmelo-Silver, C.E.; Šabanović, S. Designing human-centered robots: The role of constructive failure. *Think. Ski. Creat.* **2018**, *30*, 90–102. [CrossRef]
62. Wendell, K.B.; Wright, C.G.; Paugh, P. Reflective Decision-Making in Elementary Students' Engineering Design. *J. Eng. Educ.* **2017**, *106*, 356–397. [CrossRef]
63. Trochim, W.M.K.; Donnelly, J.P. Research Methods Knowledge Base (3rd ed.). Available online: http://www.socialresearchmethods.net/kb/ (accessed on 9 June 2022).
64. Fleer, M. The genesis of design: Learning about design, learning through design to learning design in play. *Int. J. Technol. Des. Educ.* **2022**, *32*, 1441–1468. [CrossRef]
65. Cutumisu, M.; Schwartz, D.L.; Lou, N.M. The relation between academic achievement and the spontaneous use of design-thinking strategies. *Comput. Educ.* **2020**, *149*, 103806. [CrossRef]
66. Zupan, B.; Cankar, F.; Cankar, S.S. The development of an entrepreneurial mindset in primary education. *Eur. J. Educ.* **2018**, *53*, 427–439. [CrossRef]
67. Buchanan, R. Wicked Problems in Design Thinking. *Des. Issues* **1992**, *8*, 5–21. [CrossRef]
68. Chin, D.B.; Blair, K.P.; Wolf, R.C.; Conlin, L.D.; Cutumisu, M.; Pfaffman, J.; Schwartz, D.L. Educating and Measuring Choice: A Test of the Transfer of Design Thinking in Problem Solving and Learning. *J. Learn. Sci.* **2019**, *28*, 337–380. [CrossRef]
69. Csikszentmihalyi, M. *Good Business: Leadership, Flow, and the Making of Meaning*; Viking: New York, NY, USA, 2004.
70. Wells, A. The importance of design thinking for technological literacy: A phenomenological perspective. *Int. J. Technol. Des. Educ.* **2012**, *23*, 623–636. [CrossRef]
71. International Technology Education Association. Standards for Technological Literacy. *Content for the Study of Technology (3rd ed.)*. Available online: https://www.iteea.org/42511.aspx (accessed on 9 June 2022).
72. Laal, M.; Ghodsi, S.M. Benefits of collaborative learning. *Procedia-Soc. Behav. Sci.* **2012**, *31*, 486–490. [CrossRef]
73. Wu, Q.; Lu, J.; Yu, M.; Lin, Z.; Zhan, Z. Teaching Design Thinking in a C-STEAM Project: A Case Study of developing the Wooden Arch Bridges' Intelligent Monitoring system. In Proceedings of the 13th International Conference on E-Education, E-Business, E-Management, and E-Learning (IC4E), Tokyo, Japan, 14–17 January 2022; pp. 280–285. [CrossRef]
74. De Bono, E. *New Thinking for the New Millennium*; Viking Adult: New York, NY, USA, 1999.
75. Australian Curriculum, Assessment, and Reporting Authority. The Australian Curriculum: Technology. Available online: https://www.australiancurriculum.edu.au/umbraco/Surface/Download/Pdf?subject=Digital%20Technologies&type=F10 (accessed on 9 June 2022).
76. Australian Curriculum, Assessment, and Reporting Authority. ACARA STEM Connections Project Report. Available online: https://www.australiancurriculum.edu.au/media/3220/stem-connections-report.pdf (accessed on 9 June 2022).
77. Peron, M.; Alfnes, E.; Sgarbossa, F. Learning Through Action: On the Use of Logistics4.0 Lab as Learning Developer. In Proceedings of the 7th European Lean Educator Conference, ELEC 2021, Online, 25–27 October 2021; Powell, D.J., Alfnes, E., Holmemo, M.D., Reke, E., Eds.; Springer: Trondheim, Norway, 2021; Volume 610, pp. 205–212. [CrossRef]
78. Zhan, Z.; Zhong, B.; Shi, X.; Si, Q.; Zheng, J. The design and application of IRobotQ3D for simulating robotics experiments in K-12 education. *Comput. Appl. Eng. Educ.* **2021**, *30*, 532–549. [CrossRef]
79. Simonetto, M.; Arena, S.; Peron, M. A methodological framework to integrate motion capture system and virtual reality for assembly system 4.0 workplace design. *Saf. Sci.* **2021**, *146*, 105561. [CrossRef]
80. Carlgren, L.; Rauth, I.; Elmquist, M. Framing Design Thinking: The Concept in Idea and Enactment. *Creativity Innov. Manag.* **2016**, *25*, 38–57. [CrossRef]
81. Buchanan, R. Human Dignity and Human Rights: Thoughts on the Principles of Human-Centered Design. *Des. Issues* **2001**, *17*, 35–39. [CrossRef]
82. Yan, L.; Xiaoying, M.; Jian, L.; Rui, W.; Lihong, M.; Guanxing, X.; Qiuling, G. Cultural Competence: Part I of the 5Cs Framework forTwenty-first Century Key Competences. *J. East China Norm. Univ. Educ. Sci.* **2020**, *38*, 29–44. [CrossRef]

 *applied sciences*

Article

Factors Influencing Student Satisfaction toward STEM Education: Exploratory Study Using Structural Equation Modeling

Jingbo Zhao [1,†], Tommy Tanu Wijaya [2,*,†], Mailizar Mailizar [3,*] and Akhmad Habibi [4]

[1] Key Laboratory of Data Science and Smart Education, Ministry of Education, Hainan Normal University, Haikou 571158, China
[2] School of Mathematical Sciences, Beijing Normal University, Beijing 100875, China
[3] Mathematics Education Department, Universitas Syiah Kuala, Banda Aceh 23111, Indonesia
[4] Fakultas Keguruan dan Ilmu Pendidikan, Universitas Jambi, Jambi 36122, Indonesia
* Correspondence: 202139130001@mail.bnu.edu.cn (T.T.W.); mailizar@unsyiah.ac.id (M.M.)
† These authors contributed equally to this work.

Abstract: Learning satisfaction has a relationship with student outcomes. Furthermore, this has prompted many governments to increasingly implement STEM education-based learning. Many studies have examined the improvement of STEM education by teachers. However, the studies have not analyzed STEM education's effect on students' learning satisfaction. Extending the planned behavior theory, this study aimed to predict high school students' learning satisfaction with STEM education. The questionnaire developed from the TPB model was filled out by 174 high school students in Indonesia. Furthermore, AMOS and SPSS 23 software were used for structural equation model analysis. The results showed that seven of twelve hypotheses were supported. Subjective norm and playfulness factors of STEM education positively relate to students' attitudes toward STEM education. Attitude is the most important factor influencing student satisfaction and acceptance toward STEM education. Therefore, this study provides a theoretical and practical contribution to improving learning satisfaction in technology-based STEM education.

Keywords: STEM education; senior high school; TPB; learning satisfaction

Citation: Zhao, J.; Wijaya, T.T.; Mailizar, M.; Habibi, A. Factors Influencing Student Satisfaction toward STEM Education: Exploratory Study Using Structural Equation Modeling. *Appl. Sci.* **2022**, *12*, 9717. https://doi.org/10.3390/app12199717

Academic Editor: Krzysztof Koszela

Received: 6 August 2022
Accepted: 24 September 2022
Published: 27 September 2022

Publisher's Note: MDPI stays neutral with regard to jurisdictional claims in published maps and institutional affiliations.

1. Introduction

The study of Science, Technology, Engineering, and Mathematics (STEM) education has increased in the last decade, showing its importance in improving students' abilities in the 21st century [1,2]. However, its implementation in Indonesia has many challenges. First, the teacher does not understand STEM education and how to implement it for students. Second, it combines interdisciplinary science, a challenge for schools and teachers accustomed to single-subject teaching. Other problems are class hours, communication on interdisciplinary science, teaching materials, and its relationship with Indonesia's standard curriculum [3].

Indonesia's learning curriculum is centralized, where the government monitors and evaluates teachers' school activities. The government has advised some schools in big cities to use STEM education-based learning. STEM education entered Indonesia in 2014, with Syiah Kuala University becoming one of the largest centers. In collaboration with SEAMEO and QITEP (https://www.qitepinscience.org/, accessed on 30 June 2022), the STEM education research center focuses on improving STEM education quality. They conduct various workshops and develop education-based learning to assist its implementation in schools. Examples of experiments include electroplating, aquaponics technology, electrical installation, and others related to technology and engineering from elementary to high school levels. Although STEM implementation has been running for more than five years, only a few studies measure its success [4–6]. The previous study primarily used cognitive learning performance to measure the quality of its implementation [7,8]. Additionally, several college studies have examined the effects of STEM education, though there is limited literature that focuses on K-12 students.

The success of implementing STEM education is measured in many ways. Previous studies used the perspective of student academic performance [9,10]. Other studies stated that academic performance is less effective in determining the success of implementing a new learning approach. Furthermore, several studies recommend analyzing student satisfaction as an alternative to successfully implementing a new learning model [11,12]. Student satisfaction reflects their perceptions of the learning experience with STEM education [13]. Student satisfaction is also an important outcome affecting motivation, ability, and the desire to participate in learning activities [14,15]. Furthermore, this study considered student satisfaction with STEM education as the dependent variable because it strongly relates to the perception of learning quality. Besides the importance of STEM education and its integration of various fields of science, unsatisfied students may be influenced by its difficulty to implement, impact, and inadequate facilities. Therefore, further studies should identify the predicted determinants related to student satisfaction with STEM education.

This study aimed to develop a new model to explore factors related to student satisfaction toward STEM education through attitude and acceptance. The following section elaborates on the theoretical background of predictors that may relate to student satisfaction with STEM education.

2. Literature Review

2.1. STEM Education in Indonesia

Common teaching and learning approaches used by teachers in Indonesia are problem-solving. Where learning activities are teacher-centered, students have less opportunities to participate. Furthermore, teachers rarely use technology-based learning media to explain the material. When STEM education was implemented in Indonesia, the shifting was established. Teachers with STEM education connect science, technology, engineering, and mathematics materials in instruction. The increase in the use of technology-based learning media influences pedagogical and technological knowledge.

STEM education was introduced in Indonesia by SEAQIS in 2015 through the 2015 IBSE-STEM Policy concerning cooperation agreements to organize training in STEM assisted by ATSE Australia. Cooperation between countries is encouraged to implement projects to strengthen the curriculum based on STEM education [9]. SEAQIS is one of the institutions appointed by UNESCO partners to make this program a success [16].

At the beginning of 2018, batch one STEM education training began to be pursued, containing teachers in West Java, Indonesia. In mid-2018, the achievements of technology-based experimental STEM products from batch one became the start and initial guideline for batch two training. This training is currently growing with the support and cooperation of PPPPTK IPA.

From August to October 2018, technology-based STEM experiments were implemented at various school levels. The experiments are mostly at the secondary school level, where students are more mature, careful, and knowledgeable than in elementary school. The experience of implementing teachers and product results are displayed in the STEM workshop. In 2019 and 2020, large-scale training was conducted to learn STEM-based education evenly distributed in Indonesia.

The Indonesian Ministry of Education monitors and supports collaborations and big plans to socialize STEM education [17]. It also hopes that the learning steps could be gradually combined with the national curriculum [9]. Therefore, teaching and learning activities in the future would be more scientific and fun, avoiding the lecture learning model mostly used by teachers [18,19].

Indonesia's STEM education training that produces pre-service and in-service teachers is technology- and project-based correlated with everyday life. Many studies have shown that project-based learning is student-centered and makes students more participative. This approach is suitable to be combined with STEM education, where learning is experiment-based. In line with this, several studies found that STEM education learning in the first week was filled with engineering design and theory cycles. It provided students with preparation for using STEM education experiments with technology, as well as discussing and formulating hypotheses about the product. The teacher provides brainstorming, experimental plans, and product goals before the scientific unit learning with engineering design begins.

Since 2014, several universities in Indonesia have been working to form lesson plans for experimental schools. Some studies conducted are the development of workbook-based STEM education, which improves students' abilities in every meeting, technology literacy, and outcomes. In contrast, this study focused more on the importance of student satisfaction in learning with STEM education than only technology literacy and outcomes.

After introducing the STEM education for more than five years and its implementation running in schools for approximately three years, it is necessary to understand students' acceptance, attitude, and satisfaction. Therefore, this study aimed to explore the factors related to students' attitudes and satisfaction toward STEM education. The results may help the Indonesian government develop STEM education-based learning. They could also provide deeper knowledge about important considerations when implementing learning in schools.

2.2. Theoretical Foundation and Hypotheses Development

This study proposed a comprehensive model to solve the problem and the need for further study on student satisfaction with STEM education. Figure 1 shows that the proposed model examines perceived usefulness, perceived convenience, facilitating conditions, subjective norm, playfulness, attitude, and collaborative learning. These predictors may affect student satisfaction with STEM education at the high school level. They are the combined results of previous studies derived from the theory of planned behavior (TPB) [20] and several theories about the acceptance model affecting student satisfaction [11,12,21]. The technology acceptance model theory was used because this study defined STEM education as a learning approach that integrates technology with other sciences. According to Zou [22] and Zobair [23], this theory could be used to analyze a person's intentions and acceptance of learning models related to new technology.

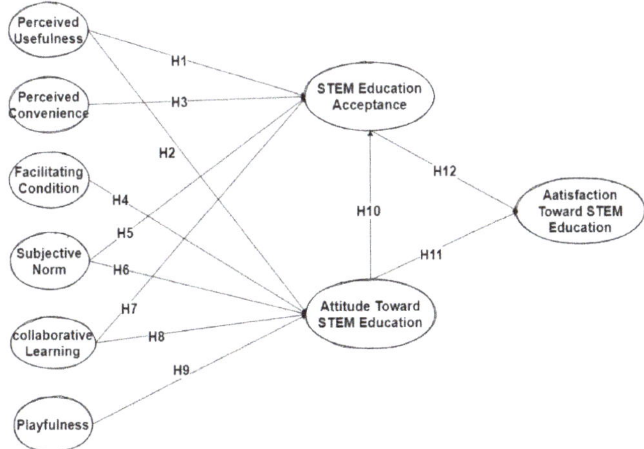

Figure 1. The proposed model with initial hypothesis.

2.3. Perceived Usefulness (PU)

PU with technology is how people believe that learning with technology-based STEM education helps them achieve learning goals. Previous studies found that PU positively and significantly relates to student attitudes and acceptance of the new learning approach [24,25]. Similarly, there is a positive relationship between PU and student satisfaction [26]. High school students may be highly satisfied with STEM education with technology-based learning when they believe technology helps them gain new knowledge.

2.4. Perceived Convenience

For students, determining convenient learning depends on time and effort [27]. Learning is convenient when it is fast, short, clear, and reduces the students' emotional and physical burden [28]. Joo [29] stated that convenience is similar to the TAM model's ease of use. It is felt by students when conducting STEM education to gain new knowledge.

From the self-determination theory perspective, perceived convenience is how people believe a model could help them achieve the desired goal [30]. Liao [31] found that perceived convenience affects one's motivation and intention to use technology-based learning models. Additionally, Yoon and Kim [31] extended TAM by adding a perceived convenience variable to analyze the acceptance of new technology.

2.5. Facilitating Conditions

Facilitating conditions (FC) are facilities and environmental factors affecting the perception of doing a task [32]. FC facilitates support, including technical assistance, teacher knowledge, and knowledge about STEM education. It could influence students' attitudes toward technology-based STEM education learning. According to Wijaya [33], technical support was the highest factor influencing teachers to implement new technology. Other studies also found FC related to the intention to use technology-based learning models [34–36]. Teo [37] stated that FC affects user satisfaction, and Ngai et al. [38] found that it affects attitudes toward computer use. Therefore, the FC factor may influence student satisfaction toward STEM education.

2.6. Subjective Norm

Subjective norm is the acceptance of social pressure to perform a behavior [39]. It is also a perception that important people support or do not support one's behavior [40]. In this study, the subjective norm is where people around students believe technology-based STEM education is important learning. Previous studies found that these norms affect technology acceptance and student attitudes [37,41]. Therefore, subjective norms relate to student attitudes and acceptance toward technology-based STEM education.

2.7. Playfulness

Perceived playfulness may also influence students' attitudes toward STEM education. Studies on technology show that user playfulness is the strongest determinant associated with using new technology-based learning approaches [42,43].

Perceived playfulness is the mindset with three dimensions, including how individuals feel their attention is focused on STEM education with technology experiments, curiosity when conducting experiments, and learning is fun and interesting. Students have a pleasant experience when conducting the experiments in groups. Experiment-based learning is sometimes liked by students and provides intrinsic motivation. Extrinsic motivation is the desire to do something because it is fun, clear, and valuable [44]. In contrast, intrinsic motivation is the desire to be seen in an activity driven by passion [45].

Perceived playfulness when using learning or objects is related to technology in previous studies. Davis [46] found that playfulness positively influences a person's attitude toward new systems and learning. Therefore, the higher the perceived playfulness students feel in learning technology-based STEM education, the better their attitude [47].

2.8. Attitude toward STEM Education

Attitude is the desire to do or use something and positive or negative feelings toward an action or system [48]. This study defined attitude toward STEM education as the level at which students have negative or positive feelings toward technology-based STEM education lessons. Previous studies found that attitude toward using embodies successful implementation [24,49,50]. Attitudes toward conformity affect behavior intention, technology acceptance, and perceived learning [51,52]. Therefore, this study hypothesized that attitude affects student satisfaction with technology-based STEM education.

2.9. Collaborative Learning

Compared to traditional classroom settings, collaborative learning in STEM education transforms the learning environment with ICT-assisted experimentation. It enables students to encourage collaboration and build higher knowledge [53]. Collaborative learning embeds the integrated power of many sciences into learning through large-scale networked education [54,55]. It also encourages students to work together to solve life problems [56,57]. In this case, the teacher is only a tutor helping students overcome difficulties conducting experiments and group discussions. Therefore, collaborative learning may relate to student acceptance of technology-based STEM education.

2.10. STEM Education Acceptance

STEM education acceptance is the willingness to learn using STEM education. Many people have used and developed acceptance, such as Venkatesh [58] and the TAM model [59], to understand the technology acceptance phenomenon. This study synthesized the technology acceptance theory in the UTAUT model [32] into technology-based STEM education. Many studies have modified the UTAUT model in psychology, information systems, marketing and banking, and education [22,60–62]. In the UTAUT model [32], technology acceptance is influenced by performance expectancy, perceived convenience, subjective norms, and facilitating conditions adopted in this study. Therefore, this study hypothesized that STEM education acceptance is influenced by PU, PC, SN, and collaborative learning, and it affects student satisfaction with technology-based STEM education.

2.11. Satisfaction toward STEM Education

Student satisfaction is the emotions after having a learning experience with technology-based STEM education. This implies the importance of helping students' post-adoption behavior [63,64]. Student satisfaction is an important factor affecting learning interest and outcomes [65,66]. It is widely used to measure the success or failure of a learning model or approach, especially in education, where student satisfaction is prioritized [67]. Student satisfaction toward technology-based STEM education strongly relates to their learning outcomes. Therefore, various studies examine the factors affecting student satisfaction [14,68,69].

This study used the perceived dimensions of internal factors and educational dimensions as variables to predict the factors related to student satisfaction with STEM education. It used six independent variables, two intermediate variables, and one dependent variable connected into twelve initial hypotheses shown in Figure 1 and Table 1.

Table 1. Initial hypotheses about the factors related to student satisfaction toward technology-based STEM education.

Hypothesis	Hypothesis Description
H1	Perceived usefulness has a relationship with STEM education acceptance.
H2	Perceived usefulness has a relationship with student attitude toward technology-based STEM education.
H3	Perceived convenience has a relationship with STEM education acceptance.
H4	Facilitating conditions have a relationship with student attitude toward technology-based STEM education.
H5	Subjective norm has a relationship with STEM education acceptance.
H6	Subjective norm has a relationship with student attitude toward technology-based STEM education.
H7	Collaborative learning has a relationship with STEM education acceptance.
H8	Collaborative learning has a relationship with student attitude toward technology-based STEM education.
H9	Perceived playfulness has a relationship with student attitude toward technology-based STEM education.
H10	Student attitude toward STEM education with technology has a relationship with STEM education acceptance.
H11	Student attitude toward STEM education with technology has a relationship with student satisfaction toward technology-based STEM education.
H12	STEM education acceptance has a relationship with student satisfaction.

2.12. Participants

This study aimed to build a model to determine the factors related to student satisfaction toward STEM education at the high school level. Based on the theoretical foundation, eight determinants were examined for their relationship with student satisfaction. All students were informed that this online questionnaire was only used for study data, and their identity would be protected. The questionnaire was distributed at the STEM education experimental school guided by the SK university. Furthermore, this study used purposive sampling and the google questionnaire platform for the online questionnaire distribution. Therefore, the identity of the questionnaire filler was ascertained anonymously.

A total of 174 valid responses were collected from May to June 2022. The respondents comprised 118 female and 56 male students participating in teaching and learning activities with STEM education at least once. Furthermore, most teachers were divided into groups of more than three for STEM education. Complete data are shown in Table 2.

Table 2. Basic student information.

Data Demographic		N	Percentage
gender	female	118	67.8%
	male	56	32.2%
class	10	70	40.2%
	11	83	47.7%
	12	21	12.1%
Ever carried out a STEM education	1 x	104	59.7%
	2–3 x	30	17.2%
	More than 3 x	40	22.9%
The number of group members when conducting the STEM education experiment	2 students	47	27.0%
	3 students	35	20.1%
	More than 3 students	92	52.9%

2.13. Instrument

This study used online google questionnaires (see Appendix A) and conducted a voluntary and anonymous survey. The first part of the questionnaire indicated students' basic information, experiences with STEM education and the predictors related to satisfaction with STEM education. It used a five-point Likert scale ranging from one (strongly disagree) to five (strongly agree) [70]. The scale indicated the student's level of agreement with the statement on the questionnaire. The original questionnaire had nine latent variables, including perceived usefulness (three items), perceived ease of use (three items), and facilitating condition (three items). Other latent variables were playfulness (three items), subjective norm (three items), attitude toward STEM education (three items), behavior intention toward STEM education (three items), collaborative learning (four items), and student satisfaction toward STEM education (three items), resulting in twenty-eight items.

The questionnaires were sent to three experts on STEM education studies to check the appropriateness and clarity of all items and constructs. They were corrected, discarding two items (play two and SN1) because they were unclear and inappropriate in the construct.

2.14. Data Analysis

An initial analysis was conducted using AMOS and SPSS software to show whether the data could answer the study objectives. Structural equation modeling (SEM) was also applied because it allows simultaneous analysis of many variables needed and tests the relationship with factor analysis [71]. Confirmatory factor analysis (CFA) testing was first performed to confirm the variables' validity and reliability, which estimated the instrument's internal consistency before processing data using SEM [72]. The Cronbach alpha value must exceed 0.6 to meet the convergent validity criteria [73]. Second, the factor loadings of observed items must exceed 0.5 [74]. Furthermore, the overall fit model in SEM is usually assessed based on the goodness-of-fit indices (GFI) value exceeding 0.90. The parsimonious goodness-of-fit index (PGFI) must exceed 0.50 [73]. RMR must be smaller than 0.08, including the minimum Chi-square value (CMIN) and the ratio of Chi-square to degrees of freedom (where 2/df must be less than 5.0) [75]. The hypothetical model is supported when all the value requirements for SEM are met.

3. Results

This section presents the proposed SEM model's verification to determine the factors related to student satisfaction with STEM education. First, it displays descriptive statistical data and normality tests, followed by the reliability and validity tests. The last section presents the results of SEM and initial hypothesis testing.

3.1. Descriptive Statistics and Normality Test

Table 3 shows descriptive statistics of all observed items in this study. Based on the definition from Kline [76], the data normality is measured based on the skewness and kurtosis, which must be at the limit of ± 3. In this study, the highest and lowest kurtosis values were 2.985 and 0.198, respectively, while the skewness ranged between -1.229 and -0.280. Therefore, the skewness and kurtosis values met the normal distribution criteria and were useful parameters in SEM.

Table 3. Descriptive statistics and normality test.

Variable	Mean	Standard Deviation	Excess Kurtosis	Skewness
PU1	4.221	0.791	2.437	−1.167
PU2	4.129	0.752	2.136	−1.090
PU3	4.202	0.800	2.113	−1.109
PEU1	3.638	0.813	1.435	−0.351
PEU2	3.798	0.800	1.832	−0.487
PEU3	4.031	0.779	2.902	−0.841
SN1	3.699	0.800	2.145	−0.491
SN2	3.681	0.757	2.214	−0.333
SN3	3.896	0.788	2.153	−0.571
FC1	3.902	0.867	1.204	−0.607
FC2	4.239	0.797	2.362	−1.192
FC3	3.712	0.804	1.582	−0.356
PLAY1	3.908	0.790	2.322	−0.664
PLAY2	3.466	0.824	1.395	−0.390
PLAY3	3.902	0.785	2.405	−0.669
CL1	4.012	0.775	2.345	−0.978
CL2	4.006	0.787	2.985	−1.154
CL3	4.037	0.782	2.326	−1.229
CL4	3.724	0.809	1.463	−0.296
ATT1	3.957	0.786	2.425	−0.689
ATT2	3.748	0.762	2.218	−0.377
ATT3	3.859	0.798	2.231	−0.690
ACC1	3.951	0.820	2.282	−0.852
ACC2	3.914	0.839	2.293	−0.841
ACC3	3.761	0.835	1.288	−0.417
SAT1	3.870	0.763	0.198	−0.280
SAT2	3.773	0.754	2.489	−0.463
SAT3	3.865	0.795	1.903	−0.491

3.2. Reliability Analysis

Data reliability was seen from the Cronbach alpha value to measure each construct's internal consistency. Table 4 shows that the Cronbach alpha value is between 0.705 and 0.889. According to Hair et al. [73], the Cronbach alpha value must be greater than 0.70. Therefore, the questionnaire items have high reliability and internal consistency between variables. The next step checked the loading factor, composite reliability (CR), and AVE value to analyze convergent validity. The loading factor, CR, and AVE must be higher than 0.5, 0.7, or 0.5 [74]. In this study, the lowest factor loading, CR, and AVE values were 0.75, 0.90, and 0.76, respectively. Therefore, the proposed model has good convergent validity.

Table 4. Measurement construct validity with the factor loading, CR, and AVE.

Construct	Items	Factor Loading	CR	AVE	Cronbach Alpha
PU	PU1	0.86	0.94	0.84	0.875
	PU2	0.91			
	PU3	0.88			
PEU	PEU1	0.86	0.93	0.81	0.853
	PEU2	0.81			
	PEU3	0.80			
SN	SN1	0.79	0.94	0.88	0.705
	SN2	0.83			
FC	FC1	0.75	0.94	0.84	0.764
	FC2	0.79			
	FC3	0.87			
PLAY	PLAY1	0.93			
			0.92	0.85	0.886
	PLAY3	0.90			
CL	CL1	0.87			
	CL2	0.94	0.94	0.83	0.853
	CL3	0.90			
ATT	ATT1	0.89	0.95	0.87	0.857
	ATT2	0.91			
	ATT3	0.90			
ACC	ACC1	0.97			
	ACC2	0.95	0.93	0.81	0.889
	ACC3	0.87			
SAT	SAT1	0.91			
	SAT2	0.83	0.90	0.76	0.887
	SAT3	0.87			

3.3. Validity Analysis

Hair [73] suggested checking convergent and discriminant validity. Convergent validity is seen in the average variance extracted (AVE) value. Table 5 shows that all AVE values exceed 0.5, indicating that all constructs are acceptable and useful for further analysis.

Table 5. Discriminant validity result and root of average variance extracted.

	PU	CL	PLAY	FC	SN	PEU	ATT	ACC	SAT
PU	**0.310**								
CL	0.235	**0.346**							
PLAY	0.275	0.313	**0.374**						
FC	0.242	0.258	0.295	**0.320**					
SN	0.222	0.259	0.269	0.249	**0.304**				
PEU	0.264	0.257	0.314	0.272	0.288	**0.344**			

Table 5. *Cont.*

	PU	**CL**	**PLAY**	**FC**	**SN**	**PEU**	**ATT**	**ACC**	**SAT**
ATT	0.276	0.269	0.338	0.278	0.278	0.315	**0.383**		
ACC	0.298	0.311	0.358	0.288	0.276	0.296	0.372	**0.442**	
SAT	0.259	0.250	0.318	0.262	0.265	0.302	0.352	0.355	**0.373**

The discriminant validity was analyzed by assessing the square root of AVE for all variables. Table 5 shows that the AVE value is greater than the inter-construct correlations, implying that discriminant validity is accepted [77].

3.4. Model Fit Assessment

The initial model tested was concluded to identify and interpret the results of the fit index from the model estimation. There are many ways to interpret the fit model tested by looking at CMIN/df, RMSEA, GFI, and AGFI [78]. The model is also tested by evaluating TLI, NFI, and CFI, as well as parsimonious fit with PNFI and PGFI reference indicators [61]. This study used the three methods to test model fit (Table 6).

The data processing results using AMOS software showed that the Chi-square value is 429,521 with 248 degrees of freedom. This indicates that the CMIN/df value is 1.75 in the limit between $1 < x < 3$. Furthermore, many studies suggest checking the goodness-of-fit index by considering the variance and covariance predicted in the reproduced matrix [65,79]. A higher GFI value of 0.80 is acceptable because it indicates a better model fit. In this study, the GFI value reached 0.88. Unlike the Chi-square, the RMSEA reference indicator considers the estimated parameters but not the sample size. Therefore, an RMSEA value less than 0.08 indicates an accepted model [80], and an RMSEA value smaller than 0.05 indicates a perfect model. This study's RMSEA value is 0.06, implying an acceptable model. For incremental fit measurement, the values of NFI, CFI, and AGFI must be greater than 0.8 to achieve an acceptable fit model [81]. In this study, the values of NFI, CFI, and AGFI exceeded 0.8. Some of the literature uses parsimonious fit measurement to determine model fit. A parsimonious model is the simplest and sharpest model to explain the analyzed phenomenon [82]. In this study, the entire parsimonious fit index exceeded 0.05, signifying the model is suitable. These three fit model tests confirm that the proposed structural model in Figure 2 is acceptable and appropriate for analyzing and interpreting the factors related to student satisfaction toward STEM education.

Table 6. Absolute, incremental, and parsimonious fit measurement.

Measurement	Indicator	*p*-Value	Recommended Criteria
Absolute fit	CMIN/df	1.75	$1 < x < 3$
	GFI	0.88	>0.8
	RMSEA	0.06	<0.08
	RMR	0.05	<0.08
	NFI	0.85	>0.8
Incremental fit	CFI	0.80	>0.8
	AGFI	0.82	>0.8
Parsimonious fit	PNFI	0.29	>0.05
	PGFI	0.60	>0.05

The linear correlation coefficient (R2) [83] is the most important value for determining whether the proposed structural model strongly explains the factors influencing student satisfaction toward STEM education. Cohen [84] stated that an R2 value greater than 0.26

(26%) was strong enough to explain a model. In this study, the R-value for acceptance toward STEM education was 78.4%, student attitude toward STEM education was 73.0%, and student satisfaction toward STEM education was 70.4%. Therefore, the proposed model explains the factors related to student satisfaction with STEM education.

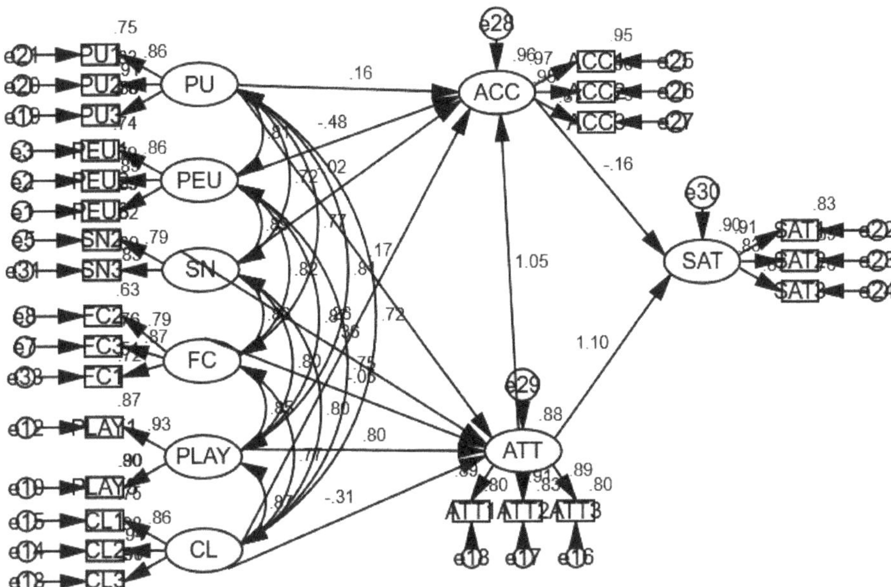

Figure 2. Whole set observation of the measurement model.

3.5. Structural Model and Hypothesis Testing

This study examined the statistical significance of the 12 initial hypotheses about the relationship with student satisfaction with STEM education. It determined the standardized regression coefficients between the dependent and independent variables. Furthermore, the study analyzed the significance of the p-value on each coefficient derived from the SEM output. The accepted hypothesis is where a statistically significant relationship in the predicted direction is confirmed. Table 7 and Figure 3 show three hypotheses with a significant level reaching 0.01 and at a significant level of 0.05 and 0.10 against seven hypotheses.

Table 7. Coefficient and hypothesis testing. ***: $p < 0.001$.

Hypothesis	Parameter			Path Coefficient (β)	SE	CR	p-Value	Interpretation		
								0.1	0.05	0.01
H1	PU	$\rightarrow$	ACC	0.190	0.137	1.385	0.166	Rejected	Rejected	Rejected
H2	PU	$\rightarrow$	ATT	0.184	0.130	1.417	0.156	Rejected	Rejected	Rejected
H3	PEU	$\rightarrow$	ACC	−0.539	0.239	−2.262	0.024	Accepted	Accepted	Rejected
H4	FC	$\rightarrow$	ATT	−0.054	0.154	−0.352	0.725	Rejected	Rejected	Rejected
H5	SN	$\rightarrow$	ACC	−0.022	0.227	−0.096	0.923	Rejected	Rejected	Rejected
H6	SN	$\rightarrow$	ATT	0.398	0.188	2.119	0.034	Accepted	Accepted	Rejected
H7	CL	$\rightarrow$	ACC	0.295	0.140	2.109	0.035	Accepted	Accepted	Rejected
H8	CL	$\rightarrow$	ATT	−0.324	0.153	−2.117	0.034	Accepted	Accepted	Rejected

Table 7. *Cont.*

Hypothesis	Parameter			Path Coefficient (β)	SE	CR	*p*-Value	Interpretation		
H9	PLAY	→	ATT	0.796	0.199	3.992	***	Accepted	Accepted	Accepted
H10	ATT	→	ACC	1.149	0.178	6.437	***	Accepted	Accepted	Accepted
H11	ATT	→	SAT	1.101	0.252	4.361	***	Accepted	Accepted	Accepted
H12	ACC	→	SAT	−0.150	0.198	−0.759	0.448	Rejected	Rejected	Rejected

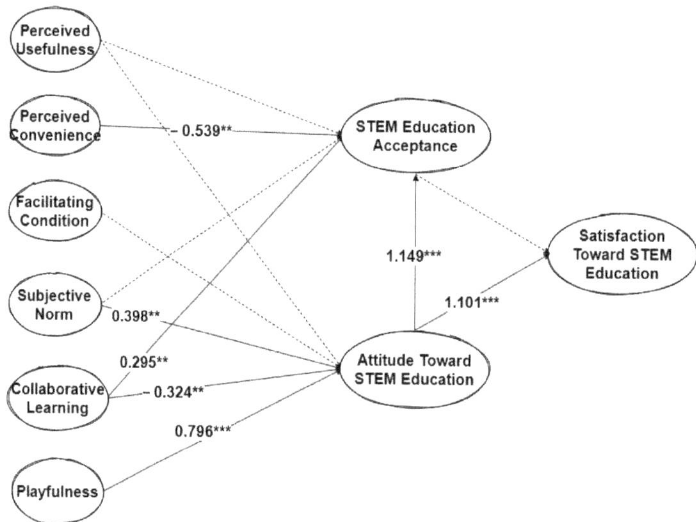

Figure 3. Final model with a coefficient (β) and *p*-value; a dotted line means an insignificant relationship. **: $p < 0.01$; ***: $p < 0.001$.

The findings showed that PU did not significantly correlate with STEM education acceptance or student attitude ($p > 0.05$), hence H1 and H2 were rejected. PEU had a relationship with STEM education acceptance (β = −0.539, $p < 0.05$), supporting H3 (PEU → ACC). Furthermore, FC had no significant relationship with attitude toward STEM education ($p > 0.05$), denoting H4 was rejected. SN does not significantly affect STEM education acceptance ($p > 0.05$), thereby H5 was rejected. In contrast, SN significantly correlates with attitude toward STEM education (β = 0.398, $p < 0.05$), supporting H6. Collaborative learning on STEM education has a relationship with STEM education acceptance (β = 0.295, $p < 0.05$) and attitude toward STEM (β = −0.324, $p < 0.05$), supporting H7 and H8. Moreover, the playfulness variable had the strongest relationship with students' attitudes toward STEM (β = 0.796, $p < 0.01$), supporting H9. The attitude had significant relationships with STEM education acceptance and student satisfaction toward STEM education (β = 1.149, $p < 0.01$) and (β = 1.101, $p < 0.01$), also supporting H10 and H11. Based on the coefficient, students' attitude strongly correlates with STEM education acceptance and satisfaction. Additionally, STEM education acceptance has no significant relationship with student satisfaction toward STEM education, meaning H12 was rejected.

4. Discussion

This study focused on integrating high school STEM education by predicting factors related to student satisfaction. It proposed a model based on the theory of planned behavior (TPB) added with predictors from the literature review.

Students might assume that when they feel the STEM education with technology experiment is relatively easy, their acceptance decreases, as shown by the PEU coefficient of acceptance of −0.184. High school students think experimental STEM education with technology should integrate several science fields with certain difficulties. When the STEM education experiment is too easy, high school students may feel they cannot improve their abilities. They are not challenged to experiment with technology-based STEM education. Moreover, students feel they are more accepting of learning commonly used by teachers. This finding supports previous studies which showed that easy activities did not improve student outcomes [85,86]. However, further understanding using qualitative studies is needed to explain these findings.

Subjective norms affect students' attitudes toward STEM education. The success of implementing learning aids is inseparable from the help of the people around the users [41,87]. Similarly, the success of technology-based STEM education depends on the role of the government, schools, and teachers. Students assume that their attitude cannot be separated from the support of the people that think it provides many opportunities to develop 21st century skills and knowledge to apply in life daily. The stronger the subjective norm, the better the student's attitude toward technology-based STEM education.

Collaborative learning in STEM education encourages students to communicate and increase their satisfaction in teaching and learning activities [88]. It allows them to talk, discuss, and convey their ideas in groups. Furthermore, collaborative learning relates to the acceptance of technology-based STEM education. Students assume that learning has a relationship with their acceptance of technology-based education. This education is a series of experiments difficult to conduct individually. Furthermore, students think that success in conducting experiments on technology-based STEM education requires collaboration. This implies they think collaborative learning is related to their acceptance of technology-based STEM education. However, students believe collaborative learning reduces their attitude toward STEM education. This is because learning in Indonesia mostly used a scientific approach or individual- and teacher-centered methods before STEM education. Therefore, students are not accustomed to working in groups and prefer individual- to experiment-based learning that requires cohesiveness. In some cases, groups with more members increase the possibility of conflicts of ideas, making the students emotional. This aspect provides input for teachers to guide and monitor experiments and student discussions. The teacher must mediate and provide a way out with deliberation when there are differences in opinions within the group during technology-based STEM education experiments.

Perceived playfulness is an advantage of technology-based STEM education. The findings showed that perceived playfulness has the strongest relationship with student attitudes. It has the largest positive indirect effect on student satisfaction toward technology-based STEM education. The model allows students to experiment inside and outside the classroom, making the learning process flexible. Unconsciously, students learn to solve complex problems related to everyday life as a team. The joy of experimenting increases their attitude toward technology-based STEM education, supporting previous studies [89,90].

Students' attitudes toward technology-based STEM education are the strongest determinants compared to acceptance and satisfaction. This implies the importance of improving the effectiveness of STEM education implementation in Indonesia. When students' attitudes improve, they easily accept STEM education-based learning. Similarly, they are more satisfied with experimental STEM education-based learning with technology. This supports a previous study finding that attitude toward a learning model or new technology significantly affected the intention to use the technology [91,92].

Students' acceptance of technology-based STEM education indirectly significantly increased their satisfaction. High school students feel that their satisfaction is more influenced by their attitude. However, further studies should examine other predictors affecting student satisfaction with technology-based STEM education.

Facilitating conditions did not significantly correlate with high school students' attitudes toward technology-based STEM education-based learning. These include teachers ready to help students overcome difficulties experimenting and schools that provide the needed facilities. The schools in question are equipped with complete laboratories to conduct STEM experiments. Furthermore, high school students feel confident and able to conduct their STEM experiments, indicating that their attitude is not influenced.

5. Theoretical and Practical Implications

This study provided theoretical and practical implications for increasing student satisfaction with technology-based STEM education. First, it is to develop a model to explain the factors related to student satisfaction with technology-based STEM education. The proposed model provides empirical evidence of student satisfaction. Second, this study provides new knowledge about the theory of student satisfaction in STEM education. The information from this new model's development contributes to STEM education's future development. This is useful for school governments and teachers to identify the factors to be considered when implementing STEM education in schools.

Practically, this study showed the relevance of student attitude and satisfaction by analyzing the factors significantly affecting the two variables. Student attitude could be improved by increasing the effectiveness of STEM education learning. This study found that subjective norms significantly affect student attitude toward STEM education. Therefore, educational institutions and teachers should create a conducive environment and spearhead the use of STEM education-based learning. The finding of playfulness implies that schools and teachers should integrate fun and exciting activities that generate pedagogically satisfying interest among students when learning with STEM education. It is also important to incorporate new innovations in experiments conducted by students. Although STEM education is rarely applied to learning activities in Indonesia, the existing experiments may always be interesting for students.

Learning using STEM education has challenges and problems. However, this approach is accepted by students because they are happy with learning activities that require teamwork. In today's era, knowledge should be balanced with the ability to work in a team. The important collaborative skills should be possessed in the 21st century. Therefore, this finding is expected to spur teachers to continue using STEM education-based learning. It could also promote students to collaborate when experimenting in STEM education than traditional or individualized teacher-centered learning.

Student attitude significantly affects their acceptance and satisfaction with STEM education. They feel that they recognize the benefits of STEM education on their learning outcomes. Furthermore, this study is beneficial to the government and academics that focus on developing STEM education in Indonesia. It could help them understand various important factors concerning student acceptance and satisfaction with STEM education. The findings provide a basis for evaluating student attitude, acceptance, and satisfaction with STEM education. Additionally, they contribute to the development of a model to analyze student acceptance.

6. Conclusions

This study tested the proposed model developed from the modified TPB model with literature review predictors to understand the determinants related to student satisfaction toward technology-based STEM education in Indonesia. Increasing student satisfaction may increase the effect of STEM education learning on student outcomes. Therefore, this study provides a better understanding of the factors to be considered to increase student satisfaction toward STEM education for governments, institutions, and teachers. Although STEM education is sustainable, it does not automatically increase student acceptance and satisfaction. The most important thing is to improve students' attitudes through support from schools and teachers. This could help students become more familiar with technology-based STEM education. The success of increasing student satisfaction is influenced by internal and external factors.

7. Limitations and Future Directions

Although this study achieves the research objectives and provides several valuable implications, this study has several limitations. First, the study was limited to the school experiment at the senior high school level in collaboration with the Syiah Kuala University in Indonesia. Therefore, the findings should be generalized with caution. Future studies should use larger samples from various countries. Second, this study also used TPB as the base model, while student satisfaction toward STEM education could still be analyzed in many models. Additionally, it was a quantitative study, implying the need for qualitative and longitudinal studies in the future.

The results showed how attitude and collaborative learning in STEM education affect student satisfaction. However, future studies could consider constructs such as student academic achievement, intention to use, and actual use of STEM education. The findings could provide additional knowledge on marginal processes and situations that describe STEM education's effect on student satisfaction.

Author Contributions: Conceptualization, T.T.W. and M.M.; methodology, T.T.W.; software, A.H.; validation M.M.; formal analysis, T.T.W. and A.H.; investigation, M.M.; funding acquisition, J.Z.; project administration, J.Z.; writing—original draft, T.T.W., M.M. and A.H.; writing—review and editing, T.T.W., M.M., A.H. and J.Z. All authors have read and agreed to the published version of the manuscript.

Funding: This research was supported by the Hainan Provincial Natural Science Foundation of China (622RC670); Project of the 13th Five Year Plan of Educational Science in Hainan Province of China (QJY20191071).

Institutional Review Board Statement: The study was conducted in accordance with the Declaration of Helsinki and approved by the Institutional Review Board (or Ethics Committee) of Syiah Kuala University (protocol code 3806/UN11.1.6/TU/01.01/2022 and 1 May 2022).

Informed Consent Statement: Not applicable.

Data Availability Statement: The data presented in this study are available on request from the corresponding author. The data are not publicly available due to privacy.

Conflicts of Interest: The authors declare no conflict of interest.

Appendix A

Table A1. Questionnaire items.

Construct	Items	Items (Indonesian Version)	References
Perceived useful	STEM learning helps me to gain more knowledge through experiments.	Pembelajaran STEM membantu saya untuk mendapatkan lebih banyak ilmu melalui experiment	[59,93]
	Learning by doing in STEM lessons is very useful for me.	Learning by doing pada pelajaran STEM sangat bermanfaat bagi saya	
	I feel STEM learning is very useful for my future.	Saya merasakan pembelajaran STEM sangat berguna bagi masa depan saya	
Perceived easy to use	I do not find it difficult to learn with STEM-based learning.	Saya tidak merasa kesulitan belajar dengan pembelajaran berbasis STEM	[59,94]
	I can follow directions and experiments easily.	Saya dapat mengikuti arahan dan melakukan experiment dengan mudah	
	I easily obtain much useful knowledge from STEM activities.	Saya dengan mudah mendapatkan banyak ilmu yang bermanfaat dari kegiatan STEM	

Table A1. *Cont.*

Construct	Items	Items (Indonesian Version)	References
Subjective norm	My teacher uses STEM-based learning when teaching.	Guru saya menggunakan pembelajaran berbasis STEM saat mengajar	[40]
	The government has a program to encourage STEM-based learning.	Pemerintah mempunyai program untuk mendorong pembelajaran berbasis STEM	
	The school has a STEM-based learning program.	Sekolah mempunyai program pembelajaran berbasis STEM	
Facilitating conditions	I do not need to provide tools for experimentation, and the school already provides them.	Saya tidak perlu menyediakan alat untuk berexperiment, sekolah sudah menyediakannya	[58,95]
	The teacher is ready to help me if I have trouble doing experiments.	Guru siap membantu saya jika saya kesulitan untuk melakukan experiment	
	My team and I have enough knowledge to experiment.	Saya dan tim mempunyai pengetahuan yang cukup untuk melakukan experiment	
Perceived playfulness	I feel happy learning at school using STEM-based learning.	Saya merasa senang belajar di sekolah dengan menggunakan pembelajaran berbasis STEM	[96]
	I spend my free time at home continuing unfinished experiments at school.	Saya menghabiskan waktu luang saya di rumah untuk melanjutkan experiment yang belum selesai di sekolah	
Collaborative learning	I feel STEM learning effectively gets students to work together while conducting experiments.	saya merasa pembelajaran STEM efektif untuk membuat para siswa bekerjasama saat melakukan experiment	[43,97]
	I feel STEM learning is effective for making students discuss conducting experiments.	saya merasa pembelajaran STEM efektif untuk membuat para siswa berdiskusi untuk melakukan experiment	
	STEM learning improves my knowledge and skills through group discussions.	Pembelajaran STEM meningkatkan pengetahuan dan kemampuan saya melalui diskusi kelompok	
	Collaboration in STEM learning is better than traditional learning.	Kerjasama pada pembelajaran STEM lebih baik dibanding pembelajaran tradisional	
attitude	I like it when teachers use STEM learning.	Saya senang jika guru menggunakan pembelajaran STEM	[92,98]
	I prefer learning that uses STEM learning.	Saya lebih menyukai pembelajaran yang menggunakan pembelajaran STEM	
	I think using STEM learning is a good idea.	Saya pikir menggunakan pembelajaran STEM adalah ide yang baik	
STEM education acceptance	I am happy to accept STEM-based learning.	Saya senang untuk menerima pembelajaran berbasis STEM	[99,100]
	I will be happy if the teacher can continue to use the STEM approach next semester.	Saya akan senang jika guru dapat terus menggunakan pendekatan STEM pada semester depan	
	I will advise teachers; therefore, we will learn with a STEM approach.	Saya akan memberikan saran kepada guru agar kita dapat belajar dengan pendekatan STEM	
Learning satisfaction	I am very satisfied with learning with STEM learning.	Saya sangat puas belajar dengan pembelajaran STEM	[21,63]
	My team is very satisfied with the experiments in STEM learning.	Tim saya sangat puas dengan experiment pada pembelajaran STEM	
	The teacher directs each team to learn to experiment well.	Guru mengarahkan setiap tim untuk belajar melakukan experiment dengan baik	
	I am very satisfied if the teacher uses STEM learning in the classroom.	Saya sangat puas jika guru menggunakan pembelajaran STEM di kelas.	

References

1. Gattullo, M.; Laviola, E.; Boccaccio, A.; Evangelista, A.; Fiorentino, M.; Manghisi, V.M.; Uva, A.E. Design of a Mixed Reality Application for STEM Distance Education Laboratories. *Computers* **2022**, *11*, 50. [CrossRef]
2. Rose, M.A.; Geesa, R.L.; Stith, K. STEM leader excellence: A modified delphi study of critical skills, competencies, and qualities. *J. Technol. Educ.* **2019**, *31*, 42–62. [CrossRef]
3. Permanasari, A.; Rubini, B.; Nugroho, O.F. STEM Education in Indonesia: Science Teachers' and Students' Perspectives. *J. Innov. Educ. Cult. Res.* **2021**, *2*, 7–16. [CrossRef]
4. Sulaeman, N.; Efwinda, S.; Putra, P.D.A. Teacher Readiness In Stem Education: Voices Of Indonesian Physics Teachers. *J. Technol. Sci. Educ.* **2022**, *12*, 68–82. [CrossRef]
5. Nugroho, O.F.; Permanasari, A.; Firman, H. The movement of stem education in Indonesia: Science teachers' perspectives. *J. Pendidik. IPA Indones.* **2019**, *8*, 417–425. [CrossRef]
6. Tupas, F.P.; Matsuura, T. Moving forward in stem education, challenges and innovations in senior high school in the Philippines: The case of Northern Iloilo polytechnic state college. *J. Pendidik. IPA Indones.* **2019**, *8*, 407–416. [CrossRef]
7. Purwaningsih, E.; Sari, S.P.; Sari, A.M.; Suryadi, A. The effect of stem-pjbl and discovery learning on improving students' problem-solving skills of the impulse and momentum topic. *J. Pendidik. IPA Indones.* **2020**, *9*, 465–476. [CrossRef]
8. Chai, C.S.; Rahmawati, Y.; Jong, M.S.-Y. Indonesian science, mathematics, and engineering preservice teachers' experiences in stem-tpack design-based learning. *Sustainability* **2020**, *12*, 9050. [CrossRef]
9. Arlinwibowo, J.; Retnawati, H.; Kartowagiran, B. How to Integrate STEM Education in The Indonesian Curriculum? A Systematic Review. *Chall. Sci.* **2021**, 18–25. [CrossRef]
10. Rusydiyah, E.F.; Indrawati, D.; Jazil, S.; Susilawati, S.; Gusniwati, G. Stem learning environment: Perceptions and implementation skills in prospective science teachers. *J. Pendidik. IPA Indones.* **2021**, *10*, 138–148. [CrossRef]
11. Azizan, S.; Lee, A.; Crosling, G.; Atherton, G.; Arulanandam, B.; Lee, C.; Rahim, R. A Online Learning and COVID-19 in Higher Education: The Value of IT Models in Assessing Students' Satisfaction. *Int. J. Emerg. Technol. Learn.* **2022**, *17*, 245–278. [CrossRef]
12. Alowayr, A. Determinants of mobile learning adoption: Extending the unified theory of acceptance and use of technology (UTAUT). *Int. J. Inf. Learn. Technol.* **2021**, *39*, 1–12. [CrossRef]
13. Zhuang, T.; Cheung, A.C.K.; Tam, W. Modeling undergraduate STEM students' satisfaction with their programs in China: An empirical study. *Asia Pacific Educ. Rev.* **2020**, *21*, 211–225. [CrossRef]
14. De, S.; López, F.; Ferrando, F.; Fabregat-Sanjuan, A. Learning/training video clips: An efficient tool for improving learning outcomes in Mechanical Engineering. *Int. J. Educ. Technol. High. Educ.* **2016**, *13*, 1–13. [CrossRef]
15. Scholtenhuis, L.O.; Vahdatikhaki, F.; Rouwenhorst, C. Flipped microlecture classes: Satisfied learners and higher performance? *Eur. J. Eng. Educ.* **2021**, *46*, 457–478. [CrossRef]
16. Rahmadi, I.F.; Lavicza, Z.; Kocadere, S.A.; Padmi, R.S.; Houghton, T. User-generated microgames for facilitating learning in various scenarios: Perspectives and preferences for elementary school teachers. *Interact. Learn. Environ.* **2021**, *10*, 1–13. [CrossRef]
17. Suwarma, I.R.; Kumano, Y. Implementation of STEM education in Indonesia: Teachers' perception of STEM integration into curriculum. *J. Phys. Conf. Ser.* **2019**, *1280*, 052052. [CrossRef]
18. Prasadi, A.H.; Wiyanto, W.; Suharini, E. The Implementation of Student Worksheet Based on STEM (Science, Technology, Engineering, Mathematics) and Local Wisdom to Improve of Critical Thinking Ability of Fourth Grade Students. *J. Prim. Educ.* **2020**, *9*, 227–237. [CrossRef]
19. Lo, J.H.; Lai, Y.F.; Hsu, T.L. The study of ar-based learning for natural science inquiry activities in taiwan's elementary school from the perspective of sustainable development. *Sustainability* **2021**, *13*, 6283. [CrossRef]
20. Ajzen, I. From Intentions to Actions: A Theory of Planned Behavior. In *Action Control*; Springer: Berlin/Heidelberg, Germany, 1985.
21. Arain, A.A.; Hussain, Z.; Rizvi, W.H.; Vighio, M.S. Extending UTAUT2 toward acceptance of mobile learning in the context of higher education. *Univers. Access Inf. Soc.* **2019**, *18*, 659–673. [CrossRef]
22. Zou, C.; Li, P.; Jin, L. Integrating smartphones in EFL classrooms: Students' satisfaction and perceived learning performance. *Educ. Inf. Technol.* **2022**, 1–22. [CrossRef]
23. Zobair, K.M.; Sanzogni, L.; Houghton, L.; Islam, M.Z. Forecasting care seekers satisfaction with telemedicine using machine learning and structural equation modeling. *PLoS ONE* **2021**, *16*, e0257300. [CrossRef]
24. Drueke, B.; Mainz, V.; Lemos, M.; Wirtz, M.A.; Boecker, M. An Evaluation of Forced Distance Learning and Teaching Under Pandemic Conditions Using the Technology Acceptance Model. *Front. Psychol.* **2021**, *12*, 1–10. [CrossRef] [PubMed]
25. Osih, S.C.; Singh, U.G.; Africa, S. Students' perception on the adoption of an e-textbook (digital) as an alternative to the printed textbook. *South Afr. J. High. Educ.* **2020**, *34*, 201–215. [CrossRef]
26. Al-Maroof, R.S.; Alnazzawi, N.; Akour, I.A.; Ayoubi, K.; Alhumaid, K.; AlAhbabi, N.M.; Alnnaimi, M.; Thabit, S.; Alfaisal, R.; Aburayya, A.; et al. The effectiveness of online platforms after the pandemic: Will face-to-face classes affect students' perception of their behavioural intention (BIU) to use online platforms? *Informatics* **2021**, *8*, 83. [CrossRef]
27. Alismaiel, O.A. Using structural equation modeling to assess online learning systems' educational sustainability for university students. *Sustainability* **2021**, *13*, 13565. [CrossRef]
28. Navarro, M.M.; Prasetyo, Y.T.; Young, M.N.; Nadlifatin, R.; Redi, A.A.N.P. The perceived satisfaction in utilizing learning management systems among engineering students during the COVID-19 pandemic: Integrating task technology fit and extended technology acceptance model. *Sustainability* **2021**, *13*, 10669. [CrossRef]

29. Joo, Y.J.; Park, S.; Shin, E.K. Students' expectation, satisfaction, and continuance intention to use digital textbooks. *Comput. Human Behav.* **2017**, *69*, 83–90. [CrossRef]

30. Jeno, L.M.; Vandvik, V.; Eliassen, S.; Grytnes, J.-A. Testing the novelty effect of an m-learning tool on internalization and achievement: A Self-Determination Theory approach. *Comput. Educ.* **2019**, *128*, 398–413. [CrossRef]

31. Kim, K.J.; Shin, D.H. An acceptance model for smart watches: Implications for the adoption of future wearable technology. *Internet Res.* **2015**, *25*, 527–541. [CrossRef]

32. Venkatesh, V.; Morris, M.G.; Davis, G.B.; Davis, F.D. User acceptance of information technology: Toward a unified view. *Manag. Inf. Syst. Q.* **2003**, *27*, 425–478. [CrossRef]

33. Wijaya, T.T.; Cao, Y.; Weinhandl, R.; Yusron, E. Applying the UTAUT Model to Understand Factors Affecting Micro-Lecture Usage by Mathematics Teachers in China. *Mathematics* **2022**, *10*, 1008. [CrossRef]

34. Zhou, Y.; Li, X.; Wijaya, T.T. Determinants of Behavioral Intention and Use of Interactive Whiteboard by K-12 Teachers in Remote and Rural Areas. *Front. Psychol.* **2022**, *13*, 1–19. [CrossRef]

35. Yoo, D.K.; Roh, J.J. Adoption of e-Books: A Digital Textbook Perspective. *J. Comput. Inf. Syst.* **2019**, *59*, 136–145. [CrossRef]

36. Šumak, B.; Šorgo, A. The acceptance and use of interactive whiteboards among teachers: Differences in UTAUT determinants between pre- and post-adopters. *Comput. Human Behav.* **2016**, *64*, 602–620. [CrossRef]

37. Teo, T. Examining the influence of subjective norm and facilitating conditions on the intention to use technology among pre-service teachers: A structural equation modeling of an extended technology acceptance model. *Asia Pacific Educ. Rev.* **2010**, *11*, 253–262. [CrossRef]

38. Ngai, E.W.; Poon, J.K.; Chan, Y.H. Empirical examination of the adoption of WebCT using TAM. *Comput. Educ.* **2007**, *48*, 250–267. [CrossRef]

39. Stössel, J.; Baumann, R.; Wegner, E. Predictors of student teachers' esd implementation intention and their implications for improving teacher education. *Sustainability* **2021**, *13*, 9027. [CrossRef]

40. Gellerstedt, M.; Babaheidari, S.M.; Svensson, L. A first step towards a model for teachers' adoption of ICT pedagogy in schools. *Heliyon* **2018**, *4*, e00786. [CrossRef]

41. Saleem, M.; Kamarudin, S.; Shoaib, H.M.; Nasar, A. Influence of augmented reality app on intention towards e-learning amidst COVID-19 pandemic. *Interact. Learn. Environ.* **2021**, *8*, 1–15. [CrossRef]

42. Akar, E.; Mardikyan, S. Analyzing factors affecting users' behavior intention to use social media: Twitter. *Int. J. Bus. Soc. Sci.* **2014**, *5*, 85–95.

43. Rabu, S.N.A.; Hussin, H.; Bervell, B. QR code utilization in a large classroom: Higher education students' initial perceptions. *Educ. Inf. Technol.* **2019**, *24*, 359–384. [CrossRef]

44. Al-Maroof, R.S.; Alhumaid, K.; Salloum, S. The continuous intention to use e-learning, from two different perspectives. *Educ. Sci.* **2021**, *11*, 6. [CrossRef]

45. Eunyoung, C. Design Framework for Math Educational Serious game Based on Cognitive Psychology. *Cartoon Animat. Stud.* **2016**, *45*, 299–320.

46. Davis, D.R.; Boone, W. Using Rasch analysis to evaluate the psychometric functioning of the other-directed, lighthearted, intellectual, and whimsical (OLIW) adult playfulness scale. *Int. J. Educ. Res. Open* **2021**, *2*, 100054. [CrossRef]

47. Wang, H.-W. An explorative study of continuance intention on cloud learning with applying social networks blended in new era of internet technology applications. *J. Internet Technol.* **2015**, *16*, 563–570. [CrossRef]

48. Han, J.; Kelley, T.; Knowles, J.G. Factors Influencing Student STEM Learning: Self-Efficacy and Outcome Expectancy, 21st Century Skills, and Career Awareness. *J. STEM Educ. Res.* **2021**, *4*, 117–137. [CrossRef]

49. Fussell, S.G.; Truong, D. Accepting virtual reality for dynamic learning: An extension of the technology acceptance model. *Interact. Learn. Environ.* **2021**, *3*, 1–18. [CrossRef]

50. Camadan, F.; Reisoglu, I.; Ursavas, Ö.F.; Mcilroy, D. How teachers' personality affect on their behavioral intention to use tablet PC. *Int. J. Inf. Learn. Technol.* **2018**, *35*, 12–28. [CrossRef]

51. Yang, H.-H.; Su, C.-H. Learner Behaviour in a MOOC Practice-oriented Course: In Empirical Study Integrating TAM and TPB. *Int. Rev. Res. OPEN Distrib. Learn.* **2017**, *18*, 35–63. [CrossRef]

52. Fang, X.; Liu, R. Determinants of teachers' attitude toward microlecture: Evidence from elementary and secondary schools. *Eurasia J. Math. Sci. Technol. Educ.* **2017**, *13*, 5597–5606. [CrossRef]

53. Fu, Y.; Zhang, D.; Jiang, H. Students' Attitudes and Competences in Modeling Using 3D Cartoon Toy Design Maker. *Sustainability* **2022**, *14*, 2176. [CrossRef]

54. Oner, D. A virtual internship for developing technological pedagogical content knowledge. *Australas. J. Educ. Technol.* **2020**, *36*, 27–42. [CrossRef]

55. Shieh, R.S.; Chang, W. Fostering student's creative and problem-solving skills through a hands-on activity. *J. Balt. Sci. Educ.* **2014**, *13*, 650–661. [CrossRef]

56. Volet, S.; Jones, C.; Vauras, M. Attitude-, group- and activity-related differences in the quality of preservice teacher students' engagement in collaborative science learning. *Learn. Individ. Differ.* **2019**, *73*, 79–91. [CrossRef]

57. Harrell, S.V.; Abrahamson, D.; Morgado, L.; Esteves, M.; Valcke, M.; Vansteenbrugge, H.; Rosenbaum, E.; Barab, S. Virtually There: Emerging designs for STEM teaching and learning in immersive online 3D microworlds. In Proceedings of the International Conference on Computer-Supported Collaborative Learning (CSCL), Utrecht, The Netherlands, 23–28 June 2008; pp. 383–391. Available online: https://www.scopus.com/inward/record.uri?eid=2-s2.0-84880418573&partnerID=40&md5=8976548d41840 1619abae2ae10e734bc (accessed on 10 June 2022).

58. Venkatesh, V.Y.L.; Thong, J.; Xu, X. Consumer acceptance and use of information Technology: Extending the unified theory of acceptance and use of technology. MIS Quarterly. *Manag. Inf. Syst. Q.* **2012**, *36*, 157–178. [CrossRef]

59. Davis, F.D. Perceived usefulness, perceived ease of use, and user acceptance of information technology. *MIS Q.* **1989**, *13*, 319–340. [CrossRef]

60. Jevsikova, T.; Stupuriene, G.; Stumbriene, D.; Juškevičiene, A.; Dagiene, V. Acceptance of Distance Learning Technologies by Teachers: Determining Factors and Emergency State Influence. *Informatica* **2021**, *32*, 517–542. [CrossRef]

61. Wijaya, T.T.; Weinhandl, R. Factors Influencing Students' Continuous Intentions for Using Micro-Lectures in the Post-COVID-19 Period: A Modification of the UTAUT-2 Approach. *Electronics* **2022**, *11*, 1924. [CrossRef]

62. Mujalli, A.; Khan, T.; Almgrashi, A. University Accounting Students and Faculty Members Using the Blackboard Platform during COVID-19; Proposed Modification of the UTAUT Model and an Empirical Study. *Sustainability* **2022**, *14*, 2360. [CrossRef]

63. Wu, C.; Gong, X.; Luo, L.; Zhao, Q.; Hu, S.; Mou, Y. Applying Control-Value Theory and Unified Theory of Acceptance and Use of Technology to Explore Pre-service Teachers' Academic Emotions and Learning Satisfaction. *Front. Psychol.* **2021**, *12*, 738959. [CrossRef] [PubMed]

64. Kosiba, J.P.B.; Odoom, R.; Boateng, H.; Twum, K.K.; Abdul-Hamid, I.K. Examining students' satisfaction with online learning during the COVID-19 pandemic-an extended UTAUT2 approach. *J. Furth. High. Educ.* **2022**, *6*, 1–18. [CrossRef]

65. Al-Rahmi, A.M.; Shamsuddin, A.; Alturki, U.; Aldraiweesh, A.; Yusof, F.M.; Al-Rahmi, W.M.; Aljeraiwi, A.A. The influence of information system success and technology acceptance model on social media factors in education. *Sustainability* **2021**, *13*, 7770. [CrossRef]

66. Liao, T.; Tang, S.; Shim, Y. The Development of a Model to Predict Sports Participation among College Students in Central China. *Int. J. Environ. Res. Public Health* **2022**, *19*, 1806. [CrossRef] [PubMed]

67. Chang, D.F.; Lee, K.Y.; Tseng, C.W. Exploring Structural Relationships in Attracting and Retaining International Students in STEM for Sustainable Development of Higher Education. *Sustainability* **2022**, *14*, 1267. [CrossRef]

68. Yuan, M.; Zeng, J.; Wang, A.; Shang, J. Would It Be Better if Instructors Technically Adjust Their Image or Voice in Online Courses? Impact of the Way of Instructor Presence on Online Learning. *Front. Psychol.* **2021**, *12*, 1–14. [CrossRef]

69. McConnell, J.R., III. A model for understanding teachers' intentions to remain in STEM education. *Int. J. STEM Educ.* **2017**, *4*, 1–21. [CrossRef]

70. Suwonjandee, N.; Mahachok, T.; Asavapibhop, B. Evaluation of Thai students and teacher's attitudes in physics using Colorado Learning Attitudes about Science Survey (CLASS). *J. Phys. Conf. Ser.* **2018**, *1144*, 012124. [CrossRef]

71. Khine, M.S.; Ali, N.; Afari, E. Exploring relationships among TPACK constructs and ICT achievement among trainee teachers. *Educ. Inf. Technol.* **2017**, *22*, 1605–1621. [CrossRef]

72. Aytekin, E.; Isiksal-Bostan, M. Middle school students' attitudes towards the use of technology in mathematics lessons: Does gender make a difference? *Int. J. Math. Educ. Sci. Technol.* **2019**, *50*, 707–727. [CrossRef]

73. Hair, J.; Black, B.; Babin, B.; Anderson, R.E.; Tatham, R.L. *Multivariate Data Analysi*, 6th ed.; Prentice Hall: Upper Saddle River, NJ, USA, 2006.

74. Doll, W.; Xia, W.; Torkzadeh, G. A confirmatory factor analysis of the end-user computing satisfaction instrument. *MIS Q.* **1994**, *18*, 357–369. [CrossRef]

75. Browne, M.W.; Cudeck, R. *Alternative Ways of Assessing Model Fit*; Sage: Beverly Hills, CA, USA, 1993.

76. Kline, R.B. *Principles and Practice of Structural Equation Modeling*, 2nd ed.; Guilford Press: New York, NY, USA, 2005.

77. Fornell, C.; Larcker, D.F. Evaluating structural equation models with unobservable variables and measurement error. *J. Mark. Res.* **1981**, *18*, 39–50. [CrossRef]

78. Bardakci, S. Exploring high school students' educational use of youtube. *Int. Rev. Res. Open Distance Learn.* **2019**, *20*, 260–278. [CrossRef]

79. Prasetyo, Y.; Roque, R.; Chuenyindee, T.; Young, M.; Diaz, J.; Persada, S.; Miraja, B.; Redi, A.P. Determining factors affecting the acceptance of medical education elearning platforms during the COVID-19 pandemic in the philippines: Utaut2 approach. *Healthcare* **2021**, *9*, 780. [CrossRef] [PubMed]

80. Hu, L.; Bentler, P.M. Fit indices in covariance structure modeling: Sensitivity to underparameterized model misspecification. *Psychol. Methods* **1998**, *3*, 424–453. [CrossRef]

81. Diván, M.J.; Sánchez-Reynoso, M.L. A Metadata and Z Score-based Load-Shedding Technique in IoT-based Data Collection Systems. *Int. J. Math. Eng. Manag. Sci.* **2021**, *6*, 363–382. [CrossRef]

82. Fennell, P.G.; Zuo, Z.; Lerman, K. Predicting and explaining behavioral data with structured feature space decomposition. *EPJ DATA Sci.* **2019**, *8*, 23. [CrossRef]

83. Sofwan, M.; Pratama, R.; Muhaimin, M.; Yusnaidar, Y.; Mukminin, A.; Habibi, A. Contribution of technology innovation acceptance and organizational innovation climate on innovative teaching behavior with ICT in Indonesian education. *Qwerty* **2021**, *16*, 33–57. [CrossRef]

84. Cohen, B. Teaching STEM after school: Correlates of instructional comfort. *J. Educ. Res.* **2018**, *111*, 246–255. [CrossRef]

85. Conrad, C.; Deng, Q.; Caron, I.; Shkurska, O.; Skerrett, P.; Sundararajan, B. How student perceptions about online learning difficulty influenced their satisfaction during Canada's COVID-19 response. *Br. J. Educ. Technol.* **2022**, *53*, 534–557. [CrossRef]

86. Yang, Q.; Zhang, Y.; Lin, Y. Study on the Influence Mechanism of Virtual Simulation Game Learning Experience on Student Engagement and Entrepreneurial Skill Development. *Front. Psychol.* **2022**, *12*, 1–11. [CrossRef]

87. Wijaya, T.T.; Zhou, Y.; Houghton, T.; Weinhandl, R.; Lavicza, Z.; Yusop, F.D. Factors affecting the use of digital mathematics textbooks in indonesia. *Mathematics* **2022**, *10*, 1808. [CrossRef]

88. Roldán-álvarez, D.; Martín, E.; Haya, P.A. Collaborative video-based learning using tablet computers to teach job skills to students with intellectual disabilities. *Educ. Sci.* **2021**, *11*, 437. [CrossRef]

89. Gómez, R.L.; Suárez, A.M. Do inquiry-based teaching and school climate influence science achievement and critical thinking? Evidence from PISA 2015. *Int. J. STEM Educ.* **2020**, *7*, 1–11. [CrossRef]

90. Park, W.; Cho, H. The interaction of history and STEM learning goals in teacher-developed curriculum materials: Opportunities and challenges for STEAM education. *Asia Pacific Educ. Rev.* **2022**, 1–18. [CrossRef]

91. Naveed, Q.N.; Alam, M.M.; Tairan, N. Structural equation modeling for mobile learning acceptance by university students: An empirical study. *Sustainability* **2020**, *12*, 8618. [CrossRef]

92. Shiferaw, K.B.; Mengiste, S.A.; Gullslett, M.K.; Zeleke, A.A.; Tilahun, B.; Tebeje, T.; Wondimu, R.; Desalegn, S.; Mehari, E.A. Healthcare providers' acceptance of telemedicine and preference of modalities during COVID-19 pandemics in a low-resource setting: An extended UTAUT model. *PLoS ONE* **2021**, *16*, e0250220. [CrossRef] [PubMed]

93. Alyoussef, I.Y. Massive open online course (Moocs) acceptance: The role of task-technology fit (ttf) for higher education sustainability. *Sustainability* **2021**, *13*, 7374. [CrossRef]

94. Fussell, S.G.; Truong, D. Using virtual reality for dynamic learning: An extended technology acceptance model. *Virtual Real.* **2022**, *26*, 249–267. [CrossRef]

95. Hu, S.; Laxman, K.; Lee, K. Exploring factors affecting academics' adoption of emerging mobile technologies-an extended UTAUT perspective. *Educ. Inf. Technol.* **2020**, *25*, 4615–4635. [CrossRef]

96. Vidergor, H.E. Effects of digital escape room on gameful experience, collaboration, and motivation of elementary school students. *Comput. Educ.* **2021**, *166*, 104156. [CrossRef]

97. Reyes, E.G.D.; Galura, J.C.; Pineda, J.L.S. C5-LMS design using Google Classroom: User acceptance based on extended Unified Theory of Acceptance and Use of Technology. *Interact. Learn. Environ.* **2022**, 1–10. [CrossRef]

98. Shao, D.; Lee, I.J. Acceptance and influencing factors of social virtual reality in the urban elderly. *Sustainability* **2020**, *12*, 9345. [CrossRef]

99. Johnston, D.J.; Berg, S.A.; Pillon, K. Ease of use and usefulness as measures of student experience in a multi-platform e-textbook pilot. *Libr. Hi Tech* **2015**, *33*, 65–82. [CrossRef]

100. Zulnaidi, H.; Oktavika, E.; Hidayat, R. Effect of use of GeoGebra on achievement of high school mathematics students. *Educ. Inf. Technol.* **2020**, *25*, 51–72. [CrossRef]

Article

How K12 Teachers' Readiness Influences Their Intention to Implement STEM Education: Exploratory Study Based on Decomposed Theory of Planned Behavior

Pengze Wu [1], Lin Yang [2,*], Xiaoling Hu [1], Bing Li [1], Qijing Liu [3], Yiwei Wang [1] and Jiayong Huang [3]

[1] School of Information Technology in Education, South China Normal University, Guangzhou 510631, China
[2] Center of Network and Modern Educational Technology, Guangzhou University, Guangzhou 510006, China
[3] School of Education, Guangzhou University, Guangzhou 510006, China
*   Correspondence: yanglin@gzhu.edu.cn

Abstract: Teachers are the key factors in ensuring the effectiveness of STEM education, and their intentions deeply influence their teaching practices. The existing research about the influencing factors of teachers' intentions to implement STEM education has some problems, such as small sample sizes, being limited to teachers of a single subject, and the need for optimization of the theoretical model relied on. This research, based on the decomposed theory of planned behavior combined with the readiness of teachers, formed an assumption model of the factors influencing teachers' STEM education intentions from the aspects of attitudes, subjective norms, and perceived behavioral control. Questionnaires were sent to 532 K12 general teachers in China. A structural equation model (SEM) was used to analyze recycled data and verify the assumption model. The results show the following: (1) The educational readiness of K12 teachers in China was at an upper–middle level. Among them, the level of emotional readiness was the highest, while the level of behavioral readiness was the lowest. (2) The STEM behavioral intention of teachers was at an upper–middle level, and attitudes and perceived behavioral control had direct significant impacts on teachers' intentions to engage in STEM education. Perceived usefulness, self-efficacy, and behavioral readiness were the three strongest indirect impact factors. (3) The emotional readiness of the teachers directly affected their intentions to implement STEM education. Behavioral readiness and cognitive readiness indirectly had an impact on teachers' intentions to implement STEM education by influencing self-efficacy.

Keywords: STEM education; teachers' readiness; teachers' intention; decomposed theory of planned behavior

Citation: Wu, P.; Yang, L.; Hu, X.; Li, B.; Liu, Q.; Wang, Y.; Huang, J. How K12 Teachers' Readiness Influences Their Intention to Implement STEM Education: Exploratory Study Based on Decomposed Theory of Planned Behavior. *Appl. Sci.* **2022**, *12*, 11989. https://doi.org/10.3390/app122311989

Academic Editor: Francesco Caridi

Received: 5 November 2022
Accepted: 20 November 2022
Published: 23 November 2022

Publisher's Note: MDPI stays neutral with regard to jurisdictional claims in published maps and institutional affiliations.

1. Introduction

STEM is the abbreviation of the initials of four disciplines, including science, technology, engineering, and mathematics. STEM education aims at solving problems such as technological competitiveness and manufacturing dilemmas [1]. STEM education, which has an interdisciplinary integration orientation, has attracted the attention of international research institutions and scholars [2]. STEM education requires students to use knowledge and skills from multiple disciplines to complete learning tasks in the background of complex thematic life [3]. Proven by many empirical studies, STEM education helps students develop creative problem-solving skills and competencies to adapt to the future [4]. These skills and competencies include research inquiry, problem solving, critical and creative thinking, entrepreneurship, collaboration, teamwork, and communication [3,5,6], which are helpful for personal study and work, as well as long-term development. After STEM education was put forward, STEAM education appeared, as well as I-STEAM, STEAMx, and so on derived from STEAM education. For ease of expression, this research uses STEM education to refer to STEM, STEAM, and so on, which are the same types of concepts as STEM education.

Under the background of fierce global competition, STEM education contributes greatly to the development of innovative talents with high-order thinking skills and 21st-century skills [7,8]. STEM education can help students acquire 21st-century skills such as problem solving [9] and creativity [10,11]. It is considered an important cornerstone for the realization of 21st-century skills [12]. Therefore, STEM education has been widely promoted by countries around the world, including China. The importance attached to STEM education has been reflected in national policies and researched in practical areas. The Ministry of Education of the People's Republic of China put forward that the country should actively explore the application of information technology in new educational models such as interdisciplinary learning (STEM education) [13]. The China STEM Education White Paper suggested that STEM education should be involved in the national strategy for training innovative talents. Teachers are a crucial medium in the implementation of STEM education in schools [14]. Their understanding of STEM, knowledge reserve of related disciplines, STEM teaching competence and experience, etc., directly determine the effectiveness of STEM teaching.

Behavioral intention is described as an important indicator of performing given behaviors. It is also the core predictive variable of behavior [15]. Unless the teacher has both the skills to use the most complex method and the desire to implement them, they cannot use the most complex method to help students learn [16]. As front-line education practitioners, teachers play a vital role in implementing STEM education [17]. Nikolopoulou et al. [18] put forth that teachers' acceptance of and disposition for instructional technology indicated their interest and deeply influenced their teaching practice. As a result, professional development programs for STEM teachers are a core issue of future STEM research [19]. Exploring the teaching intention of teachers is urgent because it directly decides their STEM teaching practice.

Relevant research has shown that the majority of teachers in China strongly agreed with the concept of STEM education. However, teachers who were willing to implement STEM education were in the minority [20] because STEM education is difficult. Li et al. [21] investigated the past behaviors and behavioral intentions of physical education teachers to incorporate STEM education into their teaching and found that the proportion of teachers who regularly incorporated STEM education into their teaching was low. Therefore, some questions remain: Why are teachers in China not so willing to implement STEM education? What are the basic factors that influence their willingness? How can we promote their willingness to implement STEM education? We need to identify which variables dominate in determining teachers' intentions to implement STEM education and put forward suggestions that may strengthen these factors more effectively. This will help to find out the factors that affect the actual performance of STEM teachers from the source and solve problems of source power that bridge the gap between reality and the theory of STEM.

2. Hypothesis Development

Research on the influencing factors of teachers' intentions to implement STEM education has some problems, such as small sample sizes, being limited to teachers of a single subject, and the need for optimization of the theoretical model relied on [22,23]. To predict an individual's behavioral intentions, a series of models have been constructed and adopted. Among them, the more impactful models include: the technology acceptance model (TAM) [24], the theory of planned behavior (TPB) [15], and the decomposed theory of planned behavior (DTPB) [25]. The DTPB was put forward by Taylor and Todd after they compared TAM theory and the TPB. In the DTPB, attitudes, subjective norms, and perceived behavioral control, the three influencing factors of the TPB, are decomposed into lower-level belief structures.

Attitudes include perceived usefulness, ease of use, and compatibility. Subjective norms include peer influence and superior influence. Perceived behavioral control includes self-efficacy, resource-facilitating conditions, and technology-facilitating conditions. The DTPB model and other models have been verified and compared in different fields, proving that the DTPB has more accurate predictions and explanatory ability by adding decomposed variables [26,27]. This model has been widely used in research on, for example, the intentions of pre-service teachers to use Web 2.0 tools [28,29] and implement ICT [30], and the intentions of junior high school teachers to use online learning platforms [31]. Therefore, this research selected the DTPB model as the basic theoretical framework to comprehensively forecast and interpret the influencing factors of the intention to implement STEM education by Chinese K12 teachers.

Readiness is the level of ability and willingness shown by an individual in a particular job [32]. The readiness of a teacher has a great impact on the quality of classroom instruction, the effectiveness of classroom instruction, and the acquisition of students' abilities [33]. Hata, Nur Fatahiyah Mohamed et al. [34] found that teachers who had more knowledge and better attitudes toward STEM education were better prepared for STEM education.

As a result, this research proposes that readiness should be included in the influencing factors of STEM educational intention; thus, readiness was divided into three dimensions: cognition, emotion, and behavior [35–37] Certain extensions of the DTPB model were established: nine independent variables, four intervening variables, and one dependent variable were used to construct a comprehensive model. Furthermore, corresponding hypotheses were formulated, as shown in Figure 1. Three factors, namely attitudes, subjective norms, and perceived behavioral control, serve as variables that predict teachers' behavioral intentions; perceived usefulness, perceived ease of use, and compatibility serve as variables that predict teachers' attitudes; superior influence, peer influence, student influence, and parental influence serve as variables that predict teachers' subjective norms; self-efficacy and facilitating conditions serve as variables that predict teachers' perceived behavioral control; and cognitive readiness and behavioral readiness serve as variables that predict teachers' self-efficacy.

The relevant hypotheses are explained as follows:

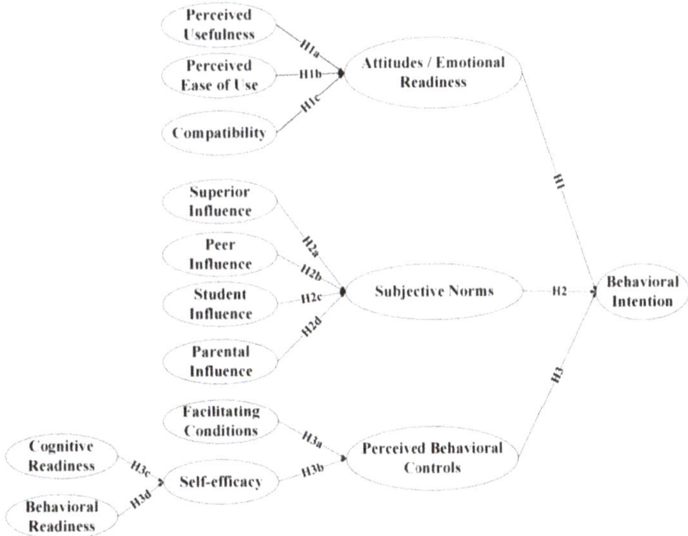

Figure 1. The proposed model with the initial hypotheses.

2.1. Attitudes

Emotional readiness explains how emotions affect teachers' achievements in performing their duties [36]. Instructional emotional readiness reflects the impact of teachers' emotions on their teaching performance. In the DTPB model, attitude is considered to be the degree to which an individual is positive or negative about a particular behavior [15,38], and it has a corresponding impact on whether to implement the target behavior [39], which is similar to the concept of emotional readiness. As a result, to reduce the conceptual overlap of influencing factors and simplify the model, attitudes and emotional preparedness were studied as the same dimension. The attitude of teachers was mainly evaluated from whether they could generate positive emotions in their implementation of STEM education and achieve good results in STEM education. Attitudes' influence on an individual's behavioral intentions has been strongly supported by previous studies [25,40]. Therefore, the following hypothesis was proposed:

Hypothesis 1: *The attitudes/emotional readiness of teachers towards STEM education have a significant impact on teachers' intentions to implement STEM courses.*

According to the deconstruction of the DTPB model for attitude, attitude was also decomposed into three aspects: perceived usefulness, perceived ease of use, and compatibility. Perceived usefulness mainly concerns the perceived STEM benefits of individual teachers when they choose to offer STEM courses. Perceived ease of use refers to the degree to which a person finds it easy to use a specific system [41], as well as the degree of difficulty for teachers who implement STEM courses. Compatibility refers to the degree of fit of technology and existing potential value and experience [42]. It also refers to the degree of adaptation between STEM education and the teacher's teaching concepts, existing teaching experience, etc.

Therefore, the following hypotheses were proposed:

Hypothesis 1a: *Perceived usefulness has a significant impact on teachers' attitudes toward engaging in STEM education.*

Hypothesis 1b: *Perceived ease of use has a significant impact on teachers' attitudes toward engaging in STEM education.*

Hypothesis 1c: *Compatibility has a significant impact on teachers' attitudes toward engaging in STEM education.*

2.2. Subjective Norms

Subjective norms refer to social pressures that lead individuals to perform specific behaviors, which reflects the cognition of important reference disciplines that want individuals to perform or not perform certain behaviors [25]. In this research, subjective norms refer to the teachers' perceptions of the related groups' encouragement and acceptance of their behaviors when they implement STEM education activities. Subjective norms of behavior are usually found to be highly accurate in predicting behavioral intentions [15].

Therefore, the following hypothesis was proposed:

Hypothesis 2: *Subjective norms have a significant impact on teachers' intentions to implement STEM courses.*

As different groups may have different opinions on adopting the same specific behaviors, Taylor and Todd believe that there are three important reference groups in an organizational environment, namely peers, superiors, and subordinates. They suggested breaking down the whole population into these three types of reference groups. In the field of education, in addition to teachers, teachers' superiors, teachers' colleagues, and students, the parents of students are also involved in primary and secondary education [43]. Equally, parents also have some influence on teachers. If teachers are not trusted and respected by the parents of their students, they may develop a sense of vulnerability [44]. Thus, whether parents influence the subjective norms of teachers is also a question worth verifying. We mainly used superiors, colleagues (other teachers), students, and parents as reference groups. Taylor considered that reference groups may influence the regulation of individual subjective norms. Therefore, the following hypotheses were proposed:

Hypothesis 2a: *Superiors' views of STEM education have a significant impact on teachers' subjective norms.*

Hypothesis 2b: *Colleagues' views of STEM education have a significant impact on teachers' subjective norms.*

Hypothesis 2c: *Students' views of STEM education have a significant impact on teachers' subjective norms.*

Hypothesis 2d: *Parents' views of STEM education have a significant impact on teachers' subjective norms.*

2.3. Perceived Behavioral Control

Perceived behavioral control reflects the beliefs about the acquisition of required abilities, resources, and opportunities, or the perception of possible internal and external factors that may hinder behavior execution [15,45]. In this research, perceived behavioral control includes the perception of the internal and external constraints of individual teachers when implementing STEM education. Research has shown that perceived behavioral control is an important determinant of intentions [25]. For example, through empirical investigations, Lin and Williams [19] verified that perceived behavioral control of a higher level was related to stronger STEM instructional intentions. Therefore, the following hypothesis was proposed:

Hypothesis 3: *Teachers' perceived behavioral control over STEM education has a significant impact on teachers' intentions to engage in STEM education.*

The DTPB model divides perceived behavioral control into internal and external constraints, which specifically indicate self-efficacy, resources, and technical factors. In this research, self-efficacy reflects teachers' self-evaluation of their abilities to implement STEM courses, indicating the degree of confidence in implementing STEM education. Greater self-efficacy leads to higher behavioral intentions [25,46]. Sadaf [28] found that self-efficacy was the strongest predictor of teachers' intentions. Convenience refers to technology, resources, and other objective factors available to teachers when implementing STEM courses. A lack of convenience may negatively affect the behavioral intentions of teachers [25].

Readiness is a significant predictor of an individual's propensity to use new technologies [47,48]. Teachers' preparation of knowledge and their self-efficacy in STEM are the key to successful STEM implementation [35]. Furthermore, for different topics such as technology applications, jobs, careers, teaching, and learning, relevant research has confirmed that there is a significant correlation between the degree of readiness and self-efficacy [49–51]. Although no research has been conducted to show that the degree of teachers' STEM readiness directly affects their teaching efficacy, it can be deduced from the relevant literature that the former could significantly predict the latter. To reduce concept repetition and simplify the models, this research chose the dimensions of cognitive and behavioral readiness from the degree of readiness to investigate and discuss whether these two could predict the self-efficacy of teachers. The cognitive readiness of teachers refers to the improvement of thinking and the ability to make cognitive choices to solve STEM problems or complete STEM teaching tasks [33]. Behavioral readiness refers to the attitudes towards STEM, changing emotional reactions, and the changes that can be seen in STEM teaching behaviors [36].

Based on this, we proposed the following hypotheses:

Hypothesis 3a: *The self-efficacy of teachers' STEM education has a significant influence on perceived behavioral control.*

Hypothesis 3b: *Facilitating conditions of teachers' STEM education have a significant influence on perceived behavioral control.*

Hypothesis 3c: *The behavioral readiness of teachers' STEM education has a significant influence on self-efficacy.*

Hypothesis 3d: *The cognitive readiness of teachers' STEM education has a significant influence on self-efficacy.*

3. Methods

3.1. Research Design

Based on the DTPB, this research constructed a model that could explain and predict teachers' STEM teaching preparation extent and behavioral intentions. Based on this model, we designed a questionnaire concerning K12 teachers' intentions to participate in STEM education and analyzed the collected questionnaires with a quantitative research method, i.e., a structural equation model, to verify the hypothetical model.

A total of 532 K12 in-service teachers of all disciplines from different regions in China were randomly contacted to take part in the survey. Informed consent was provided in the survey so that all the participants knew they were participating in an evaluation study and the data they provided were anonymous. Since some of them had selected the same options in the scale selections or wrote that they "never understood STEM education" in open questions, their questionnaires were considered invalid. Having removed invalid questionnaires, there were 479 valid questionnaires left, with an efficiency rate of about 90%. The samples covered the eastern, central, and western regions of China, as well as some other regions. From a gender perspective, more female teachers (59.9% of the total samples) took part in the survey. Young teachers (under 45 years old) accounted for a high percentage of about 73%, and middle-age teachers (over 46 years old) accounted for about 17%. Most of the teachers (78.9%) were undergraduates, and a small percentage (19.2%) had a master's degree. More than half of the teachers (57%) were fairly experienced (with a seniority of more than 10 years). The participants mainly consisted of information technology teachers (32.6%), and they mainly taught in junior and senior high schools.

3.2. Research Instrument

The questionnaire consisted of three parts with 79 questions. The first two parts were compulsory. Teachers' background information was collected in the first part with 7 questions including gender, age, and the school grades they taught. The second part was a scale for the survey of readiness and the factors that influenced their behavioral intention to implement STEM teaching in K12. The scale was mainly adapted from the DTPB scale constructed by Taylor et al. [25] and the scale constructed by Abdullah et al. to measure the readiness of STEM teachers [36]. Altogether, there were 61 adapted DTPB scale contents and 10 readiness scale contents, among which 67 were originally coded scale items and 4 were reverse-coded scale items. These 71 items constructed 15 potential variables, including behavioral intention, attitudes, perceived usefulness, perceived ease of use, compatibility, subjective norms, superior influence, peer influence, student influence, parental influence, perceived behavioral control, self-efficacy, facilitating conditions, behavioral readiness, and cognitive readiness. In order to avoid the antagonistic psychology of the participants, and to reduce the possibility of distortion of the questionnaire contents, we applied reverse-coded scale items for items of perceived ease of use. Most of these variables have been proven to be reliable and efficient in previous research on teachers' behavioral intentions [28–30,52]. At the same time, we applied a 5-point Likert scale for grading, ranging from 1 ("strongly disagree") to 5 ("strongly agree"). The third part was

1 optional item. Here, we applied open-ended questions to collect teachers' opinions about STEM education. Detailed information about the research instrument can be found in Appendix A.

In order to ensure the content validity of the instrument, we invited experts and high-performance teachers of STEM education research to check the appropriateness and clarity of the items and structure of the questionnaire. Modifications were made according to their suggestions. Cronbach's α reliability coefficient was used to evaluate the internal consistency of the research instrument, and the test results can be found in Table 1. With regards to the 15 latent variables in the scale, their Cronbach α value range was 0.848–0.947, all meeting the standard of $\alpha > 0.7$. Moreover, Cronbach's α of the whole scale was 0.980, indicating a high reliability and a reliable measurement index content. In order to verify the applicability of the factor analysis, we conducted the KMO measure test and Bartlett's test. The KMO value was 0.969 (greater than 0.9), which meant that the questionnaire data were appropriate for the factor analysis. Bartlett's test value was 0.000 (less than 0.05), which meant that it met the standard, the data were distributed spherically, and the variables were inter-independent to a certain degree.

Table 1. Reliability results.

Constructs	Number of Measurable Variables	Cronbach's Alpha
Attitudes/Emotional Readiness (AT)	5	0.923
Perceived Usefulness (PU)	5	0.944
Perceptual Ease of Use (PEU)	4	0.858
Compatibility (CO)	5	0.848
Subjective Norms (SN)	3	0.921
Superior Influence (SUI)	5	0.890
Peer Influence (PEI)	5	0.908
Student Influence (STI)	5	0.931
Parental Influence (PAI)	4	0.933
Perceived Behavioral Control (PBC)	5	0.941
Self-efficacy (SE)	6	0.947
Facilitating Conditions (FC)	4	0.919
Cognitive Readiness (CR)	5	0.881
Behavioral Readiness (BR)	5	0.893
Behavioral Intention (BI)	5	0.931
All Structures	71	0.980

4. Analysis and Results

Since the emotional readiness, behavioral readiness, and cognitive readiness in the constructed model of the research did not have a theoretical framework, we applied a method that combined both exploratory factors and confirmatory factors to test the model. We aimed to extract three main dimensions in the factor analysis, and the minimum coefficient that can be displayed was set to 0.5. The rotated component matrix is shown in Table 2. We can see that items CR1–CR5 were concentrated under the first dimension, BR2–BR5 under the second, and BI1–BI5 under the third. These items were reserved, since they accorded with the original expectation. However, since BR1 was originally anticipated to belong to the behavioral readiness dimension together with BR2–BR5, while it turned out to be concentrated under the cognitive readiness dimension, we deleted measurement item BR1 because it did not accord with the expectation.

Table 2. Rotated component matrix.

	Component		
Items	1	2	3
CR1	0.732		
CR2	0.794		
CR3	0.550		
CR4	0.823		
CR5	0.760		
BR1	0.563		
BR2			0.609
BR3			0.852
BR4			0.585
BR5			0.813
AT1		0.816	
AT2		0.867	
AT3		0.880	
AT4		0.810	
AT5		0.777	

Note: Factor loadings <0.50 are omitted, and factor loadings are sorted. CR: cognitive readiness; BR: behavioral readiness; AT: attitudes/emotional readiness.

4.1. Data Analysis of Teachers' STEM Education Readiness

The analysis results of the descriptive statistics of readiness-relevant items showed that the averages of teachers' cognitive, emotional, and behavioral readiness were 3.64, 4.03, and 3.47, respectively, and the average points of each dimension (variable) were all between 3 and 4, which means that the object groups of the research all had a cognitive, emotional, and behavioral readiness level above the medium level. The emotional readiness level was above the others, while the behavioral readiness level was the lowest. Moreover, the correlations r among the three dimensions were 0.564, 0.505, and 0.769, respectively, and all greater than 0. This means that there was a significant positive correlation among different variables.

We conducted a demography variable difference analysis through an independent *t*-test and ANOVA and found that there was no significant correlation between teachers' genders, teaching disciplines, ages, majors, school grades, seniorities, etc., and the overall level of STEM readiness and the three dimensions. Differences were found only in parts of the items. For example, this research took mathematics, physics, chemistry, geography, information technology, comprehensive practices, and scientific disciplines as STEM-relevant disciplines, while Chinese, English, politics, music, etc., were considered STEM-irrelevant disciplines. The results of the independent *t*-test show that the emotional readiness dimension items AT1, AT2, AT3, and AT4, the behavioral readiness dimension items CR4 and CR5, and the cognitive readiness dimension item BR4 all had Sig values of less than 0.05, indicating a significant difference.

4.2. Data Analysis of Teachers' STEM Education Intention

In order to analyze teachers' STEM education intention level and influence factors, we firstly applied SPSS 26.0 to obtain descriptive statistics about the behavioral intention dimension. This was carried out as part of the survey of the current level of behavioral intention of China's K12 teachers to implement STEM education. Secondly, AMOS 26.0 was applied to correct the constructed structural equation model, and to test its reliability and validity. Finally, based on this model, we tested the influence factors of the behavioral intentions of China's K12 teachers to implement STEM education.

As shown in Table 3, the total average level of teachers' behavioral intentions was 3.724, which means that the STEM behavioral intentions of the research subject groups were at a medium to upper level.

Table 3. Descriptive statistics of each item of behavioral intention.

Items	Mean	SD
BI1	3.6	0.852
BI2	3.65	0.84
BI3	3.72	0.804
BI4	3.77	0.796
BI5	3.88	0.792
OA	3.724	

Note. BI: behavioral intention.

4.2.1. Reliability and Validity Tests

This research applied AMOS 26.0 to test the original model and found that the SMC values of CO2 and CR5 were less than 0.5, indicating a weaker corresponding relation between the two measurement items and their dimensions. Therefore, we removed them. We also corrected the model according to the MI (modification indices) standard and found some unreasonable measurement items, including PU1, PEU2, SUI1, SUI3, CI3, CI4, STI4, STI5, PA4, PBC1, SE5, SE6, FC1, and BR5; therefore, they were also deleted. In order to make sure the corrected model could effectively assess teachers' behavioral intention to implement STEM education, this research first conducted the analysis at the measurement level to weigh the measurement effect of the observed variables on the latent variables, and then conducted structural model analysis to test the structural relationships among the latent variables.

1. Measurement-Level Analysis

Through our test, both parts of the research, i.e., the experiment and corrected model, had a Cronbach alpha value greater than 0.8 (Tables 1 and 4), which proves their good internal consistency. Usually, the standardized factor loading, average variance extracted (AVE), SMC, and its composite reliability estimation are used to test the convergent validity of a model [53]. With regards to the AVE, the suggested value is 0.5 or above [54]. The suggested value for the standardized factor loading is 0.50 or above; in an ideal situation, it would be 0.70 or above [54]. The suggested value for the SMC is greater than or equal to 0.5. As shown in the results in Table 4, each value reached the stipulated threshold, showing that the model had good convergent validity.

In order to test the discriminant validity of the model, we compared the index of the square root of the AVE and the absolute value of correlation coefficients among the potential variables. As shown in Table 5, the square root of the AVE of each potential variable was higher than the correlation among the latent variables, which indicates that the discriminant validity of the model was good.

Table 4. Convergent validity analysis in CFA (confirmatory factor analysis).

Constructs	Cronbach's Alpha	Composite Reliability	AVE	Items	Standardized Factor Loading	SMC
Attitudes/Emotional Readiness (AT)	0.923	0.843	0.767	AT1	0.804	0.646
				AT2	0.899	0.808
				AT3	0.916	0.839
				AT4	0.857	0.734
				AT5	0.899	0.808
Perceived Usefulness (PU)	0.939	0.944	0.809	PU2	0.843	0.711
				PU3	0.921	0.848
				PU4	0.911	0.830
				PU5	0.92	0.846
Perceived Ease of Use (PEU)	0.808	0.810	0.588	PEU1	0.713	0.508
				PEU3	0.849	0.721
				PEU4	0.732	0.536
Compatibility (CO)	0.846	0.818	0.532	CO1	0.655	0.429
				CO3	0.665	0.442
				CO4	0.868	0.753
				CO5	0.843	0.711
Subjective Norms (SN)	0.921	0.923	0.799	SN1	0.925	0.856
				SN2	0.912	0.832
				SN3	0.845	0.714
Superior Influence (SUI)	0.838	0.856	0.666	SUI2	0.751	0.564
				SUI4	0.848	0.719
				SUI5	0.901	0.812
Peer Influence (PEI)	0.864	0.902	0.756	CI1	0.931	0.867
				CI2	0.767	0.588
				CI5	0.895	0.801
Student Influence (STI)	0.933	0.935	0.827	STI1	0.889	0.790
				STI2	0.938	0.880
				STI3	0.962	0.925
Parental Influence (PAI)	0.953	0.954	0.874	PA1	0.822	0.676
				PA2	0.837	0.701
				PA3	0.939	0.882
Perceived Behavioral Control (PBC)	0.938	0.938	0.791	PBC2	0.923	0.852
				PBC3	0.901	0.812
				PBC4	0.891	0.794
				PBC5	0.818	0.669
Self-efficacy (SE)	0.937	0.935	0.782	SE1	0.921	0.848
				SE2	0.936	0.876
				SE3	0.875	0.766
				SE4	0.76	0.578

Table 4. *Cont.*

Constructs	Cronbach's Alpha	Composite Reliability	AVE	Items	Standardized Factor Loading	SMC
Facilitating Conditions (FC)	0.937	0.936	0.830	FC2	0.826	0.682
				FC3	0.775	0.601
				FC4	0.799	0.638
Cognitive Readiness (CR)	0.876	0.869	0.625	CR1	0.851	0.724
				CR2	0.78	0.608
				CR3	0.738	0.545
				CR4	0.836	0.699
Behavioral Readiness (BR)	0.822	0.833	0.626	BR2	0.825	0.681
				BR3	0.876	0.767
				BR4	0.83	0.689
Behavioral Intention (BI)	0.931	0.921	0.699	BI1	0.804	0.646
				BI2	0.899	0.808
				BI3	0.916	0.839
				BI4	0.857	0.734
				BI5	0.899	0.808

Note. AVE: average variance extracted values; SMC: squared multiple correlations.

Table 5. Discriminant validity result and root of average variance extracted.

	AT	PU	PEU	CO	SN	SUI	PEI	STI	PAI	PBC	SE	FC	CR	BR	BI
AT	0.876														
PU	0.839	0.899													
PEU	0.058	−0.007	0.767												
CO	0.636	0.608	−0.081	0.730											
SN	0.453	0.413	−0.015	0.584	0.894										
SUI	0.534	0.519	−0.05	0.611	0.734	0.816									
PEI	0.577	0.535	0.007	0.643	0.701	0.765	0.869								
STI	0.569	0.549	0.034	0.62	0.628	0.655	0.763	0.909							
PAI	0.475	0.415	0.065	0.535	0.643	0.616	0.722	0.772	0.935						
PBC	0.399	0.326	0.198	0.516	0.442	0.43	0.551	0.567	0.552	0.889					
SE	0.447	0.394	0.209	0.545	0.458	0.465	0.559	0.59	0.548	0.882	0.884				
FC	0.235	0.191	0.12	0.336	0.501	0.477	0.507	0.492	0.511	0.639	0.587	0.911			
CR	0.529	0.488	0.117	0.582	0.481	0.523	0.614	0.577	0.528	0.67	0.716	0.63	0.790		
BR	0.487	0.447	0.178	0.523	0.491	0.503	0.607	0.627	0.608	0.706	0.721	0.593	0.748	0.791	
BI	0.613	0.599	0.159	0.643	0.516	0.584	0.646	0.637	0.605	0.63	0.709	0.474	0.744	0.779	0.836

Note. AT: attitudes/emotional readiness; PU: perceived usefulness; PEU: perceived ease of use; CO: compatibility; SN: subjective norms; SUI: superior influence; PEI: peer influence; STI: student influence; PAI: parental influence; PBC: perceived behavioral control; SE: self-efficacy; FC: facilitating conditions; CR: cognitive readiness; BR: behavioral readiness; BI: behavioral intention.

2. Structural model analysis

In order to evaluate the model fit, according to the research of Hu and Bentler [55], Kashy et al. [56], Widaman [57], and Ejaz Ahmed [58], we applied the comparative fit index (CFI), incremental fit index (IFI), Tucker–53Lewis index (TLI), chi-square (x^2)-to-degree of freedom (df) ratio, root mean square error of approximation (RMSEA), and standardized root mean residual (SRMR) values. The traditional rule of thumb in CFA (confirmatory factor analysis) stipulates that when the x^2/df value is less than 3, the SRMR is less than 0.08, the IFI, TLI, and CFI are all greater than 0.9, and the RMSEA is less than 0.08, the model fit can be considered a good one. This research applied the maximum likelihood (ML) estimation technique to evaluate the model parameters, and the model fit indices we obtained are shown in Table 6, all of which met the CFA stipulated index. The x^2/df (1.954), SRMR (0.0485), and RSMEA (0.045) were all lower than the maximum requirement value, and the IFI (0.953), TLI (0.948), and CFI (0.953) were all higher than the minimum requirement value, which proves that the model fitted well with the data.

Table 6. Statistics for the final models.

	Obtained	Values for Excellent Fit	Values for Good Fit	Evaluation of Fit for Final Model
x^2/df	1.954	≤ 2	≤ 3	Excellent fit
SRMR	0.0485	<0.05	<0.08 or 0.09	Excellent fit
IFI	0.953	≥ 0.95	≥ 0.90	Excellent fit
TLI	0.948	≥ 0.95	≥ 0.90	Good fit
CFI	0.953	≥ 0.95	≥ 0.90	Excellent fit
RSMEA	0.045	<0.05	<0.08	Excellent fit

Note. The fit indices are based on the work of Hooper et al. [59], Hu and Bentler [55], and Kline [60]; x^2/df: chi-square-to-degree of freedom ratio; SRMR: standardized root mean square residual; IFI: incremental fit index; TLI: Tucker–Lewis index; CFI: cumulative fit index; RMSEA: root mean square error of approximation.

4.2.2. Assessment of Hypothesized Relations

According to the model fit test, the fit of this model was good. Through the path coefficients and hypothesis testing, we obtained the model of teachers' intentions to implement STEM education, as shown in Table 7. When the value of the critical ratio (C.R.) was 1.96 or higher, the value of the coefficient (P) was considered significant at the level of 0.05 [53]. In this research, we used the path coefficients (β) of the fitted model and the p-value in the analysis and verified that 12 hypotheses, including H1, H3, H1a, H1b, H1c, H2a, H2b, H2d, H3a, H3b, H3c, and H3d were valid, while two hypotheses, namely H2 and H2c, were rejected. Therefore, we constructed the model of teachers' intentions to implement STEM education and its factor path coefficients (as shown in Figure 2). Every path constructed from one latent variable to another showed their relationships, and the path coefficient points of the valid hypotheses were between 0.093 and 0.890, which indicates a positive correlation among the exogenous variables and endogenous variables of the twelve valid hypotheses. In other words, the paths had statistical significance and supported the model.

The value of the SMC (squared multiple correlations) represents the strength of the structural relationships [61]. The value of the SMC ranges between 0 and 1, where stronger relationships are indicated by values close to 1 [62]. The explained maximum variance of the three exogenous variables (perceived behavioral control, subjective norms, and attitudes) to behavioral intention was 0.90. The SMC value of the main endogenous latent construct, i.e., behavioral intention, was 0.53. That is, 53% of the variance among behavioral intention, attitudes, subjective norms, and perceived behavioral control could be explained by behavioral intention.

Table 7. Path coefficients and hypothesis testing; *** $p < 0.001$.

Hypothesis	Parameter	Path Coefficient (β)	S.E.	C.R.	*p*-Value	Whether the Hypothesis Is Established
H1	AT→BI	0.331	0.049	7.144	***	Yes
H2	SN→BI	0.002	0.033	0.056	0.955	No
H3	PBC→BI	0.880	0.068	12.657	***	Yes
H1a	PU→AT	0.671	0.046	15.698	***	Yes
H1b	PEU→AT	0.079	0.026	3.067	0.002	Yes
H1c	CO→AT	0.265	0.038	7.058	***	Yes
H2a	SUI→SN	0.550	0.079	7.670	***	Yes
H2b	PEI→SN	0.198	0.102	2.428	0.015	Yes
H2c	STI→SN	0.016	0.066	0.237	0.813	No
H2d	PAI→SN	0.141	0.056	2.281	0.023	Yes
H3a	FC→PBC	0.093	0.020	3.517	***	Yes
H3b	SE→PBC	0.890	0.036	21.615	***	Yes
H3c	CR→SE	0.270	0.128	2.571	0.01	Yes
H3d	BR→SE	0.594	0.120	5.340	***	Yes

Note. S.E.: standard error; C.R.: critical ratio.

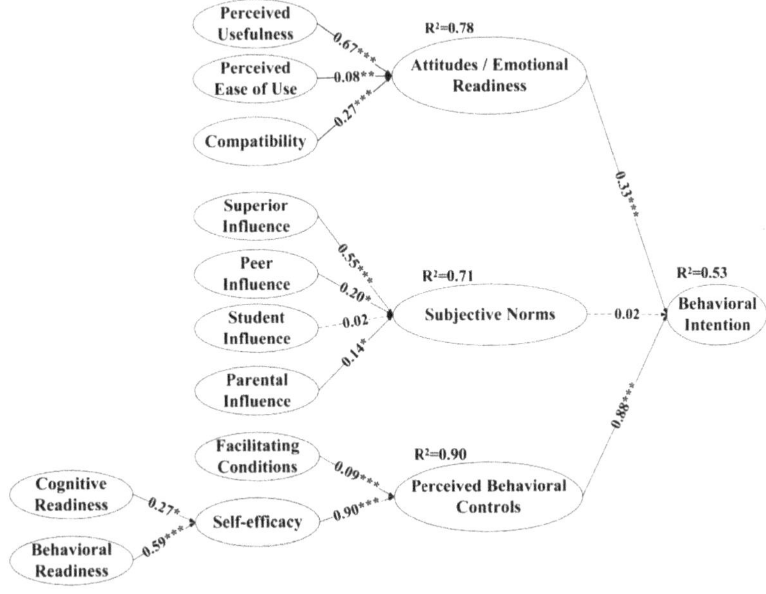

Figure 2. Final model with the coefficients (β) and *p*-values. Note. Dotted lines indicate an insignificant relationship; * $p < 0.05$, ** $p < 0.01$, and *** $p < 0.001$.

5. Discussion

The data reported here indicate that teachers' readiness for STEM education was generally in the high to average range, among which emotional readiness was the highest. This finding demonstrates that K12 teachers have positive attitudes towards STEM education, and that they were satisfied with the effects of STEM education and were willing to apply

STEM education in their classes. Research conducted in Indonesia evaluating teachers' attitudes towards STEM education found similar results [63]. From a cognitive aspect, the teachers' moderate level of readiness indicates that they had a basic understanding of STEM education. Though the teachers had a certain understanding of STEM education in terms of knowledge and methods as well as the teachers' role, their understanding of comparatively sophisticated STEM programs was still scant. Research conducted by DanyiZheng [33] obtained similar results. The level of teachers' readiness from a behavioral aspect was relatively low compared to that of cognitive readiness and emotional readiness. According to the questionnaire data, teachers' behavioral readiness focused on making elaborate plans before STEM education, but most teachers lacked enough time to implement STEM education in practice. This finding is similar to that of Abdullah [36], which suggested that teachers' behavioral readiness was at an intermediate level and most teachers did not have ample time to implement STEM education. Variables including the teachers' genders, disciplines, ages, majors, teaching stages, and seniorities did not have a significant correlation with the overall level of teachers' readiness and its three aspects, namely cognitive, behavioral, and emotional readiness, for STEM education. More relevant variables should be further explored.

The statistics of intention demonstrated that teachers' behavioral intentions were in the high to average range. This finding, although different from that of Peng [20] and Li [21], is consistent with that of Tse [22]. It was shown that with the introduction of relevant policies and the development of STEM education practices, many K12 teachers realized the importance of interdisciplinary teaching and were willing to implement STEM education in practice.

The data collected from SEM indicated that teachers' behavioral intentions to implement STEM education were mainly impacted by their attitudes and perceived behavioral control, with strong positive effects, while subjective norms did not generate prominent effects. Relevant studies have verified that attitudes, subjective norms, and perceived behavioral control are impact factors on behavioral intention [64–67]. Furthermore, some researchers have also found that subjective norms were normally a weak predictor of behavioral intention [28,68,69]. In addition, research by Atsoglou [70] and Haya Ajjan [71] argued that attitudes and perceived behavioral control were comparatively strong positive impact factors, while subjective norms had no impact on behavioral intention. Haya Ajjan explained that this was because teachers had a high degree of independence in improving the classroom environment.

Negative attitudes may cause teachers to be reluctant to work in STEM education. Three factors, namely ease of use, perceived usefulness, and compatibility, all had impacts on teachers' attitudes, with perceived usefulness having the strongest. Research by Ayesha Sadaf [28,72,73] and Mdutshekelwa Ndlovu [30] pointed out that perceived usefulness, compared with ease of use and compatibility, exerted the greatest impacts on teachers' attitudes. Consequently, to improve teachers' attitudes towards STEM, corresponding interventions should mainly focus on enhancing teachers' awareness that teachers' actions contribute to results, and on guiding teachers to positively evaluate results. Only when teachers realize that STEM education is helpful to their work and can improve students' abilities in all aspects will they be willing to adopt STEM education.

The greater the sense of control teachers have over STEM education, the more willing they are to apply STEM education. This study found that both self-efficacy and facilitating conditions had positive impacts on perceived behavioral control for teachers in STEM education, of which self-efficacy was the strongest impact factor. In previous studies, researchers such as Ayesha [28,73], Mdutshekelwa Ndlovu [30], and Kuan-Chuan Tao [31] all confirmed that self-efficacy and facilitating conditions were impact factors of perceived behavioral control, and most of these studies [28,30,73] showed that, compared with facilitating conditions, self-efficacy had the most significant impacts on behavioral control, which is consistent with the conclusions drawn in this study. This indicates that if schools can provide effective support with resources, it will be easier and more convenient for teachers to carry out teaching activities, and, more importantly, it will improve teachers' confidence in teaching. Confidence is built on success, and thus, a reliable way to improve self-efficacy is to repeatedly experience a sense of success in the middle of completing a certain task. Therefore, to improve their proficiency in STEM education, teachers should strive for successful experiences and belief in conducting STEM education, and thus believe that they are qualified for success. After that, teachers should have high "self-efficacy" with confidence in their capabilities of conducting STEM education and consequently be willing to implement STEM education.

The significant impacts of teachers' readiness on their intentions to implement STEM education were tested in the hypotheses established by this model. Currently, there is no research exploring the impacts of teachers' readiness on their intentions to implement STEM education, but some studies have confirmed that teachers' readiness for AI significantly influenced their intentions to implement AI teaching [48]. Highly consistent with other studies [21,22], this study found that emotional readiness had a significant correlation with teachers' intentions to implement STEM education. In addition, both behavioral readiness and cognitive readiness had positive impacts on the self-efficacy of teachers in implementing STEM education, among which behavioral readiness was the strongest predictor. This shows that the more teachers understand and prepare for STEM education, the more they will enhance their self-efficacy, which has been verified to be an effective factor in improving teachers' perceived behavioral control in STEM education. Although there are no studies directly exploring the impact of teachers' cognitive readiness and behavioral readiness on teachers' self-efficacy, researchers have found that TPACK (technological pedagogical content knowledge) had an indirect impact on teachers' STEM educational intention [22]. In addition, teachers who fail to prepare themselves well with knowledge of STEM [74] such as science, mathematics, and engineering [75] may be afraid of teaching design-related engineering content to students [76], thus having a low self-efficacy in STEM education. For many primary and secondary school teachers in China, STEM education is still a relatively unfamiliar teaching method. For example, in this study, some primary and secondary school teachers mentioned that they did not understand "what STEM education is". Though the Chinese government encourages and guides colleges and enterprises to commence STEM training for teachers, such training still hardly attracts teachers to participate and receives unsatisfactory results due to the incomplete training mechanism. Teachers' self-efficacy in teaching is usually highly intertwined with the corresponding knowledge and positive experience [77–79]. Insufficient cognitive and behavioral preparation would naturally undermine teachers' confidence in implementing STEM education. Therefore, with the aim of promoting STEM education, the top-level design of STEM professional development for teachers should be strengthened from an emotional aspect (attitudes), cognitive aspect, and behavioral aspect.

6. Conclusions and Further Developments

6.1. Conclusions and Suggestions

This study established and tested a research model of the behavioral intention of K12 teachers to implement STEM education based on the DTPB (decomposed theory of planned behavior) and teachers' readiness. The results of this study demonstrate that the model fit the data well and was suitable for the research of teachers' behavioral intentions to implement STEM education. Constructs including self-efficacy, perceived behavioral control, perceived usefulness, behavioral readiness, attitudes, cognitive readiness, compatibility, facilitating conditions, and perceived ease of use had direct or indirect significant impacts on K12 teachers' behavioral intentions to implement STEM education. Amid the above-mentioned constructs, self-efficacy, perceived behavioral control, perceived usefulness, behavioral readiness, attitudes, and cognitive readiness had comparatively stronger impacts. However, there was no evidence that subjective norms had significant impacts on teachers' intentions to implement STEM education. In addition, based on the analysis of Chinese K12 teachers' readiness from emotional, cognitive, and behavioral aspects, this study found that teachers' readiness for STEM education was a strong predictor of teachers' behavioral intentions, which is consistent with the results of Ayanwale et al. [48]. This study also indicated that Chinese K12 teachers' readiness for and intentions to implement STEM education were both in the high to average range. "Behavioral readiness" was a strong predictor of teaching intention to implement STEM education, but it maintained at a relatively minimal level.

The findings of this study have certain reference significance for policymakers and educators, and serve to promote STEM education and truly improve students' STEM literacy and problem-solving abilities in an interdisciplinary and innovative manner [2]. Compared with studies of its kind, this study achieved breakthroughs in terms of the sample size, the scope of the research objects, and the basic theoretical model and creatively proposed adding teachers' readiness into the DTPB model. Through analyses of K12 teachers' readiness and impact factors of behavioral intentions to implement STEM education, this study provides a reference for the development of STEM education in primary and secondary schools. Policymakers can follow the findings of this study to formulate practical top-level strategic plans and effective STEM teacher professional development programs around key impact factors to strengthen teachers' interdisciplinary teaching literacy.

Given the results of this study, how to improve teachers' teaching intention to implement STEM education can be discussed from two aspects: attitudes and perceived behavioral control. Firstly, only when teachers hold positive attitudes towards the practicability of a certain action can they adopt positive attitudes towards such behavior. Perceived usefulness was the most significant predictor of attitudes, which demonstrates that the teachers paid great attention to the effects of the implementation of teaching behavior; thus, it was crucial to improving teachers' sense of value in teaching. This inspires us to encourage more teachers to develop STEM projects and implement STEM education in real practice so that teachers can compare their original teaching methods with practical cases and truly experience the value of STEM education. Secondly, self-efficacy was the strongest predictor of perceived behavioral control, meaning it is a necessity to enhance teachers' self-efficacy in STEM education. The authors suggest that corresponding interventions should be implemented based on cognitive readiness and behavioral readiness. In terms of cognitive readiness, teachers are expected to pay more attention to relevant requirements, learn from excellent cases, and spend time analyzing them in detail to deepen their understanding of STEM education and their cognition of teachers' roles. In addition, schools can organize relevant lectures and competitions for teachers. With regard to behavioral readiness, most teachers noted that they lack enough time to implement STEM education. Thus, the workload of teachers on non-teaching tasks should be reduced to ensure that they have enough time to prepare for STEM courses. Teachers should also improve teaching strategies, integrate STEM education into discipline education, and determine specific teaching contents tailored to students with different characteristics to make the class more interesting. Considering that support, guidance, and leadership are critical for teachers to shift from traditional teaching styles [80], schools should strengthen their pedagogic support for STEM education and provide opportunities for cooperation among teachers of different disciplines. Finally, favorable facilitating conditions can also improve teachers' perceived behavioral control, thereby enhancing teachers' teaching intentions. The state and government can provide certain funds or formulate policies so as to provide corresponding educational spaces, resources, technologies, and expert guidance, among others, for conducting STEM education. By establishing a cross-school and cross-regional STEM community of practice, a balanced layout of "facilitating conditions" can be encouraged.

6.2. Limitations and Further Development

Despite the contributions of this study, some limitations remain to be addressed. Firstly, since the data collected in this questionnaire were not balanced from different regions, i.e., there were relatively small sample sizes from central and western China, this study did not explore the impact of regions on teachers' readiness for and intentions to implement STEM education. Previous studies have proven that data from different regions could predict behavior more accurately and meaningfully [81]. Therefore, more studies investigating whether the educational development of regions has impacts on teachers' readiness and teaching intentions to implement STEM education should be carried out. Secondly, the significant impacts of subjective norms on teachers' intentions to implement STEM education have yet to be tested, but this does not indicate that there is no significant impact between them. Research on non-STEM education has proven that subjective norms are among the impact factors of behavioral intention [67,82]. Further studies can focus on the correlation between subjective norms and teachers' intentions to implement STEM education. Thirdly, most of the previous studies focused on explaining and predicting behavioral intention and actions. Since the DTPB model can intervene in behavior, further studies can explore how to intervene in STEM teaching behavior based on the DTPB.

Author Contributions: Conceptualization, P.W. and L.Y.; methodology, P.W. and L.Y.; software, B.L. and Y.W.; validation, L.Y.; formal analysis, B.L. and Y.W.; investigation, P.W. and X.H.; project administration, P.W. and X.H.; writing—original draft preparation, B.L., Q.L., Y.W., and J.H.; writing—review and editing, P.W., L.Y., and X.H. All authors have read and agreed to the published version of the manuscript.

Funding: This research was financially supported by the Development Planning Project of Philosophy and Social Science in Guangzhou (2020GZYB27 and 2022GZYB50).

Institutional Review Board Statement: Not applicable.

Informed Consent Statement: The informed consent was obtained in the survey.

Data Availability Statement: Not applicable.

Conflicts of Interest: The authors declare no conflict of interest.

Appendix A

Attitudes/Emotional Readiness (AT)

AT1: I enjoy implementing STEM education approach in my lesson.

AT2: I am satisfied with the implementation of STEM education approach as it is able to stimulate students' insterest in the class.

AT3: I am satisfied with the implementation of STEM education approach as it enables students to use knowledge of different disciplines to solve real-life problems.

AT4: I think it is interesting to implement STEM education.

AT5: I think it is a good idea for students to carry out STEM courses.

Perceived Usefulness (PU)

PU1: I think STEM education is an important driving force for the rapid development of the country

PU2: I think STEM education can improve students' employment competitiveness

PU3: I think STEM education helps to cultivate students' comprehensive quality and innovative ability to solve problems

PU4: I think STEM education helps to cultivate students' cooperation ability, sense of responsibility and team spirit

PU5: I think STEM education is helpful to cultivate students' scientific inquiry ability and practical ability

Perceptual Ease of Use (PEU)

PEU1: I think it is difficult for me to implement STEM education

PEU2: I think the implementation of STEM education will make me feel stressed and anxious

PEU3: I think it is difficult to become a skilled STEM teacher

PEU4: I hesitate to implement STEM education because it requires a lot of extra preparation and effort

Compatibility (CO)

CO1: The interdisciplinary project-based teaching method of STEM education can be used with my original teaching method

CO2: The teaching method of STEM education is similar to my original teaching method

CO3: My original idea of educating people is similar to that of STEM education for cultivating compound innovative talents

CO4: My original evaluation concept is similar to STEM education's multiple evaluation concept based on real situations

CO5: Carrying out STEM education is in line with my current teaching needs.

Subjective Norms (SN)

SN1: People who affect my teaching behavior (e.g., superiors, peers, students, students' parents, etc.) think STEM education is an important way of education

SN2: People who affect my teaching behavior (e.g., superiors, peers, students, students' parents, etc.)will think that I should carry out STEM education

SN3: People who affect my teaching behavior (e.g., superiors, peers, students, students' parents, etc.) think that carrying out STEM education benefits me a lot

Superior Influence (SUI)

SUI1: Superiors believe that STEM education should be carried out

SUI2: Superiors believe that STEM education is an important way of education

SUI3: If superiors advocate STEM education, I will try this education method

SUI4: The requirements and suggestions of superiors on STEM education are very important to me

SUI5: Superiors believe that STEM education is conducive to teachers' professional development

Peer Influence (PEI)

PEI1: My colleagues think that STEM education should be carried out

PEI2: My colleagues think STEM education is an important way of education

PEI3: If my colleagues invite me to carry out STEM education, I will actively cooperate with them

PEI4: If my colleagues are conducting STEM education, I will do the same

PEI5: Colleagues' opinions and suggestions on STEM education are very important to me

Student Influence (STI)

STI1: Students believe that STEM education should be carried out

STI2: Students believe that STEM education is an important way of education

STI3: Students strongly support me to carry out STEM education in the class

STI4: The actual needs of students for STEM education are very important to me

STI5: Students' opinions and suggestions on STEM education are very important to me

Parental Influence (PAI)

PAI1: Students' parents believe that STEM education should be carried out

PAI2: Students' parents believe thinks STEM education is an important way of education

PAI3: Students' parents strongly support me to carry out STEM education in the class

PAI4: The parents' opinions and suggestions on STEM education are very important to me

Perceived Behavioral Controls (PBC)

PBC1: I can carry out STEM education activities
PBC2: I can control the whole process of STEM education activities
PBC3: I have enough resources to carry out STEM education activities
PBC4: I have sufficient knowledge to carry out STEM education activities
PBC5: I have sufficient teaching skills to carry out STEM education activities

Self-efficacy (SE)

SE1: I can properly assess students using various assessment strategies in STEM education
SE2: I can solve problems raised by students during STEM activities
SE3: I can establish rules for activities based on the characteristics of STEM education to keep them running smoothly
SE4: I can get students to follow the rules of STEM education activities
SE5: I can motivate students who have a low interest in STEM activities
SE6: I can help students to innovate during STEM activities

Facilitating Conditions (FC)

FC1: When I encounter difficulties in STEM education, I can easily find relevant people for help
FC2: My school has teachers for STEM education
FC3: My school has the necessary laboratory conditions for STEM education
FC4: My school has carried out teacher training of STEM education

Cognitive Readiness (CR)

CR1: I understand the objectives of promoting STEM education drawn up by Ministry of Education.
CR2: I understand the teacher's role in implementing STEM education at school.
CR3: I understand the more mature STEM programs at the moment.
CR4: I am responsible for ensuring that students have fun and meaningful learning experiences during STEM activities
CR5: I need to understand and master various knowledge contents and implementation methods of STEM education

Behavioral Readiness (BR)

BR1: I am prepared to attend STEM education training courses to enhance my skills and knowledge.
BR2: I always analyze the existing personality characteristics and cognitive levels of students in order to carry out STEM education.
BR3: I have enough time to implement STEM education after my subject teaching work.
BR4: I do rigorous preparations before implementing STEM education approach in the classroom
BR5: I'm going to work with my colleagues to organize STEM education.

Behavioral Intention (BI)

BI1: I plan to carry out STEM education in the future subject teaching
BI2: I will keep abreast of the latest STEM education information
BI3: I plan to spend time in the future to learn how to carry out STEM education
BI4: I will encourage and be willing to collaborate with colleagues in STEM education
BI5: I will encourage students to participate in STEM project activities

Answer the following question

Please talk about your views on STEM education (Optional):

References

1. Ross, P.M.; Scanes, E.; Poronnik, P.; Coates, H.; Locke, W. Understanding STEM Academics' Responses and Resilience to Educational Reform of Academic Roles in Higher Education. *Int. J. STEM Educ.* **2022**, *9*, 11. [CrossRef] [PubMed]
2. Bybee, R.W. *STEM Education: Challenges and Opportunities*; National Science Teachers Association Press: Arlington, TX, USA, 2013.
3. Honey, M.; Pearson, G.; Schweingruber, H.A. (Eds.) *STEM Integration in K-12 Education*; National Academies Press: Washington, DC, USA, 2014.
4. Fioriello, P. Understanding the Basics of STEM Education. Available online: https://drpfconsults.com/understanding-the-basics-of-stem-education/ (accessed on 17 November 2022).
5. English, L.D. STEM Education K-12: Perspectives on Integration. *Int. J. STEM Educ.* **2016**, *3*, 3. [CrossRef]
6. Madden, M.E.; Baxter, M.; Beauchamp, H.; Bouchard, K.; Habermas, D.; Huff, M.; Ladd, B.; Pearon, J.; Plague, G. Rethinking STEM Education: An Interdisciplinary STEAM Curriculum. *Procedia Comput. Sci.* **2013**, *20*, 541–546. [CrossRef]
7. Thomas, T.A. Elementary Teachers' Receptivity to Integrated Science, Technology, Engineering, and Mathematics (STEM) Education in the Elementary Grades. Ph.D. Thesis, University of Nevada, Reno, NV, USA, 2014.
8. Ercan, S.; Şahin, F. Fen Eğitiminde Mühendislik Uygulamalarının Kullanımı: Tasarım Temelli Fen Eğitiminin Öğrencilerin Akademik Başarıları Üzerine Etkisi. *Necatibey Eğitim Fakültesi Elektron. Fen Ve Mat. Eğitimi Derg.* **2015**, *9*, 128–164. [CrossRef]
9. National Science Foundation. Shaping the Future: New Expectations for Undergraduate Education in Science, Mathematics, Engineering, and Technology. Available online: https://files.eric.ed.gov/fulltext/ED404158.pdf (accessed on 4 October 2022).
10. Pinasa, S.; Srisook, L. STEM Education Project-Based and Robotic Learning Activities Impacting on Creativity and Attitude of Grade 11 Students in Khon Kaen Wittayayon School. *J. Phys. Conf. Ser.* **2019**, *1340*, 1–6. [CrossRef]
11. Altan, E.B.; Tan, S. Concepts of creativity in design based learning in STEM education. *Int. J. Technol. Des. Educ.* **2021**, *31*, 503–529.
12. Committee on Successful Out-of-School STEM Learning. Identifying and Supporting Productive STEM Programs in Out-of-School Settings. Available online: https://edc.org/identifying-and-supporting-productive-stem-programs-out-school-settings (accessed on 4 October 2022).
13. Ministry of Education of the People's Republic of China. Notice of the Ministry of Education on the 13th Five-Year Plan Issued by Education. Available online: http://www.moe.gov.cn/srcsite/A16/s3342/201606/t20160622_269367.html (accessed on 4 October 2022).
14. Rahman, N.A.; Rosli, R.; Rambley, A.S. Mathematical Teachers' Knowledge of STEM-Based Education. *J. Phys. Conf. Ser.* **2021**, *1806*, 1–5. [CrossRef]
15. Ajzen, I. The Theory of Planned Behavior. *Organ. Behav. Hum. Decis. Process.* **1991**, *50*, 179–211. [CrossRef]
16. Steyn, G.M. Exploring Factors That Influence the Effective Implementation of Professional Development Programmes on Invitational Education. *J. Invit. Theory Pract.* **2005**, *11*, 7–34. [CrossRef]
17. Kurup, P.M.; Li, X.; Powell, G.; Brown, M. Building Future Primary Teachers' Capacity in STEM: Based on a Platform of Beliefs, Understandings and Intentions. *Int. J. STEM Educ.* **2019**, *6*, 1–14. [CrossRef]
18. Nikolopoulou, K.; Gialamas, V.; Lavidas, K.; Komis, V. Teachers' Readiness to Adopt Mobile Learning in Classrooms: A Study in Greece. *Technol. Knowl. Learn.* **2021**, *26*, 53–77. [CrossRef]
19. Lin, K.-Y.; Williams, P.J. Taiwanese Preservice Teachers' Science, Technology, Engineering, and Mathematics Teaching Intention. *Int. J. Sci. Math. Educ.* **2016**, *14*, 1021–1036. [CrossRef]
20. Peng, M.; Zhu, D. Chinese Teachers' Perception of STEM Education:Based on the Qualitative Analysis of 52 STEM Teachers by NVivo11. Software. *Educ. Dev. Res.* **2020**, *40*, 60–65.
21. Li, C.; Kevin Kam, W.K.; Zhang, M. Physical Education Teachers' Behaviors and Intentions of Integrating STEM Education in Teaching. *Phys. Educ.* **2019**, *76*, 1086–1101. [CrossRef]
22. Cheung, H.C.; Tse, A.W.C. Hong Kong Science In-Service Teachers' Behavioural Intention towards STEM Education and Their Technological Pedagogical Content Knowledge (TPACK). In Proceedings of the 2021 IEEE International Conference on Engineering, Technology & Education (TALE), Wuhan, China, 5–8 December 2021; pp. 630–637. Available online: https://ieeexplore.ieee.org/document/9678933 (accessed on 5 October 2022).
23. Günbatar, M.S.; Bakırcı, H. STEM Teaching Intention and Computational Thinking Skills of Pre-Service Teachers. *Educ. Inf. Technol.* **2019**, *24*, 1615–1629. [CrossRef]
24. Davis, F.D.; Bagozzi, R.P.; Warshaw, P.R. User Acceptance of Computer Technology: A Comparison of Two Theoretical Models. *Manag. Sci.* **1989**, *35*, 982–1003. [CrossRef]
25. Taylor, S.; Todd, P.A. Understanding Information Technology Usage: A Test of Competing Models. *Inf. Syst. Res.* **1995**, *6*, 144–176. [CrossRef]
26. Smarkola, C. Efficacy of a Planned Behavior Model: Beliefs That Contribute to Computer Usage Intentions of Student Teachers and Experienced Teachers. *Comput. Hum. Behav.* **2008**, *24*, 1196–1215. [CrossRef]
27. Giovanis, A.; Athanasopoulou, P.; Assimakopoulos, C.; Sarmaniotis, C. Adoption of Mobile Banking Services. *Int. J. Bank Mark.* **2019**, *37*, 1165–1189. [CrossRef]
28. Sadaf, A.; Newby, T.J.; Ertmer, P.A. An Investigation of the Factors That Influence Preservice Teachers' Intentions and Integration of Web 2.0 Tools. *Educ. Technol. Res. Dev.* **2016**, *64*, 37–64. [CrossRef]
29. Alkhayat, L.; Ernest, J.; LaChenaye, J. Exploring Kuwaiti Preservice Early Childhood Teachers' Beliefs about Using Web 2.0 Technologies. *Early Child. Educ. J.* **2020**, *48*, 715–725. [CrossRef]

30. Ndlovu, M.; Ramdhany, V.; Spangenberg, E.D.; Govender, R. Preservice Teachers' Beliefs and Intentions about Integrating Mathematics Teaching and Learning ICTs in Their Classrooms. *ZDM* **2020**, *52*, 1365–1380. [CrossRef]

31. Tao, K.-C.; Hsieh, T.-F.; Hsu, C.-Y.; Yang, J.-J.; Sia, W.Y. A Study on the Intention of Using Online E-Learning Platform to Assist in Teaching of Junior High School Teachers. In Proceedings of the 2019 3rd International Conference on Education and Multimedia Technology, Nagoya, Japan, 22–25 July 2019; Volume 7, pp. 41–44.

32. Hersey, P. *The Situational Leader*; Center For Leadership Studies: Escondido, Calif, 2008.

33. Zheng, D.; Tse, A. The Readiness of Shenzhen Primary Science Teachers Using Problem-Based Learning to Implement STEM Education. In Proceedings of the 2021 IEEE International Conference on Engineering, Technology & Education (TALE), Wuhan, China, 5–8 December 2021; pp. 609–614. Available online: https://ieeexplore.ieee.org/document/9678733 (accessed on 10 October 2022).

34. Hata, N.F.M.; Mahmud, S.N.D. Kesediaan guru Sains dan Matematik dalam melaksanakan pendidikan STEM dari aspek pengetahuan, sikap dan pengalaman mengajar (Teachers' readiness in implementing STEM education from knowledge, attitude and teaching experience aspects). *Akademika* **2020**, *90*, 85–101.

35. Adams, D.; Mohamed, A.; Moosa, V.; Shareefa, M. Teachers' readiness for inclusive education in a developing country: Fantasy or possibility? *Educ. Stud.* **2021**, *48*, 1–18. [CrossRef]

36. Abdullah, A.H.; Hamzah, M.H.; Hussin, R.H.S.R.; Kohar, U.H.A.; Rahman, S.N.S.A.; Junaidi, J. Teachers' Readiness in Implementing Science, Technology, Engineering and Mathematics (STEM) Education from the Cognitive, Affective and Behavioural Aspects. In Proceedings of the 2017 IEEE 6th International Conference on Teaching, Assessment, and Learning for Engineering (TALE), Hong Kong, China,, 12–14 December 2017; pp. 6–12. Available online: https://ieeexplore.ieee.org/document/8252295 (accessed on 10 October 2022).

37. Mustafa, N.; Ismail, Z.; Tasir, Z.; Said, M.N.H.M. Teacher Readiness Towards Integrating Stem Education into Teaching and Learning. In Proceedings of the Asia International Multidisciplinary Conference (AIMC 2017), Johor Bahru, Malaysia, 1–2 May 2017; pp. 333–345. Available online: https://www.europeanproceedings.com/article/10.15405/epsbs.2018.05.27 (accessed on 10 October 2022).

38. Francis, J.; Johnston, M.; Eccles, M.; Walker, A.; Grimshaw, J.M.; Foy, R.; Kaner, E.F.S.; Smith, L.; Bonetti, D. *Constructing Questionnaires Based on the Theory of Planned Behaviour: A Manual for Health Services Researchers*; Quality of Life and Management of Living Resources, Centre for Health Services Research: Newcastle upon Tyne, UK, 2004.

39. Ham, M.; Jeger, M.; Frajman Ivković, A. The Role of Subjective Norms in Forming the Intention to Purchase Green Food. *Econ. Res. -Ekon. Istraživanja* **2015**, *28*, 738–748. [CrossRef]

40. Ajzen, I.; Fishbein, M. *Understanding Attitudes and Predicting Behavior*, 1st ed.; Prentice Hall: Upper Saddle River, NJ, USA, 1980.

41. Davis, F.D. Perceived Usefulness, Perceived Ease of Use, and User Acceptance of Information Technology. *MIS Q.* **1989**, *13*, 319–340. [CrossRef]

42. Everett, M.R. *Diffusion of Innovations*, 5th ed.; Free Press: Glencoe, IL, USA, 2003.

43. Hoover-Dempsey, K.V.; Sandler, H.M. Why Do Parents Become Involved in Their Children's Education? *Rev. Educ. Res.* **1997**, *67*, 3–42. [CrossRef]

44. Day, C.; Kington, A.; Stobart, G.; Sammons, P. The Personal and Professional Selves of Teachers: Stable and Unstable Identities. *Br. Educ. Res. J.* **2006**, *32*, 601–616. [CrossRef]

45. Ajzen, I. From Intentions to Actions: A Theory of Planned Behavior. In *Action Control: From Cognition to Behavior*; Kuhl, J., Beckmann, J., Eds.; Springer: Berlin/Heidelberg, Germany, 1985; pp. 11–39, ISBN 978-3-642-69746-3.

46. Compeau, D.R.; Higgins, C.A. Computer Self-Efficacy: Development of a Measure and Initial Test. *MIS Q.* **1995**, *19*, 189–211. [CrossRef]

47. Parasuraman, A.; Colby, C.L. An Updated and Streamlined Technology Readiness Index. *J. Serv. Res.* **2014**, *18*, 59–74. [CrossRef]

48. Ayanwale, M.A.; Sanusi, I.T.; Adelana, O.P.; Aruleba, K.D.; Oyelere, S.S. Teachers' Readiness and Intention to Teach Artificial Intelligence in Schools. *Comput. Educ. Artif. Intell.* **2022**, *3*, 100099. [CrossRef]

49. Makki, B.I.; Salleh, R.; Harun, H. Work Readiness, Career Self-Efficacy and Career Exploration: A Correlation Analysis. In Proceedings of the 2015 International Symposium on Technology Management and Emerging Technologies (ISTMET), Langkawi, Malaysia, 25–27 August 2015; pp. 427–431. Available online: https://ieeexplore.ieee.org/document/7359072/ (accessed on 11 October 2022).

50. Birzina, R.; Cedere, D. Students' readiness for massive open online courses (moocs). In Proceedings of the International Scientific Conference, Rezekne, Latvia, 22–23 May 2020; pp. 403–413. Available online: http://journals.rta.lv/index.php/SIE/article/view/4957 (accessed on 14 October 2022).

51. Okuonghae, O.; Igbinovia, M.O.; Adebayo, J.O. Technological Readiness and Computer Self-Efficacy as Predictors of E-Learning Adoption by LIS Students in Nigeria. *Libri* **2021**, *72*, 13–25. [CrossRef]

52. Teo, T.; Huang, F.; Hoi, C.K.W. Explicating the Influences That Explain Intention to Use Technology among English Teachers in China. *Interact. Learn. Environ.* **2017**, *26*, 460–475. [CrossRef]

53. Hair, J.; Black, W.C.; Babin, B.J.; Anderson, R.E. *Multivariate Data Analysis: A Global Perspective*, 7th ed.; Pearson Education, Cop: Upper Saddle River, NJ, USA, 2010.

54. Bagozzi, R.P.; Yi, Y. On the Evaluation of Structural Equation Models. *J. Acad. Mark. Sci.* **1988**, *16*, 74–94. [CrossRef]

55. Hu, L.; Bentler, P.M. Cutoff Criteria for Fit Indexes in Covariance Structure Analysis: Conventional Criteria versus New Alternatives. *Struct. Equ. Model. A Multidiscip. J.* **1999**, *6*, 1–55. [CrossRef]
56. Kashy, D.A.; Donnellan, M.B.; Ackerman, R.A.; Russell, D.W. Reporting and Interpreting Research in PSPB: Practices, Principles, and Pragmatics. *Personal. Soc. Psychol. Bull.* **2009**, *35*, 1131–1142. [CrossRef]
57. Widaman, K.F. Multitrait-multimethod analysis. In *The Reviewer's Guide to Quantitative Methods in the Social Sciences*, 2nd ed.; Routledge: London, UK, 2018; pp. 331–347. ISBN 978-1-315-75564-9.
58. Ahmed, E.; Ward, R. Analysis of Factors Influencing Acceptance of Personal, Academic and Professional Development E-Portfolios. *Comput. Hum. Behav.* **2016**, *63*, 152–161. [CrossRef]
59. Hooper, D.; Coughlan, J.; Mullen, M.R. Structural Equation Modelling: Guidelines for Determining Model Fit. *Electron. J. Bus. Res. Methods* **2008**, *6*, 53–60.
60. Markus, K.A. Principles and Practice of Structural Equation Modeling by Rex B. Kline. *Struct. Equ. Model. A Multidiscip. J.* **2012**, *19*, 509–512. [CrossRef]
61. Schumacker, R.E. *A Beginner's Guide to Structural Equation Modeling*; Routledge: London, UK, 2012.
62. Tabachnick, B.G.; Fidell, L.S. *Using Multivariate Statistics*, 6th ed.; Pearson Education: Boston, MA, USA, 2013.
63. Wahono, B.; Chang, C.-Y. Assessing Teacher's Attitude, Knowledge, and Application (AKA) on STEM: An Effort to Foster the Sustainable Development of STEM Education. *Sustainability* **2019**, *11*, 950.
64. Ajzen, I. *Attitudes, Personality and Behavior*; Open University Press: Maidenhead, UK, 2005.
65. Ajzen, I. The Theory of Planned Behaviour: Reactions and Reflections. *Psychol. Health* **2011**, *26*, 1113–1127. [CrossRef] [PubMed]
66. Armitage, C.J.; Conner, M. Efficacy of the Theory of Planned Behaviour: A Meta-Analytic Review. *Br. J. Soc. Psychol.* **2010**, *40*, 471–499. [CrossRef] [PubMed]
67. McEachan, R.R.C.; Conner, M.; Taylor, N.J.; Lawton, R.J. Prospective Prediction of Health-Related Behaviours with the Theory of Planned Behaviour: A Meta-Analysis. *Health Psychol. Rev.* **2011**, *5*, 97–144. [CrossRef]
68. Zint, M. Comparing Three Attitude-Behavior Theories for Predicting Science Teachers' Intentions. *J. Res. Sci. Teach.* **2002**, *39*, 819–844. [CrossRef]
69. Alhendal, D.; Marshman, M.; Grootenboer, P. Kuwaiti Science Teachers' Beliefs and Intentions Regarding the Use of Inquiry-Based Instruction. *Int. J. Sci. Math. Educ.* **2015**, *14*, 1455–1473. [CrossRef]
70. Atsoglou, K.; Jimoyiannis, A. Teachers' Decisions to Use ICT in Classroom Practice. *Int. J. Digit. Lit. Digit. Competence* **2012**, *3*, 20–37. [CrossRef]
71. Ajjan, H.; Hartshorne, R. Investigating Faculty Decisions to Adopt Web 2.0 Technologies: Theory and Empirical Tests. *Internet High. Educ.* **2008**, *11*, 71–80. [CrossRef]
72. Sadaf, A.; Newby, T.J.; Ertmer, P.A. Exploring Factors That Predict Preservice Teachers' Intentions to Use Web 2.0 Technologies Using Decomposed Theory of Planned Behavior. *J. Res. Technol. Educ.* **2012**, *45*, 171–196. [CrossRef]
73. Sadaf, A.; Gezer, T. Exploring Factors That Influence Teachers' Intentions to Integrate Digital Literacy Using the Decomposed Theory of Planned Behavior. *J. Digit. Learn. Teach. Educ.* **2020**, *36*, 124–145. [CrossRef]
74. Banilower, E.R.; Smith, P.S.; Weiss, I.R.; Malzahn, K.A.; Campbell, K.M.; Weis, A.M. *Report of the 2012 National Survey of Science and Mathematics Education*; Horizon Research, Inc.: Chapel Hill, NC, USA, 2013. Available online: http://www.horizon-research.com/2012nssme/wp-content/uploads/2013/02/2012-NSSME-Full-Report1.pdf (accessed on 17 October 2022).
75. Ryu, M.; Mentzer, N.; Knobloch, N. Preservice Teachers' Experiences of STEM Integration: Challenges and Implications for Integrated STEM Teacher Preparation. *Int. J. Technol. Des. Educ.* **2018**, *29*, 493–512. [CrossRef]
76. Cunningham, C.M. Engineering is elementary. *Bridge* **2009**, *30*, 11–17.
77. Leader-Janssen, E.M.; Rankin-Erickson, J.L. Preservice Teachers' Content Knowledge and Self-Efficacy for Teaching Reading. *Lit. Res. Instr.* **2013**, *52*, 204–229. [CrossRef]
78. Swackhamer, L.E.; Koellner, K.; Basile, C.; Kimbrough, D. Increasing the Self-Efficacy of Inservice Teachers through Content Knowledge. *Teach. Educ. Q.* **2009**, *36*, 63–78.
79. Hoy, A.W.; Spero, R.B. Changes in Teacher Efficacy during the Early Years of Teaching: A Comparison of Four Measures. *Teach. Teach. Educ.* **2005**, *21*, 343–356. [CrossRef]
80. NRC [National Research Council]. *Inquiry and the National Science Education Standards: A Guide for Teaching and Learning*; The National Academies Press: Washington, DC, USA, 2000.
81. Horng, J.-S.; Liu, C.-H.; Chou, S.-F.; Huang, Y.-C. The Roles of University Education in Promoting Students' Passion for Learning, Knowledge Management and Entrepreneurialism. *J. Hosp. Tour. Manag.* **2020**, *44*, 162–170. [CrossRef]
82. Ursavaş, Ö.F.; Bahçekapılı, T.; Camadan, F.; İslamoğlu, H. Teachers' behavioural intention to use ICT: A structural equation model approach. In Proceedings of the Society for Information Technology & Teacher Education International Conference, Las Vegas, NV, USA, 2 March 2015; pp. 2875–2880. Available online: https://www.learntechlib.org/primary/p/150400/ (accessed on 17 October 2022).

MDPI

St. Alban-Anlage 66

4052 Basel

Switzerland

Tel. +41 61 683 77 34

Fax +41 61 302 89 18

www.mdpi.com

Applied Sciences Editorial Office

E-mail: applsci@mdpi.com

www.mdpi.com/journal/applsci